Y0-BSD-001

PROBABILITY AND STATISTICS

Essays in Honor of Franklin A. Graybill

PROBABILITY AND STATISTICS

Essays in Honor of Franklin A. Graybill

edited by

Jaya N. SRIVASTAVA
Department of Statistics
Colorado State University
Fort Collins, Colorado, U.S.A.

1988

NORTH-HOLLAND
AMSTERDAM • NEW YORK • OXFORD • TOKYO

ISBN: 0 444 70457 4

Publishers:

ELSEVIER SCIENCE PUBLISHERS B.V.
P.O. Box 1991
1000 BZ Amsterdam
The Netherlands

Sole distributors for the U.S.A. and Canada:

ELSEVIER SCIENCE PUBLISHING COMPANY, INC.
52 Vanderbilt Avenue
New York, N.Y. 10017
U.S.A.

LIBRARY OF CONGRESS
Library of Congress Cataloging-in-Publication Data

Probability and statistics : essays in honor of Franklin A. Graybill / edited by Jaya N. Srivastava.
p. cm.
"Publications of Franklin A. Graybill": p.
ISBN 0-444-70457-4
1. Probabilities. 2. Mathematical statistics. 3. Graybill, Franklin A. I. Graybill, Franklin A. II. Srivastava, J. (Jaya)
QA273.18.P748 1988
519.2--dc19 88-16463
CIP

PRINTED IN THE NETHERLANDS

PREFACE

This collection contains a series of interesting articles, dedicated to our beloved colleague, Professor Franklin A. Graybill.

I have been assisted in the development of this volume by an Editorial Committee, whose members included Professors Duane Boes, David Bowden, Paul Mielke, Moinuddin Siddiqui, and James Williams, all of my department; I am very thankful to each among them for their help.

Each paper in this volume has been refereed. Each paper was handled by one member of the Editorial Committee, who also, in many cases, asked other person(s) (not on the Committee) to act as referee(s). The number of submissions was larger, only those papers which were acceptable to the Committee members are included here.

My sincere thanks are due to all authors, for their efforts are truly responsible for the publication of this volume.

Jaya N. Srivastava

Franklin A. Graybill

FRANKLIN A. GRAYBILL
A BRIEF LIFE SKETCH

Professor Franklin A. Graybill was born to Arno and Lula Graybill in Carson, Iowa, a town of 600 inhabitants in a farming community in southwestern Iowa, on September 23, 1921. His family consisted of two brothers and four sisters. He attended all twelve years of public school in Carson. After graduation from high school in 1939 he attended college for two years at Tabor Junior College in Tabor, Iowa. In September 1941 he enrolled at William Penn College in Oskaloosa, Iowa and during his first semester there the United States entered World War II. A few weeks later he enlisted in the Army Air Corps and later went through cadet training in Santa Anna, California. After graduation he started pilot flight training at Twenty Nine Palms, California. His flight training terminated after a few weeks due to a physical impairment; his depth perception was defective. He eventually received training in a special troops unit and was sent to the Phillipine Islands.

When the war was over, Graybill returned to his home in Iowa and then spent the summer of 1946 at Greeley College in Greeley, Colorado. In September 1946, he returned to William Penn College where he received a B.S. degree in mathematics in May 1947. It was then that he met and married Jeanne Bunting.

In September 1947 they went to Stillwater, Oklahoma where Graybill attended Oklahoma State University. He received a Master's degree in mathematics in May 1949. His Master's paper was under the direction of Professor Carl Marshall.

In September 1949, Frank and Jeanne went to Iowa State University where Frank started working towards a Ph.D. degree in statistics. He received this degree in the spring of 1952. His dissertation topic was on Variance Components, and his advisor was Professor Oscar Kempthorne.

in July 1952, Frank, Jeanne and their two children, Dan and Kathryn, returned to Stillwater, Oklahoma where Frank joined the statistics group as an Assistant Professor. He remained there for eight years where his duties were teaching, research and consultant in the Oklahoma Agricultural Experiment Station.

In August 1960, Frank accepted a position at Colorado State University as Professor of Statistics. He was hired to develop a graduate program in Statistics. He served as the first chairman of the Department of Statistics that was created in 1971. In 1981, Frank retired at age 60, but continued on a 'transitional' appointment in which he teaches during the fall semester each year.

In 1968, Frank served as an Associate Editor of the Annals of Mathematical Statistics. From 1972–1974, he was Editor of Biometrics. He served on the Panel on Statistics for the Committee on the Undergraduate Program in Mathematics. He served on several committees in the various statistical societies. He was President of WNAR of the International Biometric Society in 1967, and was President of the American Statistical Association in 1976.

Frank is the author or co-author of over 50 published papers and eight textbooks, but in his own words, "the most satisfying results of the past twenty years are our three grandchildren, Tonia, Jules, and Emily".

Indeed, besides his professional accomplishments, Frank will be particularly remembered for his keen sense of humor.

PUBLICATIONS OF FRANKLIN A. GRAYBILL

Books

Linear Statistical Models, Vol. I., McGraw-Hill, February 1961

An Introduction to the Theory of Statistics (with A. M. Mood), Revised edition published by McGraw-Hill Book Company, 1963. (Printed in Japanese, French, and Spanish).

An Introduction to Statistical Models in Geology (with W. Krumbein), McGraw-Hill Book Company, 1965.

An Introduction to Matrices that have Application in Statistics, Wadsworth Publishing Company, 1969.

Introduction to the Theory of Statistics (with A. M. Mood and D. C. Boes), Third Edition, McGraw-Hill, 1974, 564 pages.

Introduction to the Theory and Application of the Linear Model. Duxbury Press, 1976, 704 pages.

Matrices That Have Applications in Statistcs. Brooks/Cole Advanced Series, 1983.

Papers

"Effect of lindane on egg productioon," (with George F. Godfrey and D. E. Howell) Poultry Science, Vol. XXXII, No. 1, pp. 183-184.

"On quadratic estimates of variance components," Annals of Mathematical Statistics, Vol. 25, No. 2, June 1954, pp. 367-372.

"Hot weather shelters for livestock," (with G. L. Nelson, W. A. Mahoney and E. R. Berousek) Agriculture Engineering, September 1954, pp. 638-645.

"Variance heterogeneity in a randomized block design," Biometrics, Vol. 10, No. 4, Dec. 1954, pp. 516-520.

"Factors affecting results of grazing trials with yearling steers," (with J. Q. Lynd and R. Totusek) Agronomy Journal, Vol. 48, 1956, pp. 352-355.

"Sample size in food-habits analyses," (with William Hanson) The Journal of Wildlife Management, Vol. 20, No. 1, Jan. 1956, pp. 64-68.

"A note on uniformly best unbiased estimators for vor variance components," (with A. W. Wortham) Journal of the American Statistical Association.

"Confidence intervals for variance ratios specifying genetic heritability," (with Frank Martin and George Godfrey) Biometrics, June 1956, pp. 99-109.

"Weight losses of packaged ice cream and mellorine-type products during storage," (with J. V. Mickle) Journal of Dairy Science, Sept. 1956, Vol. XXXIX, No. 9, pp. 1251-1259.

"Grazing trial evaluations using paired pastures with yearling steers," (with J. Q. Lynd and R. Totusek) Agronomy Journal, Vol. 49, 1947, pp. 488-492.

"Methods of determining the total solids in fluid milk," (with J. B. Mickle, S. D. Musgrave, R. E. Walton, M. Loewenstein, and P. E. Johnson) Oklahoma A & M Experiment Station Technical Bulletin, No. T-67, Feb. 1957.

"Calculating confidence intervals for genetic heritability," (with W. H. Robertson) Poultry Science, Vol. XXXVI, No. 2, March 1957, pp. 261-265.

"Heterogeneity of error variances in a randomized block design," (with John Leroy Folks) Biometrika, Vol. 44, Parts 1 and 2, June 1957, pp. 275-277.

"Idempotent matrices and quadratic forms in the general linear hypothesis," (with George Marsaglia) The Annals of Mathematical Statistics, Vol. 28, No. 3, Sept. 1957, pp. 678-686.

"Determining sample size for a specified width confidence interval," The Annals of Mathematical Statistics, Vol. 29, No. 1, March 1958, pp. 282-287.

"The staircase design: theory," (with William Pruitt) The Annals of Mathematical Statistics, Vol. 29, No. 2, June 1958, pp. 523-533.

"Evaluation of formulas for predicting various components of mixed herd milk," (with J. B. Mickle, E. R. Egermeier, S. D. Musgrave, R. D. Morrison, and R. MacVicar) Oklahoma State University Experiment Station Technical Bulletin, No. T-74, Sept. 1958.

"Determining minimum populations for initial evaluation of breeding material," (with William R. Kneebone) Agronomy Journal, Vol. 51, 1959, pp. 4-6.

"Combining inter-block and intra-block information in balanced incomplete blocks," (with David L. Weeks) The Annals of Mathematical Statistics, Vol. 30, No. 3, Sept. 1959, pp. 799-805.

"Combining unbiased estimators," (with R. B. Deal) Biometrics, Vol. 15, No. 4, Dec. 1959, pp. 543-550.

"On the unbiasedness of Yates' method of estimation using interblock information," (with V. Seshadri) The Annals of Mathematical Statistics, Vol. 31, No. 3, Sept. 1960, pp. 786-787.

"Sample size for a specified width confidence interval on the variance of a normal distribution," (with Robert D. Morrison) Biometrics, Vol. 16, No. 4, Dec. 1960, pp. 636-641.

"Minimal sufficient statistics for the balanced incomplete block design under an Eisenhart Model II," (with David L. Weeks) Sankhya: The Indian Journal of Statistics, Series A., Vol. 23, Part 3, 1961, pp. 261-269.

"Theorems concerning Eisenhart's Model II," (with Robert A. Hultquist) The Annals of Mathematical Statistics, Vol. 32, No. 1,March 1961, pp. ? ? ?

"A minimum sufficient statistic for a general class of designs," (with David L. Weeks) Sankhya: The Indian Journal of Statistics, Series A., Vol. 24, Part 4, 1962, pp. 339-354.

"Sample size required to estimate the ratio of variances with bounded relative error," (with Terrence L. Connell) Journal of the American Statistical Association, Vol. 58, Dec. 1963, pp. 1044-1047.

"Sample size for estimating the variance within d units of the true value," (with Terrence L. Connell) The Annals of Mathematical Statistics, Vol. 35, No. 1, March 1965, pp. 438-440.

"Sample size required to estimate the parameter in the uniform density within d units of the true value," (with Terrence L. Connell) Journal of the American Statistical Association, Vol. 59, June 1964, pp. 550-556.

"Minimal sufficient statistics for the two-way classification mixed model design," (with R. A. Hultquist) Journal of the American Statistical Association, Vol. 60, No. 309, March 1965, pp. 182-192.

"Confidence bands of uniform and proportional width for linear models," (with D. C. Bowden) Journal of the American Statistical Association, Vol. 61, No. 313, March 1966.

"Linear segment confidence bands for simple linear models," (with D. C. Bowden) Journal of the American Statistical Association, June 1967, Vol. 62.

"Note on the computation of the generalized inverse of a matrix," (with D. C. Meyer and R. J. Painter) SIAM Review , Vol. 8, No. 4, October 1966.

"Quadratic forms and idempotent matrices with random elements," (with G. A. Milliken) Annals of Mathematical Statistics, Vol. 40, No. 4, August 1969.

"Extensions of the general linear hypothesis model," (with G. A. Milliken) Journal of the American Statistical Association, June 1970, Vol. 65, No. 330.

"A non-linear characterization of the normal distribution," (with A. Kingman) Annals of Mathematical Statistics, Vol. 41, No. 6, Dec. 1970.

"Tests for intereaction in a two-way model with missing data," (with A. Kingman) Biometrics, Vol. 27, No. 4, December 1971.

"Preparation for graduate work in statistics." CUPM Panel (F. A. Graybill, Chairman)

"Introductory statistics without calculus." CUPM Panel (F. A. Graybill, Chairman) The American Statistician, Feb. 1972.

"Estimating σ^2 in a two-way classification model with interaction," (with D. Johnson) Journal of the American Statistical Association, June 1972, Vol. 67, No. 338.

"Interaction models for the Latin square," (with G. Milliken) Australian Journal of Statistics, Vol. 14, No. 2, Aug. 1972.

"Analysis of a two-way model with interaction and no replication," (with D. E. Johnson) Journal of the American Statistical Association, Dec. 1972, Vol. 67, No. 340, pp. 862-868.

"Least squares programs--a look at the square root procedure," (with R. Kopitzke and T. Boardman) The American Statistician, Vol. 29, 1975.

"What lies ahead?," Journal of the American Statistical Association, Mar. 1977, Vol. 72, No. 357, pp. 7-10. Presidential Address, American Statistical Association.

"Confidence intervals for proportions of variability in two-factor nested variance component models," Journal of the American Statistical Association, June 1979, Vol. 74, No. 366.

"Evaluation of a Method for setting confidence intervals on the common mean of two normal populations," Communications in Statistics--Simulation Computation, B8 (1), 53-60, 1979.

"Small samples confidence intervals on common mean of two normal distributions with unequal variances," Communications in Statistics--Theory and Methods, A8(13), 1255-1269, 1979.

"Confidence intervals on variance components in three-factor cross-classification models," (with S. Jeyaratnam), Technometrics, Aug. 1980.

"Confidence intervals on nonnegative linear combinations of variance," (with C. M. Wang), Journal of the American Statistical Association, 1980, Vol. 75, pp. 863-869.

"Confidence interval estimation in heteroscedastic linear models with replicates," (with A. K. Mohmand), Journal of Statistical Research, Vol. 14. Nos. 1 & 2, 1980.

"Confidence intervals on a ratio of variances in the two-factor nested components of variance model," (with C. M. Wang), Communications in Statistics--Theory and Methods, A10(14), 1357-1368, 1981.

"Confidence intervals on a ratio of variances in the two-factor nested components-of-variance model," (with C. Arteaga), Communications in Statistics, 1982.

Panel discussion on graduate programs in statistics. Teaching of Statistics and Statistical Consulting, 1982. Academic Press.

"A nonparametric procedure for confidence intervals in linear models with unequal error variances," (with A. K. Mohmand), The Arabian Journal for Science and Engineering, Vol. 7, No. 1, 1982.

"Confidence intervals on variance components in the unbalanced one-way classification," (with R. Burdick), Technometrics, Vol. 26, No. 2, 1984.

"Confidence intervals on the total variance in an unbalanced two-fold nested classification with equal subsampling," (with R. K. Burdick). Communications in Statistics, 1985.

"Confidence intervals for variance components in the balanced two way model with interaction," (with Ricardo A. Leiva) Commun. Statist.-Simula., 1986.

"Confidence intervals on the intraclass correlation in the unbalanced one-way classification," (with Richard k. Burdick, and Farooq Maqsood) Commun. Statist.-Theory Meth., 1986.

CONTENTS

PROBABILITY AND STATISTICS
Essays in Honor of Franklin A. Graybill
J.N. Srivastava (Editor)

ADMISSIBILITY IN LINEAR MODELS WITH POLYHEDRAL COVARIANCE STRUCTURE

Abdul-Mordy Azzam

Faculty of Commerce
University of Alexandria, Egypt

David Birkes and Justus Seely

Department of Statistics
Oregon State University, Corvallis, USA

Consider the problem of choosing a linear unbiased estimator for a given linear function of the fixed effects in the usual mixed linear model. Within this class of estimators, those that are admissible under squared error loss are characterized.

AMS Subject Classification: Primary 62J99; Secondary 62C15

Key Words: Admissible linear unbiased estimator, mixed linear model, closed convex cone

1. Introduction.

In the linear model $E(Y) = \theta$ with $Cov(Y) = \sigma^2 I$, consider the problem of choosing a linear estimator of θ. Using the criterion of squared error loss, Cohen (1966, Remark 3) found necessary and sufficient conditions for a linear estimator $H'Y$ to be admissible: H must be symmetric and have all its eigenvalues in the closed interval $[0,1]$. Admissibility here is only within the class of linear estimators. Thus, Y is admissible within this linear class even though Stein (1956) showed it to be inadmissible within the larger class of all estimators when Y has an n-dimensional normal distribution with $n \geq 3$. Cohen's conditions can be generalized to apply to the problem of finding admissible linear estimators of a linear parametric vector in the model $E(Y) = X\beta$ with $Cov(Y) = \sigma^2 V$ (Rao, 1976, Theorem 6.6).

Taking a different approach, Olsen, Seely and Birkes (1976) confined their attention to linear unbiased estimators of a real-valued linear parametric function, but allowed $Cov(Y) \in \mathcal{V}$, where $\mathcal{V}$ is an arbitrary set of covariance matrices. They found that a necessary condition for a linear unbiased estimator to be admissible is for it to be best at some nonzero covariance matrix in the closed convex cone generated by $\mathcal{V}$. LaMotte (1982), by introducing the notion of trivial points and allowing 'covariance' matrices that were not nonnegative definite, obtained a necessary and sufficient condition for admissibility in terms of bestness. LaMotte treated the problem of estimating a linear parametric vector in a linear model $E(Y) = X\beta$ with $Cov(Y) \in \mathcal{V}$, and characterized admissibility within a certain type of affine set of linear estimators (including all affine sets in the case of a real-valued linear parametric function).

In this paper we consider admissiblity within the class of linear unbiased estimators of a real-valued linear parametric function when $\mathcal{V}$ generates a polyhedral cone. This is the type of covariance structure encountered in the usual fixed, mixed, and random linear models. The theory of LaMotte is applicable here, but in our special setting we are able to provide a more specific recipe for expressing admissible estimators as best estimators (see Theorem 5.3). Our setting also allows a relatively direct attack on the problem, which avoids some of the complications encountered in LaMotte's general approach.

2. The model and problem.

Throughout this paper we will be concerned with an $n \times 1$ random vector Y whose distributional assumptions fall within a general linear model framework. In particular, we suppose that the expectation of Y is of the form $E(Y) = X\beta$ where X is a known $n \times p$ matrix and β is an unknown $p \times 1$ vector of parameters. We also suppose that the covariance matrix of Y is of the form $Cov(Y) = V$, $V \in \mathscr{V}$, where $\mathscr{V}$ is a given set of nonnegative definite (nnd) matrices.

Suppose that $\lambda'\beta$ is an estimable parametric function. That is, λ is an element of $\underline{R}(X')$. (For a matrix A, the notation A', $\underline{R}(A)$, $\underline{r}(A)$, and $\underline{R}(A)^{\perp}$ is used to denote its transpose, range, rank, and the orthogonal complement of its range.) Our object is to select a subset of estimators for $\lambda'\beta$ which in some sense have optimal properties. We do this through the decision-theoretic notion of admissibility when the candidate estimators are confined to the linear unbiased estimators (lue's) for $\lambda'\beta$ and when they are compared according to their possible variances.

Let $\mathbf{H}$ denote the set of all vectors h for which $h'Y$ is an lue for $\lambda'\beta$. We adopt the usual terminology of decision theory with the convention that an lue $h'Y$ (decision rule) is to be identified via its coefficient vector h and that the loss function is squared error. Since we are confining attention to unbiased estimators, our risk function reduces to the variance. For elements d and h in $\mathbf{H}$, we say d is <u>as good as</u> h if

$$d'Vd = Var(d'Y|V) \leq Var(h'Y|V) = h'Vh$$

for all $V \in \mathscr{V}$. We say d is <u>better than</u> h if it is as good as h and there is a strict inequality for at least one $V \in \mathscr{V}$. An element $h \in \mathbf{H}$ is said to be <u>admissible</u> if there is no $d \in \mathbf{H}$ that is better than it.

Let the vector space $\mathscr{S}$ of $n \times n$ symmetric matrices be endowed with the usual Euclidean topology. A convex cone in $\mathscr{S}$ is any nonempty convex set that is closed under nonnegative scalar multiplication. The smallest closed convex cone containing a subset $\mathscr{A}$ of $\mathscr{S}$ is denoted $[\mathscr{A}]$. If $\mathscr{C}$ is a closed convex cone, we say that $\mathscr{G}$ is a generator set for $\mathscr{C}$ provided $[\mathscr{G}] = \mathscr{C}$. Olsen et al. (1976) showed that the definitions of 'as good as', 'better than', and 'admissible' remain the same if $\mathscr{V}$ is replaced by any set that generates $[\mathscr{V}]$. They also showed that the class $\mathbf{A}$ of admissible lue's is a minimal complete class. It is the class $\mathbf{A}$ which we wish to characterize when there is a generator set for $[\mathscr{V}]$ that consists of a finite number of elements.

3. Best sets.

Let $\mathbf{L}$ be a nonempty affine set in Euclidean n-space and let W be an $n \times n$ nonnegative definite matrix. A vector d is said to be <u>best</u> in $\mathbf{L}$ at W if $d \in \mathbf{L}$ and $d'Wd \leq h'Wh$ for all $h \in \mathbf{L}$. The set $\mathbf{B}(\mathbf{L},W)$ consisting of all vectors that are best in $\mathbf{L}$ at W is called the <u>best set</u> in $\mathbf{L}$ at W. Best sets have been extensively studied in the theory of best linear unbiased estimation. For future use, we list two basic known facts about these sets.

<u>Lemma 3.1</u>. Let T be any matrix such that $\underline{R}(T)^{\perp}$ is the subspace parallel to the affine set $\mathbf{L}$. Then:

(a) $\mathbf{B}(\mathbf{L},W) = \{h \in \mathbf{L} : Wh \in \underline{R}(T)\}$.

(b) $\mathbf{B}(\mathbf{L},W)$ is a nonempty affine set and its parallel subspace is $\underline{R}(T,W)^{\perp}$.

Proof: Part (a) can be found in Zyskind (1967, Theorem 3); and (b) can be derived from (a). []

In generalizing the results of Olsen et al. (1976), both Azzam (1981) and LaMotte (1982) applied the idea of best sets in a sequential fashion. For a sequence of nnd matrices $W_1, \ldots, W_t$ define

$$\mathbf{B}(\mathbf{L}, W_1, \ldots, W_m) = \mathbf{B}(\mathbf{B}(\mathbf{L}, W_1, \ldots, W_{m-1}), W_m) \ , \quad m = 2, \ldots, t \ .$$

Some useful summary facts about these sets are given next.

Lemma 3.2. Let T be as in Lemma 3.1 and let $\mathbf{B}_t$ denote the sequential best set $\mathbf{B}(\mathbf{L}, W_1, \ldots, W_t)$. Then:

(a) $\mathbf{B}_t = \{h \varepsilon \mathbf{L} : W_j h \ \varepsilon \ \underline{R}(T, W_1, \ldots, W_{j-1}), \ j = 1, \ldots, t\}$.

(b) $\mathbf{B}_t$ is a nonempty affine set and $\underline{R}(T, W_1, \ldots, W_t)^{\perp}$ is its parallel subspace.

(c) $\mathbf{B}_t$ is a singleton set if and only if $\underline{r}(T, W_1, \ldots, W_t) = n$.

(d) If $\underline{R}(W_t) \subset \underline{R}(T, W_1, \ldots, W_{t-1})$, then $\mathbf{B}_t = \mathbf{B}(\mathbf{L}, W_1, \ldots, W_{t-1})$.

Proof: Set $\mathbf{L}_1 = \mathbf{B}(\mathbf{L}, W_1)$ and $T_1 = (T, W_1)$. Lemma 3.1b implies that $\underline{R}(T_1)^{\perp}$ is the subspace parallel to $\mathbf{L}_1$. Use $\mathbf{L}_1$ and T_1 in Lemma 3.1 to see that

$$\mathbf{B}(\mathbf{L}, W_1, W_2) = \mathbf{B}(\mathbf{L}_1, W_2) = \{h \varepsilon \mathbf{L}_1 : W_2 h \ \varepsilon \ \underline{R}(T_1)\}$$

$$= \{h \varepsilon \mathbf{L} : W_1 h \ \varepsilon \ \underline{R}(T) \ , \ W_2 h \ \varepsilon \ \underline{R}(T, W_1)\} \ .$$

Its parallel subspace is $\underline{R}(T_1, W_2)^{\perp}$. Thus, (a) and (b) follow by induction. Now $\mathbf{B}_t$ is a singleton set if and only if its parallel subspace is $\{0\}$. This is equivalent to the condition in part (c). Part (d) is immediate from (a). []

4. Admissibility of uniquely best estimators.

To discuss the ideas in this and the following sections, it is convenient to extend the decision theory terms beyond the classes $\mathbf{H}$ and $\mathscr{V}$. Let $\mathbf{L}$ be a nonempty affine set in Euclidean n-space and let $\mathscr{G}$ be a nonempty set of nnd matrices. A vector d is said to be as good as a vector h in $\mathbf{L}$ wrt $\mathscr{G}$ provided d and h are in $\mathbf{L}$ and $d'Gd \leq h'Gh$ for all $G \ \varepsilon \ \mathscr{G}$. The terms better than and admissible in $\mathbf{L}$ wrt $\mathscr{G}$ are defined in a similar manner. The set of all vectors that are admissible in $\mathbf{L}$ wrt $\mathscr{G}$ is denoted by $\mathbf{A}(\mathbf{L}, \mathscr{G})$. Thus, $\mathbf{A}$ of Section 2 can also be expressed as $\mathbf{A}(\mathbf{H}, \mathscr{V})$.

Lemma 4.1. Let $G \ \varepsilon \ [\mathscr{G}]$. Then

$$\mathbf{A}\{\mathbf{B}(\mathbf{L}, G), \mathscr{G}\} = \mathbf{A}(\mathbf{L}, \mathscr{G}) \cap \mathbf{B}(\mathbf{L}, G) \ .$$

Proof: Set $\mathbf{C} = \mathbf{B}(\mathbf{L}, G)$. Take any $h \ \varepsilon \ \mathbf{A}(\mathbf{C}, \mathscr{G})$. Suppose $h \not\varepsilon \ \mathbf{A}(\mathbf{L}, \mathscr{G})$. Then there is some $d \ \varepsilon \ \mathbf{L}$ that is better than h wrt $[\mathscr{G}]$. In particular, $d'Gd \leq h'Gh$ so that $d \ \varepsilon \ \mathbf{C}$. This contradicts $h \ \varepsilon \ \mathbf{A}(\mathbf{C}, \mathscr{G})$. This shows that $\mathbf{A}(\mathbf{C}, \mathscr{G}) \subset \mathbf{A}(\mathbf{L}, \mathscr{G}) \cap \mathbf{C}$. The reverse containment follows directly by noting that $\mathbf{C} \subset \mathbf{L}$. []

Recall that $\mathbf{H} = \{h : X'h = \lambda\}$ is the affine set of lue's for $\lambda'\beta$. It is nonempty because $\lambda'\beta$ is estimable. An lue h is said to be uniquely best in $\mathbf{H}$ if $h \ \varepsilon \ \mathbf{H}$ and if there exist $W_1, \ldots, W_t \ \varepsilon \ [\mathscr{V}]$ such that $\mathbf{B}(\mathbf{H}, W_1, \ldots, W_t) = \{h\}$. Let $\mathbf{B}$ denote the set of lue's that are uniquely best in $\mathbf{H}$. By Lemma 3.2c, $\mathbf{B}$ is the union of all $\mathbf{B}(\mathbf{H}, W_1, \ldots, W_t)$ for which $W_1, \ldots, W_t \ \varepsilon$

$[\mathcal{V}]$, $t \geq 1$, and $\underline{r}(X,W_1,\ldots,W_t) = n$. Both Azzam (1981) and LaMotte (1982) realized that uniquely best lue's are admissible.

Lemma 4.2. $\mathbf{B} \subset \mathbf{A}$.

Proof: Take $b \in \mathbf{B}$. Then $\{b\} = \mathbf{B}(\mathbf{H},W_1,\ldots,W_t)$ for matrices $W_1,\ldots,W_t \in [\mathcal{V}]$. For $k = 1,\ldots,t$, set $\mathbf{L}_k = \mathbf{B}(\mathbf{H},W_1,\ldots,W_k)$ and $\mathbf{A}_k = \mathbf{A}(\mathbf{L}_k,[\mathcal{V}])$. By definition, $\mathbf{L}_k = \mathbf{B}(\mathbf{L}_{k-1},W_k)$ for $k \geq 2$. By successively applying Lemma 4.1, we get $\mathbf{A} \supset \mathbf{A}_1 \supset \ldots \supset \mathbf{A}_t$. Since $\{b\} = \mathbf{A}_t$, the result follows. []

For some covariance structures the set $\mathbf{B}$ describes the entire set $\mathbf{A}$. Olsen et al. (1976) established that $\mathbf{B} = \mathbf{A}$ when $\underline{r}(X,V) = n$ for all nonzero $V \in [\mathcal{V}]$. Another case when equality holds is when $[\mathcal{V}]$ is equal to the set of all nnd matrices. To see this take any $h \in \mathbf{H}$. Let $W = I - hh'/h'h$. Then W is nnd, h is best at W , and $\underline{r}(X,W) = n$. This shows that $\mathbf{H} = \mathbf{B} = \mathbf{A}$. However, LaMotte's (1982) Example 1 shows that sometimes $\mathbf{B} \neq \mathbf{A}$.

5. Polyhedral cones of covariance matrices.

Let $\mathbf{L}$ and $\mathcal{G}$ be defined as in Section 4. A partial converse to Lemma 4.2 is given by Proposition 3.6 in Olsen et al. (1976), which states that if $h \in \mathbf{A}(\mathbf{L},\mathcal{G})$, then $h \in \mathbf{B}(\mathbf{L},G)$ for some nonzero $G \in [\mathcal{G}]$. This fact together with Lemma 4.1 gives:

Lemma 5.1. If $h \in \mathbf{A}(\mathbf{L},\mathcal{G})$, then $h \in \mathbf{A}\{\mathbf{B}(\mathbf{L},G),\mathcal{G}\}$ for some nonzero $G \in [\mathcal{G}]$.

Suppose $h \in \mathbf{A}$. Then the lemma implies that $h \in \mathbf{A}\{\mathbf{B}(\mathbf{H},W_1),\mathcal{V}\}$ for some nonzero W_1 in $[\mathcal{V}]$. Apply the lemma again with $\mathbf{L} = \mathbf{B}(\mathbf{H},W_1)$ to obtain $h \in \mathbf{A}\{\mathbf{B}(\mathbf{H},W_1,W_2),\mathcal{V}\}$ for some nonzero W_2 in $[\mathcal{V}]$. We can continue this procedure in hope of finding a sequence $W_1,\ldots,W_t$ for which h is uniquely best in $\mathbf{H}$. However, there is no guarantee that the uniqueness condition $\underline{r}(X,W_1,\ldots,W_t) = n$ will be achieved. LaMotte (1982) found a way around this by introducing the notion of trivial matrices and extending the set of possible 'covariance' matrices from $[\mathcal{V}]$ to a set including matrices that are not nnd. When $[\mathcal{V}]$ is polyhedral, we will see that this complication can be avoided and that a more explicit representation of $\mathbf{A}$ can be obtained.

The basic problem with the procedure in the previous paragraph occurs at the second step. It may happen that $\underline{R}(W_2) \subset \underline{R}(X,W_1)$, in which case $\underline{r}(X,W_1)$ is not increased by the addition of W_2 . If $[\mathcal{V}]$ can be appropriately decomposed, however, this problem can be solved. For a set of matrices $\mathcal{F}$, a matrix M is said to be a <u>maximal element</u> for $\mathcal{F}$ if $\underline{R}(F) \subset \underline{R}(M)$ for all $F \in \mathcal{F}$.

Lemma 5.2. Let $\mathcal{F}$ and $\mathcal{Q}$ be nonempty sets of nnd matrices. Assume that M is nnd and is a maximal element for $\mathcal{F}$. Then

$$\mathbf{A}\{\mathbf{B}(\mathbf{L},M),\mathcal{F}+\mathcal{Q}\} = \mathbf{A}\{\mathbf{B}(\mathbf{L},M),\mathcal{Q}\} .$$

Proof: Take d and h in $\mathbf{B}(\mathbf{L},M)$. By Lemma 3.1d, $\mathbf{B}(\mathbf{L},M,F) = \mathbf{B}(\mathbf{L},M)$ for all $F \in \mathcal{F}$, and so $d'Fd = h'Fh$. Thus, for all $F+Q \in \mathcal{F}+\mathcal{Q}$,

$$d'(F+Q)d - h'(F+Q)h = d'Qd - h'Qh .$$

Therefore, d is better than h wrt $\mathcal{F}+\mathcal{Q}$ if and only if d is better than h wrt $\mathcal{Q}$. []

Lemma 5.2 provides the possibility of sequentially applying Lemma 5.1 in such a way that eventually one arrives at the conclusion that an admissible lue is a uniquely best lue. To do this, $[\mathcal{V}]$ must allow decomposition into the summands $\mathscr{F}$ and $\mathscr{Q}$ required for Lemma 5.2. Polyhedral structure allows this.

A closed convex cone that can be generated by a finite number of elements is called a <u>polyhedral</u> convex cone. That is, a polyhedral convex cone consists of all nonnegative combinations $W = \rho_1 D_1 + \ldots + \rho_k D_k$, $\rho_1,\ldots,\rho_k \geq 0$, of a fixed finite set $\mathscr{D} = \{D_1,\ldots,D_k\}$. If all the ρ_j are positive, we call W a positive combination of $\mathscr{D}$. Note that $[\mathscr{D}]$ allows the decomposition $[\mathscr{D}] = [\mathscr{E}] + [\mathscr{F}]$ whenever $\mathscr{D} = \mathscr{E} \cup \mathscr{F}$.

<u>Theorem 5.3</u>. Suppose $[\mathscr{V}]$ is polyhedral with a generator set $\mathscr{D} = \{D_1,\ldots,D_m\}$ satisfying $\underline{r}(X,D_1,\ldots,D_m) = n$. If $h \in \mathbf{A}$, then there exist mutually disjoint and nonempty subsets $\mathscr{K}_1,\ldots,\mathscr{K}_t$ of $\mathscr{D}$ such that $\{h\} = \mathbf{B}(\mathbf{H},W_1,\ldots,W_t)$, where $\underline{r}(X,W_1,\ldots,W_t) = n$ and W_j is a positive combination of $\mathscr{K}_j$ for $j = 1,\ldots,t$.

<u>Proof</u>: Let $h \in \mathbf{A}$. By Lemma 5.1, $h \in \mathbf{A}(\mathbf{B}_1,\mathscr{V})$ where $\mathbf{B}_1 = \mathbf{B}(\mathbf{H},W_1)$ for some nonzero $W_1 \in [\mathscr{V}]$. Write $W_1 = \sum_j \rho_j D_j$ where $\rho_1,\ldots,\rho_m \geq 0$. Let $\mathscr{K}_1$ be the set of $D_j \in \mathscr{D}$ such that $\rho_j > 0$. Note that W_1 is a maximal element for $[\mathscr{K}_1]$ because the range of a sum of nnd matrices is the sum of their ranges. If $\underline{r}(X,W_1) = n$, then we are finished. Otherwise, let $\mathscr{D}_1$ be the set of all elements in $\mathscr{D}$ which are not in $\mathscr{K}_1$. Note $[\mathscr{V}] = [\mathscr{D}] = [\mathscr{K}_1] + [\mathscr{D}_1]$. From Lemma 5.2 we get

$$\mathbf{A}(\mathbf{B}_1,\mathscr{V}) = \mathbf{A}(\mathbf{B}_1,[\mathscr{K}_1]+[\mathscr{D}_1]) = \mathbf{A}(\mathbf{B}_1,[\mathscr{D}_1]) .$$

Apply Lemma 5.1 to $\mathbf{A}(\mathbf{B}_1,[\mathscr{D}_1])$ to obtain $h \in \mathbf{A}(\mathbf{B}_2,[\mathscr{D}_1])$ where $\mathbf{B}_2 = \mathbf{B}(\mathbf{H},W_1,W_2)$ with W_2 nonzero in $[\mathscr{D}_1]$. Write W_2 as a nonnegative combination of the elements in $\mathscr{D}_1$; and let $\mathscr{K}_2$ consist of those elements in $\mathscr{D}_1$ with positive coefficients in this expression. Then W_2 is a maximal element for $[\mathscr{K}_2]$. If $\underline{r}(X,W_1,W_2) = n$, we are finished. Otherwise, proceed as above with $\mathscr{D}_2$ the set of all elements in $\mathscr{D}_1$ which are not in $\mathscr{K}_2$. Since $\mathscr{D}$ has only a finite number of elements and $\underline{r}(X,D_1,\ldots,D_m) = n$, we must eventually arrive at the desired conclusion. []

Combining this theorem with Lemma 4.2, we obtain:

<u>Corollary 5.4</u>. Under the hypotheses of Theorem 5.3, $\mathbf{B} = \mathbf{A}$.

Theorem 5.3 gives a more concise description of $\mathbf{A}$ than does the mere statement $\mathbf{B} = \mathbf{A}$. Suppose the hypotheses of Theorem 5.3 are satisfied. Then according to the theorem, the admissible class can be described as the class of uniquely best elements in $\mathbf{H}$; and all uniquely best elements can be found in the following systematic way. First write down all ordered partitions of $\mathscr{D}$, that is, all ordered sequences $(\mathscr{K}_1,\ldots,\mathscr{K}_t)$ of disjoint nonempty subsets of $\mathscr{D}$ whose union is $\mathscr{D}$. For example, when $m = 2$, these are the three ordered partitions: $(\{D_1,D_2\})$, $(\{D_1\},\{D_2\})$ and $(\{D_2\},\{D_1\})$. Then for each ordered partition $(\mathscr{K}_1,\ldots,\mathscr{K}_t)$, form the best sets $\mathbf{B}(W_1,\ldots,W_t)$ for all sequences $W_1,\ldots,W_t$ where W_j is a positive combination of the elements of $\mathscr{K}_j$.

There are some redundancies in the procedure of the previous paragraph. For instance, we can divide any W_j by a positive constant and not change the best set. This means we need only consider W_j's that are convex combinations (with positive coefficients) of $\mathscr{K}_j$. Or alternatively, we could take all positive combinations with the coefficient of a particular element of $\mathscr{K}_j$ set equal to one. Also, if the first s elements in a sequence $W_1,\ldots,W_t$ give a uniquely best element, then the remaining elements in the sequence can be discarded.

6. Two variance components.

Suppose Y has the mixed linear model structure

$$Y = X\beta + Bb + e$$

where β denotes the fixed effects and b , e denote the random effects. The random vectors b , e are independent with zero mean vectors and covariance matrices $\tau^2 I$ and $\sigma^2 I$ respectively. This model has the structure assumed in Section 2 with $E(Y) = X\beta$ and $Cov(Y) = \tau^2 V + \sigma^2 I$ where $V = BB'$. Suppose the parameters β , τ^2 and σ^2 are all unknown. Then $[\mathcal{V}]$ is polyhedral with generator set $\mathcal{D} = \{I,V\}$.

We illustrate the results in Section 5 by characterizing the admissible class $\mathbf{A}$. Recall that $\mathbf{H}$ is the set of coefficient vectors of the lue's for a given estimable parametric function $\lambda'\beta$. For conciseness, we suppress $\mathbf{H}$ in the best set notation.

According to the discussion following Theorem 5.3, the admissible class $\mathbf{A}$ can be determined in the following way. First list the three ordered partitions of $\mathcal{D}$. From the partition $\{I,V\}$, take $W_1 = D_\alpha = \alpha I + (1-\alpha)V$ for $\alpha \in (0,1)$. From the partition $(\{I\},\{V\})$, take $W_1 = I$ (V can be discarded); and from the partition $(\{V\},\{I\})$ take the sequence $W_1 = V$, $W_2 = I$. Then

$$\mathbf{A} = [\cup_{\alpha\in(0,1)}\mathbf{B}(D_\alpha)] \cup \mathbf{B}(I) \cup \mathbf{B}(V,I)$$

is a characterization of the admissible lue's.

With the exception of $\mathbf{B}(V,I)$, it is well known how to calculate the vectors in $\mathbf{A}$. Let F^- denote any g-inverse for a matrix F . One way of expressing the element in $\mathbf{B}(V,I)$ is $t = GX(X'GX)^-\lambda$ where G is any g-inverse of $XX' + V$ having the property that $\underline{R}(G) = \underline{R}(X,V)$. For example, the Moore-Penrose inverse would be a satisfactory choice. This follows from Lemma 3.2a because $t \in \mathbf{B}(V)$ by Section 2.1 of Rao (1984) and $t \in \underline{R}(X,V)$ by the condition on G .

<u>Remark 6.1</u>. The admissible lue's can be described via normal equations. Let $G(\alpha)$ denote the inverse of $\alpha I + (1-\alpha)V$ for $\alpha \in (0,1]$ and let $G(0)$ be the transpose of G defined in the previous paragraph. Let $\beta(\alpha)$ be any random vector satisfying $X'G(\alpha)X\hat{\beta}(\alpha) = X'G(\alpha)Y$. Then $\{\lambda'\hat{\beta}(\alpha) : \alpha \in [0,1]\}$ describes the entire set of admissible lue's for any estimable parametric function $\lambda'\beta$.

<u>Remark 6.2</u>. Sometimes in mixed models one treats the random effects as fixed and uses least squares for estimation purposes. That is, one assumes the artificial model with mean vector $X\beta + B\theta$ and covariance matrix $\sigma^2 I$. It is interesting to note that if $\lambda'\beta$ is estimable wrt this artificial model, then the best linear unbiased estimator for $\lambda'\beta$ computed from the artificial model coincides with the estimator in $\mathbf{B}(V,I)$.

7. Three variance components.

Suppose Y has the mixed linear model structure

$$Y = X\beta + Bb + Cc + e$$

where β denotes the fixed effects and b , c , e denote the random effects. With the usual assumptions about the random effects, we have the structure assumed in Section 2 with $E(Y) = X\beta$ and $Cov(Y) = \tau^2 V + \pi^2 W + \sigma^2 I$ where $V =$

BB′ and $W = CC'$. Suppose the parameters are all unknown. Then $[\mathscr{V}]$ is polyhedral with generator set $\mathscr{D} = \{I,V,W\}$.

Let us characterize **A** via the remarks following Theorem 5.3. As in the previous section, we suppress **H** in the best set notation. There are thirteen ordered partitions of $\mathscr{D}$. Setting one of the coefficients equal to one in the positive combinations, we can express **A** as the union of:

$$\mathbf{B}(I+\gamma V+\delta W)\ ,\quad \mathbf{B}(V+\delta W,I)\ ,\quad \mathbf{B}(V,I+\delta W)\ ,$$

$$\mathbf{B}(W,I+\gamma V)\ ,\quad \mathbf{B}(V,W,I)\quad \text{and}\quad \mathbf{B}(W,V,I)$$

for all γ , $\delta \geq 0$. Notice here that γ , δ can be zero. This allows the best sets from several partitions to be described simultaneously. For example, the best sets $\mathbf{B}(I+\gamma V+\delta W)$ come from the six partitions that have $I \in \mathscr{K}_1$; and the best sets $\mathbf{B}(V+\delta W,I)$ come from the two partitions $(\{V,W\},\{I\})$ and $(\{V\},\{I\},\{W\})$.

For any nnd matrix V , an expression for the element in $\mathbf{B}(V,I)$ is given in the previous section. For any two nnd matrices V and W , an expression for an element in $\mathbf{B}(V,W)$ is $t = GX(X'GX)^{-}\lambda$ where G can be any g-inverse of $XX' + V + WNW$ and N is the orthogonal projection operator on $\underline{R}(X,V)^{\perp}$. To see this, we require the following facts.

<u>Lemma 7.1</u>. Set $T = XX' + V + WNW$. Then:

(a) $\underline{R}(T) = \underline{R}(X,V,W)$.

(b) $\underline{R}(X'GX) = \underline{R}(X')$.

(c) $\underline{R}(VGX) \subset \underline{R}(X)$.

(d) $\underline{R}(WGX) \subset \underline{R}(X,V)$.

<u>Proof</u>: Let $A = XX' + V$ and $U = WNW$. First note that

$$(*)\qquad \underline{R}(A) \cap \underline{R}(U) = \underline{R}(N)^{\perp} \cap \underline{R}(WN) = \{0\}$$

because W is nnd. For (a) use $\underline{r}(T) = \underline{r}(A) + \underline{r}(U)$ and

$$\underline{r}(U) = \underline{r}(NW) = \underline{r}(W) - \dim[\underline{R}(W) \cap \underline{R}(N)^{\perp}] = \underline{r}(W) - \dim[\underline{R}(W) \cap \underline{R}(A)]$$

to conclude that $\underline{r}(T) = \underline{r}(X,V,W)$. For (b) write $X = TC$. Then

$$\underline{R}(X'GX) = \underline{R}(C'TGTC) = \underline{R}(C'TC) = \underline{R}(C'T) = \underline{R}(X') \ .$$

Because of (*), G is a g-inverse for A . This implies $AGX = X$. So, $VGX = X - XX'GX$, which shows (c). Also, (*) implies $UGA = 0$; hence $NWGX = 0$. So, $\underline{R}(WGX) \subset \underline{R}(N)^{\perp} = \underline{R}(X,V)$, which establishes (d). []

Using (b)-(d), we see that $X't = \lambda$, $Vt \in \underline{R}(X)$, and $Wt \in \underline{R}(X,V)$. Hence $t \in \mathbf{B}(V,W)$ by Lemma 3.2a. This lemma also indicates how an element in $\mathbf{B}(V,W,I)$ can be characterized. In particular, we need $t \in \mathbf{B}(V,W)$ and $t \in \underline{R}(X,V,W)$. The vector t already satisfies the first condition and if we select G so that $\underline{R}(G) = \underline{R}(X,V,W)$, then t will satisfy both conditions. Lemma 7.1a implies that one such choice for G is the Moore-Penrose inverse.

REFERENCES

[1] Azzam, Abdul-Mordy (1981). Admissibility in Linear Models. PhD Thesis, Oregon State University.

[2] Cohen, Arthur (1966). All admissible linear estimates of the mean vector. Annals of Math. Statist. 37, 458-463.

[3] LaMotte, Lynn Roy (1982). Admissibility in linear estimation. Annals of Statistics 10, 245-255.

[4] Olsen, A., Seely, J. and Birkes, D. (1976). Invariant quadratic unbiased estimation for two variance components. Annals of Statistics 4, 878-890.

[5] Rao, C. R. (1976). Estimation of parameters in a linear model. Annals of Statistics 4, 1023-1037.

[6] Rao, C. R. (1984). Inference from linear models with fixed effects: recent results and some problems. In: H.A. David and H.T. David, Eds., Statistics: An Appraisal, The Iowa State University Press, pp. 345-369.

[7] Stein, Charles (1956). Inadmissibility of the usual estimator for the mean of a multivariate normal distribution. Third Berkeley Symposium, Vol. I, pp. 197-206.

[8] Zyskind, George (1967). On canonical forms, nonnegative covariance matrices and best and simple least squares linear estimators in linear models. Annals of Math. Statist. 38, 1092-1109.

Original paper received: 25.03.85
Final paper received: 21.05.86

Paper recommended by J.S. Williams

PROBABILITY AND STATISTICS
Essays in Honor of Franklin A. Graybill
J.N. Srivastava (Editor)

SIMULATED POWER COMPARISONS OF THE ASYMPTOTIC AND NONASYMPTOTIC GOODMAN AND KRUSKAL TAU TESTS FOR SPARSE R BY C TABLES

Kenneth J. Berry
Department of Sociology

and

Paul W. Mielke, Jr.
Department of Statistics
Colorado State University
Fort Collins, CO 80523 USA

ABSTRACT

Simulation methods are used to assess the relative power of asymptotic and nonasymptotic Goodman and Kruskal tests for contingency tables based on tau-a and tau-b statistics. Four different table configurations, two different sample sizes, and a variety of alternative hypotheses are evaluated at six significance levels. Results of the study indicate that the nonasymptotic tau-a and tau-b tests yield probability values very close to those expected under the model specified by the null hypothesis, and the nonasymptotic tau-a and tau-b tests can be substantially more powerful than the comparable asymptotic tests, for sparse contingency tables.

1. INTRODUCTION

It is often necessary to test a null hypothesis of association for two categorical variables, given a sample of n observations arranged in a sparse r by c contingency table. Goodman and Kruskal (1954) present asymmetrical proportional prediction measures, τ_a and τ_b, for the analysis of two cross-classified categorical variables, A and B. While τ_a and τ_b norm properly from 0 to 1, possess a clear and meaningful proportional-reduction-in-error interpretation (Costner, 1965), and are characterized by high intuitive and factorial validity (Hunter, 1973), the sample versions of τ_a and τ_b, t_a and t_b, pose difficulties whenever the null hypothesis under test posits either H_0: τ_a = 0 or H_0: τ_b = 0 (Margolin and Light, 1974). Because in practice, the vast majority of null hypotheses posit either H_0: $\tau_a = 0$ or H_0: $\tau_b = 0$, the problem is pervasive and the applicability of these measures is severely circumscribed. In this paper we report results of Monte Carlo power comparisons of the Margolin and Light (1974) asymptotic test for Goodman and Kruskal's (1954) t_a and t_b and the nonasymptotic test of Berry and Mielke (1985), for sparse r by c contingency tables sampled under restricted multinomial sampling.

2. DEVELOPMENT OF THE PROBLEM

For the sake of brevity, the following remarks are restricted to τ_b and t_b; the parallel argument can easily be developed for τ_a and t_a. Although τ_b was developed by Goodman and Kruskal in 1954, the asymptotic normality was established and an asymptotic variance was given for t_b in 1963 when $0 < \tau_b < 1$

(Goodman and Kruskal, 1963). Unfortunately, the asymptotic variance for t_b given in 1963 was incorrect and it was not until 1972 that the correct asymptotic variance for t_b was reported (Goodman and Kruskal, 1972). Light and Margolin (1971) developed R^2, an analysis of variance technique for categorical response variables. They apparently were unaware that R^2 was identical to t_b and that they had solved the longstanding problem of testing H_0: $\tau_b = 0$. The identity between R^2 and t_b was first recognized by Sarndal (1974) and discussed by Margolin and Light (1974) where they showed that $t_b(n-1)(r-1)$ has an approximate χ^2 distribution with $(r-1)(c-1)$ degrees of freedom under H_0: $\tau_b = 0$, where n is the number of cases in the table, c is the number of categories of the independent variable, and r is the number of categories of the dependent variable. The Goodman and Kruskal (1963, 1972) test when $0 < \tau_b < 1$ and the Margolin and Light (1974) test for $\tau_b = 0$ are both asymptotic tests. A nonasymptotic test under H_0: $\tau_b = 0$ is given by Berry and Mielke (1985).

3. THE NONASYMPTOTIC TEST

Goodman and Kruskal's t_b test statistic for the analysis of r by c contingency tables belongs to a class of statistics known as proportional-reduction-in-error (PRE) statistics. Suppose n objects are cross-classified by two categorical variables, A and B, with A indicating an independent column variable, B indicating a dependent row variable, and n_{ij} indicating the observed frequency of objects in row i and column j, $1 \leq i \leq r$ and $1 \leq j \leq c$. The marginal frequency totals of the r by c cross classification are given by $n_{i\bullet} = \Sigma_{j=1}^{c} n_{ij}$, $n_{\bullet j} = \Sigma_{i=1}^{r} n_{ij}$, and $n = \Sigma_{i=1}^{r} n_{i\bullet} = \Sigma_{j=1}^{c} n_{\bullet j}$. The probability of error in assigning any one of the n objects to the ith category of the dependent row variable is $(n-n_{i\bullet})/n$. Since there is a total of $n_{i\bullet}$ objects in that category, the expected error for the ith row category is $n_{i\bullet}(n-n_{i\bullet})/n$. Summed over all r row categories, the error, conditioned on the dependent variable marginal frequency totals, is given by $\Sigma_{i=1}^{r} n_{i\bullet}(n-n_{i\bullet})/n$.

Now consider the joint distribution of A and B. The probability of error in assigning any of the $n_{\bullet j}$ objects in column j to the n_{ij}th joint category of the independent and dependent variables is $(n_{\bullet j}-n_{ij})/n_{\bullet j}$. Since there is a total of n_{ij} objects in that joint category, the expected error for the n_{ij}th cell category is $n_{ij}(n_{\bullet j}-n_{ij})/n_{\bullet j}$. Summed over all r row categories in the jth column and summed over all c columns, the error, conditioned on the independent and dependent joint frequency distribution, is given by
$\Sigma_{i=1}^{r}\Sigma_{j=1}^{c} n_{ij}(n_{\bullet j}-n_{ij})/n_{\bullet j}$.
Goodman and Kruskal's (1954) t_b test statistic is then defined as the proportional reduction in error conditioned on the marginal frequency totals of the dependent variable by knowledge of the error conditioned on the joint frequency totals of the independent and dependent variables. Thus

$$t_b = \left[\sum_{i=1}^{r} n_{i\cdot}(n-n_{i\cdot})/n - \sum_{i=1}^{r} \sum_{j=1}^{c} n_{ij}(n_{\cdot j}-n_{ij})/n_{\cdot j} \right] / \left[\sum_{i=1}^{r} n_{i\cdot}(n-n_{i\cdot})/n \right]$$

which reduces to the conventional computing formula,

$$t_b = \left(\sum_{i=1}^{r} \sum_{j=1}^{c} n_{ij}^2/n_{\cdot j} - \sum_{i=1}^{r} n_{i\cdot}^2/n \right) / \left(n - \sum_{i=1}^{r} n_{i\cdot}^2/n \right)$$

(Mueller, Schuessler, and Costner, 1977).

For the nonasymptotic test, consider both $n_{i\cdot} \geq 2$ and $n_{\cdot j} \geq 2$, $i = 1,\ldots,r$ and $j = 1,\ldots,c$. If each of the n objects is represented by a binary column vector v of dimension r with a single unit value indicating the row placement of the object and zero otherwise, then an r by n matrix V may be defined by

$$V = [v_1,\ldots,v_n].$$

If $\Delta_{I,J} = 1 - v_I'v_J$, then $\Delta_{I,J}$ is 1 if v_I and v_J are orthogonal and 0 otherwise. Thus, $\Delta_{I,J}$, the difference between objects I and J, is 1 if I and J occur in different categories of the dependent variable (i.e., are different on B) of the realized cross-classification. The variation for categorical responses in the jth category of A is given by

$$\xi_j = \binom{n_{\cdot j}}{2}^{-1} \sum_{I<J} \Delta_{I,J}\psi_j(v_I)\psi_j(v_J),$$

where $\Sigma_{I<J}$ is the sum over all I and J such that $1 \leq I < J \leq n$ and $\psi_j(v_I)$ is an indicator function that is 1 if v_I belongs to the jth category of A and 0 otherwise for $j = 1,\ldots,c$. Thus defined, the value of ξ_j represents the average between-object difference for all $\binom{n_{\cdot j}}{2}$ differences within the jth category of the independent variable and is zero when all objects in the jth category are concentrated in a single category of the dependent variable; that is, the ith category of
B $(i = 1,\ldots,r)$.

The test statistic of interest is the weighted average of the
ξ_j $(j = 1,\ldots,c)$ and is given by

$$\delta_b = \sum_{j=1}^{c} W_j\xi_j,$$

where $\Sigma_{j=1}^{c} W_j = 1$ and $W_j = (n_{\cdot j}-1)/(n-c)$ for $j = 1,\ldots,c$ is, in an analysis variance context, the proportion of the within degrees of freedom contributed by the jth category of the independent variable.

An efficient computation form for δ_b is

$$\delta_b = 1 - \left[\sum_{i=1}^{r} \sum_{j=1}^{c} (n_{ij}^2/n_{\cdot j}) - c \right]/(n-c)$$

and t_b, defined in terms of δ_b, may be obtained from

$$t_b = 1 - (n-c)\delta_b/\left[n - \sum_{i=1}^{r} (n_{i\cdot}^2/n)\right].$$

Under the null hypothesis H_0: $\tau_b = 0$, each of the

$$M_b = n!/(\prod_{j=1}^{c} n_{.j}!)$$

possible permutations of the n objects over the c categories of the independent variable is equally probable with $n_{.j}$ fixed ($j = 1,...,c$). The exact probability of the realized t_b value is the proportion of M_b possible values of t_b equal to or greater than the realized t_b value or, equivalently, the proportion of δ_b values equal to or less than the realized δ_b value.

4. MOMENT APPROXIMATION PROCEDURE

Although a permutation test provides an exact probability value, each application of the test requires the complete enumeration of all possible permutations of the realized data set (Berry, 1982). Consequently, the exact test necessitates unacceptable amounts of computation except for very small samples. An alternative method that avoids the excessive computational demands of an exact test is a moment approximation procedure. A moment approximation procedure approximates the discrete probability distribution of an exact test with a continuous distribution that fits the lower order exact moments of the discrete probability distribution.

If δ_k denotes the kth value among the M_b possible values of δ_b, then, under the null hypothesis, the mean, variance, skewness, and kurtosis of δ_b are given by

$$\mu_b = M_b^{-1} \sum_{k=1}^{M_b} \delta_k,$$

$$\sigma_b^2 = M_b^{-1} \sum_{k=1}^{M_b} \delta_k^2 - \mu_b^2,$$

$$\gamma_{1b} = (M_b^{-1} \sum_{k=1}^{M_b} \delta_k^3 - 3\mu_b\sigma_b^2 - \mu_b^3)/\sigma_b^3,$$

and

$$\gamma_{2b} = (M_b^{-1} \sum_{k=1}^{M_b} \delta_k^4 - 4\mu_b\sigma_b^3\gamma_{1b} - 3\sigma_b^4 - 6\mu_b^2\sigma_b^2 - 4\mu_b^4)/\sigma_b^4,$$

respectively. Efficient computational techniques for obtaining μ_b, σ_b^2, γ_{1b}, and γ_{2b} are contained in a program described by Berry and Mielke (1986). The moment approximation is based on the Pearson type I, III, and VI distributions as described by Berry, Mielke, and Wong (1986). If $T_o = (\delta_o - \mu_b)/\sigma_b$ denotes the observed value of the standardized test statistic, then the approximated probability value is given by

$$\int_{-\infty}^{T_o} f(x)dx$$

and attained with numerical integration using the Newton-Cotes formula for a tenth-order polynomial (Abramowitz and Stegun, 1964; see formula 25.4.20 on page 887) where f(x) is the appropriate standardized density function.

Incidentally, the asymmetric t_a and t_b test statistics are functionally based on $\Sigma_{i=1}^{r}\Sigma_{j=1}^{c}n_{ij}^2/n_{i.}$ and $\Sigma_{i=1}^{r}\Sigma_{j=1}^{c}n_{ij}^2/n_{.j}$, respectively. For comparison, the symmetric χ^2 test statistic for an r by c contingency table is functionally based on $\Sigma_{i=1}^{r}\Sigma_{j=1}^{c}n_{ij}^2/(n_{i.}n_{.j})$.

5. POWER COMPARISONS

Power comparisons between the asymptotic and nonasymptotic t_a and t_b tests were conducted on four different table configurations (r×c = 2×4, 3×4, 4×4, and 5×4) with two sample sizes (n = 20 and 50). Each of the eight combinations of r by c and n was analyzed under both the null hypothesis, where each cell proportion p_{ij}, was set equal to 1/(rc), and a selected alternative hypothesis designed to produce probability values within the range of the significance levels of interest (i.e., 0.10 to 0.0005). Given below are the p_{ij} values for the alternative hypotheses. For n = 20, the 2×4, 3×4, 4×4, and 5×4 tables are, respectively:

.205	.045	.205	.045
.045	.205	.045	.205

.190	.060	.060	.023
.037	.130	.130	.037
.023	.060	.060	.190

.150	.040	.040	.020
.040	.150	.020	.040
.040	.020	.150	.040
.020	.040	.040	.150

.110	.030	.050	.010
.030	.140	.020	.010
.090	.020	.010	.080
.010	.020	.150	.020
.010	.040	.020	.130

For n = 50, the 2×4, 3×4, 4×4, and 5×4 tables are, respectively:

.175	.075	.175	.075
.075	.175	.075	.175

.130	.060	.060	.083
.037	.130	.130	.037
.083	.060	.060	.130

.120	.040	.050	.040
.040	.120	.040	.050
.050	.040	.120	.040
.040	.050	.040	.120

.080	.020	.090	.010
.040	.060	.030	.070
.020	.080	.020	.080
.090	.010	.080	.020
.020	.080	.030	.070

6. SAMPLING PROCEDURE

Each cell in a given table was assigned an integer number between 1 and rc and a cumulative proportion distribution was constructed from the cell proportions, yielding an exhaustive set of bounds for the rc cells: $[0, p_1)$, $[p_1, p_1+p_2)$, ..., $[1-p_{rc}, 1)$, where $[a, b)$ indicates $[a \leq x < b]$ and x is a continuous variable. A sequence of n uniform unit-interval pseudorandom numbers (x_i, i = 1,...,n) was generated on a CDC CYBER-205 using the FORTRAN-77 library function RANF. Since the cell bounds provide a disjoint and exhaustive covering of [0, 1), each x_i falls within precisely one of the rc cell bounds, whereupon the corresponding cell frequency was increased by one. This procedure was repeated to create 1000 random contingency tables, each generated under the same hypothesis, with fixed r, c, and n, but with varying marginal frequency totals. This multinomial sampling procedure models the most common application of the t_a and t_b test statistics to r by c contingency tables, where n objects are randomly sampled from a population and classified on each of two polytomies, yielding an r by c table in which the row and column marginal frequency totals vary from sample to sample. Since the definitions of the t_a and t_b test statistics require $n_{i.} \geq 2$ and $n_{.j} \geq 2$ (i = 1,...,r and j = 1,...,c), any Monte Carlo table not satisfying this restriction was discarded and a new table was generated to replace it.

For each random table, asymptotic and nonasymptotic probability values were calculated for both the t_a and t_b test statistics. The probability values were ordered from low to high and the power of the asymptotic and nonasymptotic tests was determined by the proportion of probability values less than the selected significance levels (i.e., 0.10, 0.05, 0.01, 0.005, 0.001, and 0.0005). Table 1 documents the fit between the model specified by the null hypothesis and the method, where the values for power are combined for both sample sizes, over all four table configurations, and for t_a and t_b considered together. An examination of Table 1 reveals that the 16,000 nonasymptotic t_a and t_b probability values provide an excellent approximation to the model, within the significance levels of interest (i.e., 0.10 to 0.0005). As expected, the asymptotic t_a and t_b probability values are less satisfactory.

Table 1

Power comparisons of the nonasymptotic and asymptotic t_a and t_b tests, under the null hypothesis, combined over four table configurations and two sample sizes

Significance level	Nonasymptotic test	Asymptotic test
.10	.0945	.0826
.05	.0464	.0375
.01	.0097	.0058
.005	.0044	.0021
.001	.0008	.0003
.0005	.0003	.0001

7. DISCUSSION

Table 2 contains the asymptotic and nonasymptotic power comparisons for t_a, under the null hypothesis; Table 3 contains the same power comparisons for t_b. Examination of Tables 2 and 3 reveals that, although the asymptotic t_a and t_b tests approximate the model quite well for r by c tables with relatively large sample sizes, with rc relatively small, and with relatively high significance levels (e.g., the 2×4 tables with n = 50 and $\alpha = 0.10$), they quickly generate as n decreases, rc increases, and/or the significance level decreases. For the results summarized in Tables 2 and 3, the asymptotic tests have generally underestimated the significance levels over all table configurations and at both sample sizes. On the other hand, the nonasymptotic tests provide a close approximation to the model specified by the null hypothesis in all instances and are well behaved even with small sample sizes, various configurations of r and c, and low significance levels.

Table 2

Power comparisons of the nonasymptotic (N) and asymptotic (A) t_a tests, under the null hypothesis

Table/test	Significance level .10	.05	.01	.005	.001	.0005
n = 20						
2 × 4 N	.100	.052	.014	.006	.001	.001
A	.094	.049	.009	.005	.001	.001
3 × 4 N	.087	.044	.015	.010	.003	.001
A	.070	.035	.011	.004	.001	.000
4 × 4 N	.097	.051	.011	.006	.001	.001
A	.078	.040	.006	.002	.000	.000
5 × 4 N	.110	.048	.013	.008	.002	.001
A	.074	.023	.006	.004	.001	.000
n = 50						
2 × 4 N	.096	.046	.007	.001	.000	.000
A	.095	.044	.006	.000	.000	.000
3 × 4 N	.106	.046	.008	.003	.000	.000
A	.099	.044	.008	.003	.000	.000
4 × 4 N	.086	.037	.006	.002	.000	.000
A	.076	.030	.007	.002	.000	.000
5 × 4 N	.084	.048	.003	.002	.000	.000
A	.077	.041	.002	.001	.000	.000

Table 3

Power comparisons of the nonasymptotic (N) and asymptotic (A) t_b tests, under the null hypothesis

Table/test	Significance level .10	.05	.01	.005	.001	.0005
n = 20						
2 × 4 N	.093	.053	.010	.004	.001	.000
A	.088	.035	.002	.001	.000	.000
3 × 4 N	.090	.043	.013	.008	.001	.000
A	.075	.034	.006	.002	.000	.000
4 × 4 N	.094	.050	.014	.006	.002	.001
A	.072	.036	.005	.003	.001	.000
5 × 4 N	.100	.049	.012	.004	.002	.000
A	.078	.031	.005	.002	.000	.000
n = 50						
2 × 4 N	.098	.047	.006	.000	.000	.000
A	.096	.043	.003	.000	.000	.000
3 × 4 N	.097	.054	.011	.007	.000	.000
A	.089	.047	.009	.003	.000	.000
4 × 4 N	.082	.034	.008	.003	.000	.000
A	.076	.032	.006	.002	.000	.000
5 × 4 N	.092	.040	.004	.001	.000	.000
A	.084	.036	.002	.000	.000	.000

Tables 4 and 5 provide the same summary information as Tables 2 and 3, except that the power comparisons were run under the alternative hypotheses. In these cases, attempts were made to detect actual differences in the populations. The specific alternative hypotheses were chosen to provide a large number of probability values between 0.10 and 0.0005. Too extreme alternative hypotheses yield uninformative probability values of 1.0 for the larger significance levels (e.g., 0.10 and 0.05), while too moderate alternative hypotheses yield uninformative probability values of 0.0 for the smaller significance levels (e.g., 0.001 and 0.0005). An examination of Tables 4 and 5 reveals that, while the differences between the asymptotic and nonasymptotic t_a and t_b tests are very small for r by c tables with relatively large sample sizes, with rc relatively small, and with relatively large significance levels (e.g., the 2×4 tables with n = 50 and $\alpha = 0.10$), they quickly increase as n increases, rc increases, and/or the significance level decreases. While the nonasymptotic t_a and t_b tests provide more power than the asymptotic t_a and t_b tests for the comparisons in Tables 4 and 5, the important distinction lies in the substantial magnitude of many of the differences.

Table 4

Power comparisons of the nonasymptotic (N) and asymptotic (A) t_a tests, under the alternative hypotheses

Table/test	Significance level .10	.05	.01	.005	.001	.0005
n = 20						
2 × 4 N	.799	.697	.432	.345	.164	.125
A	.796	.678	.389	.295	.111	.055
3 × 4 N	.738	.613	.361	.278	.154	.116
A	.704	.565	.294	.219	.106	.065
4 × 4 N	.784	.699	.459	.372	.230	.185
A	.755	.635	.363	.280	.139	.086
5 × 4 N	.859	.783	.543	.452	.287	.229
A	.827	.706	.394	.312	.146	.109
n = 50						
2 × 4 N	.782	.672	.426	.330	.168	.115
A	.781	.666	.417	.316	.143	.106
3 × 4 N	.725	.593	.327	.245	.115	.081
A	.723	.576	.312	.231	.098	.068
4 × 4 N	.809	.710	.484	.405	.250	.205
A	.800	.696	.463	.376	.226	.180
5 × 4 N	.910	.826	.597	.471	.262	.199
A	.899	.811	.542	.424	.205	.149

8. CONCLUSION

The Monte Carlo power simulations reported here show that for many applications likely to arise in practice, the nonasymptotic t_a and t_b tests (Berry and Mielke, 1985) provide substantial advantages over the Margolin and Light (1974) asymptotic t_a and t_b tests. Nonasymptotic tests of the null hypothesis should be applied to those cross-classification tables in which the asymptotic tests would be questionable. The simulations have demonstrated the considerable power of the nonasymptotic t_a and t_b tests for combinations of four table configurations, two sample sizes, six significance levels, and a variety of alternative hypotheses. Moreover, they have shown that the nonasymptotic t_a and t_b tests provide an excellent approximation to the model specified by the null hypothesis, under the same conditions. The use of the nonasymptotic t_a and t_b tests is, of course, not limited only to tables with small n, small rc, and small significance levels; the nonasymptotic t_a and t_b tests may be routinely employed in any application where the asymptotic t_a and t_b tests would ordinarily be utilized.

Table 5

Power comparisons of the nonasymptotic (N) and asymptotic (A) t_b tests, under the alternative hypotheses

Table/test	Significance level .10	.05	.01	.005	.001	.0005
n = 20						
2 × 4 N	.795	.711	.466	.352	.218	.160
A	.774	.643	.323	.203	.020	.017
3 × 4 N	.736	.620	.358	.278	.156	.113
A	.700	.553	.258	.179	.076	.051
4 × 4 N	.774	.676	.452	.380	.221	.177
A	.743	.610	.362	.270	.133	.081
5 × 4 N	.854	.757	.526	.442	.279	.228
A	.811	.694	.427	.332	.183	.138
n = 50						
2 × 4 N	.773	.666	.435	.342	.177	.125
A	.773	.652	.406	.293	.119	.090
3 × 4 N	.711	.575	.331	.252	.118	.080
A	.700	.561	.303	.219	.089	.059
4 × 4 N	.815	.716	.487	.406	.258	.203
A	.806	.700	.464	.380	.224	.178
5 × 4 N	.888	.804	.577	.477	.271	.202
A	.883	.789	.542	.433	.241	.177

REFERENCES

[1] Abramowitz, M. and I. A. Stegun (1964). Handbook of Mathematical Functions. National Bureau of Standards, Washington, D.C.

[2] Berry, K. J. (1982). Enumeration of all permutations of multi-sets with fixed repetition numbers. Appl. Statist. 31, 169-173.

[3] Berry, K. J. and P. W. Mielke (1985). Goodman and Kruskal's tau-b statistic: a nonasymptotic test of significance. Sociological Meth. Res. 13, 543-550.

[4] Berry, K. J. and P. W. Mielke (1986). Goodman and Kruskal's tau-b statistic: a FORTRAN-77 subroutine. Educ. Psych. Meas. 46, 645-649.

[5] Berry, K. J., P. W. Mielke and R. K. W. Wong (1986). Approximate MRPP p-values obtained from four exact moments. Commun. Statist.-Simulation Comput. 15, 581-589.

[6] Costner, H. L. (1965). Criteria for measures of association. Amer. Sociological Rev. 30, 341-353.

[7] Goodman, L. A. and W. H. Kruskal (1954). Measures of association for cross classifications. J. Amer. Statist. Assoc. 49, 732-764.

[8] Goodman, L. A. and W. H. Kruskal (1963). Measures of association for cross classifications, III: approximate sampling theory. J. Amer. Statist. Assoc. 58, 310-364.

[9] Goodman, L. A. and W. H. Kruskal (1972). Measures of association for cross classifications, IV: simplification of asymptotic variances. J. Amer. Statist. Assoc. 67, 415-421.

[10] Hunter, A. A. (1973). On the validity of measures of association: the nominal-nominal, two-by-two case. Amer. J. Sociology 79, 99-109.

[11] Light, R. J. and B. H. Margolin (1971). An analysis of variance for categorical data. J. Amer. Statist. Assoc. 66, 534-544.

[12] Margolin, B. H. and R. J. Light (1974). An analysis of variance for categorical data, II: small sample comparisons with chi square and other competitors. J. Amer. Statist. Assoc. 69, 755-764.

[13] Mueller, J. H., K. F. Schuessler, and H. L. Costner (1977). Statistical Reasoning in Sociology. 3rd ed. Houghton-Mifflin, Boston.

[14] Särndal, C. E. (1974). A comparative study of association measures. Psychometrika 39, 165-187.

Original paper received: 03.06.86
Final paper received: 17.04.87

Paper recommended by M.M. Siddiqui

PROBABILITY AND STATISTICS
Essays in Honor of Franklin A. Graybill
J.N. Srivastava (Editor)

SCHEMES EXHIBITING HURST BEHAVIOR

Duane C. Boes
Colorado State University

Abstract

An exposition of several schemes capable of mimicking the growth behavior, which has been observed in the data of various geophysical phenomena, of the range of partial sums is given.

Keywords: *Hurst phenomenon, range of partial sums, fractional Brownian noise, stable domain of attraction, shifting level processes.*

1. Introduction

This paper is basically expository and by no means exhaustive. Its purpose is to describe the so-called Hurst phenomenon and several models/interpretations that have the potential to exhibit/explain such.

For the most part only discrete-time indexed series will be considered. In order to provide a link to the real world, net inflows of water into a reservoir will be used as a generic example. Let X_t denote the net inflow during time increment t for $t = 1,2,\ldots$. Define $S_0 = 0$ and

$$S_n = \sum_1^n X_t, \tag{1.1}$$

the partial sum process associated with $\{X_t\}_{t=1}^{\infty}$. S_n can be thought of as the reservoir level at time epoch n if the reservoir neither spills nor goes dry during $t = 1,\ldots,n$. Adjusted partial sums are defined as

$$S^*_{j,n} = S_j - j\,\overline{X}_n \tag{1.2}$$

for $j = 0,1,\ldots,n$ and $n = 1,2,\ldots$ where the $\overline{X}_n = (1/n)\sum_1^n X_t$. The effect of adjusting is to return the partial sums to zero at time epoch n. Various ranges can be defined in terms of these partial sums.

Definition 1.1 The *raw range* is defined to be

$$R_n = \max(S_0,\ldots,S_n) - \min(S_0,\ldots,S_n), \tag{1.3}$$

the *adjusted range* as

$$R^*_n = \max(S^*_{0,n},\ldots,S^*_{n,n}) - \min(S^*_{0,n},\ldots,S^*_{n,n}) \tag{1.4}$$

and the *rescaled adjusted range* as

$$R^{**}_n = R^*_n/D_n \tag{1.5}$$

where $D^2_n = (1/n)\sum_1^n (X_i - \overline{X}_n)^2$.

These ranges are sometimes used in reservoir sizing. For example, if a reservoir at time zero is at level equal to $-\min(S_0,\ldots,S_n)$ and has size R_n then the reservoir would neither spill nor go dry during time $0,1,\ldots,n$. Other quantities, such as the maximum accumulated deficit, defined to be $\max(0, M_1-S_1, M_2-S_2,\ldots,M_n-S_n)$ are also used in reservoir design, but they will not be considered here, where $M_j = \max(S_0,\ldots,S_j)$.

The Hurst phenomenon, defined precisely below, deals with the growth relative to n, of one of the ranges defined above. Theoretically at least, by modeling the background X_t, $t = 1,2,\ldots$, the growth behavior of these ranges can be derived.

Definition 1.2. A stochastic process $\{X_t\}_{t=1}^{\infty}$ is said to exhibit *Hurst behavior* (possess a *Hurst effect* or satisfy the *Hurst phenomenon*) for the rescaled adjusted range if the sequence $\{R_n^{**}/n^H\}$, where $.5 < H < 1$, converges in law (distribution) as $n \to \infty$ to a limit random variable which may be degenerate but not almost surely zero. H is known as the *Hurst exponent*.

In Definition 1.2, growth of the rescaled adjusted range is expressed in terms of convergence in distribution; such growth could alternatively be expressed in terms of convergence of the mean.

Definition 1.3 A stochastic process $\{X_t\}_{t=1}^{\infty}$ is said to satisfy the *Hurst phenomenon in the mean* for the rescaled adjusted range if $\mathcal{E}[R_n^{**}/n^H]$ converges to a positive constant as $n \to \infty$.

A Hurst effect for either the raw range or adjusted range can be similarly defined by replacing the rescaled adjusted range in Definitions 1.2 and 1.3 by the appropriate alternate range.

The concept of a Hurst effect was engendered by studies of Harold Edwin Hurst reported in Hurst (1951), and in Hurst, Black, and Simaika (1965), where it was noted that for many different geophysical time series, including streamflow, rainfall, temperature, atmospheric pressure, tree rings, and mud varves, the rescaled adjusted range seemed to grow like n^H with H larger than .5. Hurst (1951) plotted log R_n^{**} versus log n, for 75 different phenomena, with various n values, and the 690 computed Hurst slopes H, yielded a mean of .729, standard deviation of .092, and range .46 to .96. Hurst's empirical work, coupled with that of Feller (1951), where he showed for independent and identically distributed $\{X_j\}$ that $R_n^{**}/n^{1/2}$ converged in distribution to a non-zero random variable as $n\to\infty$, suggesting that theory supports $n^{1/2}$ growth rather than n^H, $H > 1/2$, sparked practitioners and theoreticians alike to seek models/interpretations that would exhibit/explain such behavior.

Several possible explanations that have been posed will be discussed. These are:

(i) the $\{X_t\}$ have heavy-tailed marginal distributions,

(ii) the $\{X_t\}$ are appropriately nonstationary,

(iii) the n^H growth is transcient or preasymptotic with ultimate asymptotic growth of $n^{1/2}$,

(iv) the $\{X_t\}$, though stationary, possess an appropriate strong serial dependence, and

(v) various combinations of these.

Another possible explanation that has been considered, but not to be explored herein, is that of highly skewed marginal distributions for the $\{X_t\}$.

The remainder of the paper is partitioned into sections, one for each of the above first four possible explanations, preceded by a section that commences with a discussion of Hurst's studies and presents several theoretical results where the growth of the range is $n^{1/2}$.

2. Empirical Evidence versus Early Analytics

Hurst's work was motivated by his long term study of the Nile river basin including his involvement in the planning and operation of the Aswan Dam. A delightful mix of theoretical, Monte Carlo, and empirical results appears in Hurst (1951). On the theoretical side, he derived the expected value of the raw range, conditioned on a return to zero at time n, when the partial sums are a simple symmetric random walk. Specifically, if the $\{X_t\}$ are assumed to be independent and identically distributed with distribution given by

$$P[X_t = -1] = P[X_t = +1] = 1/2,$$

he showed that

$$\mathcal{E}[R_{2n} | S_{2n} = 0] = \frac{2^{2n}}{\binom{2n}{n}} - 1 \qquad (2.1)$$

using a path counting technique. He further noted that

$$\mathcal{E}[R_{2n} | S_{2n} = 0] \sim \sqrt{\pi n} - 1 \sim \sqrt{\pi n} \qquad (2.2)$$

where "~" reads "behaves like" and means that the ratio of the two sides is one in the limit as $n \to \infty$. Equation (2.2) represents Hurst's analytical foray into the growth behavior of a range. One suspects that Hurst considered this conditional range believing that its behavior would be close to that of the adjusted range, which it is (see Equation (2.11) below). Using a similar approach, it can be shown that

$$\mathcal{E}[R_{2n}] = (4n+1)\binom{2n}{n}\left(\frac{1}{2}\right)^{2n} - 1 \qquad (2.3)$$

when the partial sums are a simple symmetric random walk. Here

$$\mathcal{E}[R_{2n}] \sim \frac{(4n+1)}{\sqrt{\pi n}} - 1 \sim 4\sqrt{\frac{n}{\pi}} \qquad (2.4)$$

and again the $n^{1/2}$ growth appears.

Hurst supplemented his analytical work with Monte Carlo experiments involving tossing sixpences and cutting cards from a probability pack of cards composed of sixty-six cards, where each card was labelled with one of the following numbers: -9, -7, -5, -3, -1, +1, +3, +5, +7, or +9; there were thirteen of each of the ones, ten of the threes, six of the fives, three of the sevens and one of the nines. The distribution of the cards in the pack provided an approximation to a normal distribution with mean zero. He found close agreement between his Monte Carlo experiments and his analytical derivations and proceeded to look at natural phenomena with river discharges his primary concern. He noted that natural data had a greater tendency to have clusters of high or low values than his Monte Carlo experiments exhibited and that this was the main difference between the two. For his empirical studies Hurst collected a large number of geophysical time series such as river discharges, lake and river levels, rainfall, temperature, pressure, tree rings, layers of mud (varves), sunspot numbers, and wheat prices, which altogether constituted 75 time series of varying length which were divided into 690 cases of lengths ranging from 30 to 2,000. Later Hurst et al. (1965, 1966) extended his experiments to 120 time series with a total of 872 cases of lengths varying from

10 to 2,000. Since Hurst's analytical and Monte Carlo studies had shown that for large n the range grew like $n^{1/2}$, he conjectured n^K behavior for natural time series, possibly expecting to find similar $n^{1/2}$ growth behavior. Toward this end, from the data he computed a value of the rescaled adjusted range, say r_n^{**}, and set

$$r_n^{**} = (\tfrac{n}{2})^K. \tag{2.5}$$

He used Equation (2.5) rather than the more general fit equation $r_n^{**} = c\, n^K$ since for n = 2, $r_n^{**} = 1$. He actually estimated the exponent K via

$$K = \log(r_n^{**})/\log(n/2) \tag{2.6}$$

for each of the 690 (Hurst 1951) individual values of r_n^{**} (837 cases in Hurst et.al. 1965) obtaining an arithemetic mean value of K = 0.729, a standard deviation of 0.092 and a range of individual values of 0.46 to 0.96. By specific phenomenon, he got K = 0.75 for stream levels and discharges, K = 0.70 for rainfall and temperature, and K = 0.80 for tree rings. Plots on log-log paper with varying n within phenomenon actually looked quite straight and provided rather convincing evidence that natural phenomenon behaved differently than what Hurst's analytics and Monte Carlo experiments had indicated. The pursuit of possible explanations dominates the remaining sections, after a brief review of some of the early results indicating $n^{1/2}$ growth of the various ranges.

Feller (1951), aware of Hurst's results, derived the asymptotic distributions of R_n and R_n^* under the assumption that the net inputs $\{X_t\}$ are independent and identically distributed with mean zero and finite variance σ^2. In particular, he found that for the raw range

$$\mathcal{E}[R_n] \sim 2\sigma(2/\pi)^{1/2} n^{1/2}, \tag{2.7}$$

$$\mathrm{var}[R_n] \sim 4\sigma^2[\ell n2 - (2/\pi)]n, \tag{2.8}$$

and

$$\lim_{n\to\infty} P[R_n/\sigma n^{1/2} \le r] = F_R(r), \tag{2.9}$$

where $F_R(\cdot)$ has density

$$f_R(z) = [8 \sum_{j=1}^{\infty}(-1)^{j+1} j^2\, \phi(jz)] I_{(0,\infty)}(z), \tag{2.10}$$

and $\phi(\cdot)$ is the standard normal density. And for the adjusted range he found that

$$\mathcal{E}[R_n^*] \sim \sigma(\pi/2)^{1/2} n^{1/2}, \tag{2.11}$$

$$\mathrm{var}[R_n^*] \sim \sigma^2[(\pi^2/6) - (\pi/2)]n, \tag{2.12}$$

and

$$\lim_{n\to\infty} P[R_n^*/\sigma n^{1/2} \le r] = F_{R*}(r), \tag{2.13}$$

when $F_{R*}(\cdot)$ has density

$$f_{R*}(r) = \sum_{j=1}^{\infty}[8j^2 r(4j^2r^2 - 3) \exp(-2\, j^2r^2)] I_{(0,\infty)}(r). \tag{2.14}$$

Equation (2.14) is the version given by Gomide (1978). Feller gave a longer different expression. These limiting distribution results for the raw range were proved by observing that the partial sums S_n behave like W(t) for t = n and large n, where $W(t)$, $t \geq 0$, is a Wiener process or Brownian motion. Similarly, the results for the adjusted range were obtained by comparing the adjusted partial sums to the tied-down Wiener process or Brownian bridge.

It is now well known that the above results for the independently and identically distributed case can be extended to a rather broad class of stationary processes via weak convergence theory and functional central limit theorems. A popular exposition of such notions is that of Billingsley (1968). Troutman (1976) and (1978) and Siddiqui (1976) consider such results in the language of storage theory employed here.

The basic results are very similar to those given in Equations (2.7) through (2.14). Assume now that the net inputs $\{X_t\}$ are stationary, dependent belonging to the Brownian domain of attraction then Equations (2.7), (2.8), (2.9), (2.11), 2.12) and (2.13) remain intact with σ replaced by γ, where

$$\gamma^2 = \gamma_o + 2 \sum_{k=1}^{\infty} \gamma_k, \tag{2.15}$$

and γ_k is the lag k covariance of the process $\{X_t\}$. If, in addition to $\{X_t\}$ belonging to the Brownian domain of attraction, D_n^2 converges in probability to $\gamma_o = \sigma^2 = \text{var}[X_t]$ then

$$\lim_{n\to\infty} P[R_n^{**}/\beta n^{1/2} \leq r] = F_{R*}(r) \tag{2.16}$$

where $F_{R*}(\cdot)$ is as in Equation (2.13) and

$$\beta^2 = \gamma^2/\sigma^2 = 1 + 2 \sum_{k=1}^{\infty} \rho_k \tag{2.17}$$

and ρ_k is the lag k correlation of the process $\{X_t\}$. Troutman (1978) extends these results to certain periodic dependent processes $\{X_t\}$ that are capable of modeling seasonality. The crux of the above asymptotics is that for net inputs $\{X_t\}$ belonging to the Brownian domain of attraction, the growth behavior is necessarily $n^{1/2}$ and not the n^H for $H > 1/2$ that Hurst had observed in natural phenomena.

There are several exact formulas for the expected value of the various ranges that show $n^{1/2}$ growth behavior; some are given below. Chronologically, the first of these is

$$\mathcal{E}[R_n] = \sqrt{\frac{2}{\pi}} \sum_{i=1}^{n} i^{-1/2}, \tag{2.18}$$

given by Anis and Lloyd (1953), for $\{X_t\}$ independent standard normal random variables. The adjusted range counterpart is

$$\mathcal{E}[R_n^*] = \sqrt{\frac{n}{2\pi}} \sum_{i=1}^{n-1} [i(n-i)]^{-1/2}, \tag{2.19}$$

given by Solari and Anis (1957), again for independent standard normal random variables.

Boes and Salas (1973), capitalizing on an identity of Spitzer (1956), gave

$$\mathcal{E}[R_n] = \sum_{i=1}^{n} \mathcal{E}[|\bar{X}_i|] \tag{2.20}$$

and

$$\mathcal{E}[R_n^*] = \sum_{i=1}^{n} \mathcal{E}[|\bar{X}_i - \bar{X}_n|] \tag{2.21}$$

where $\bar{X}_i = (1/i) \sum_{t=1}^{i} X_t$, for $X_1,\ldots,X_n$ any set of exchangeable random variables for which the indicated expectations exist. Utilizing the same technique, Anis and Lloyd (1976) gave

$$\mathcal{E}[R_n^{**}] = \frac{\Gamma[(n-1)/2]}{\Gamma[n/2]\pi^{1/2}} \sum_{i=1}^{n-1} [(n-i)/i]^{1/2} \tag{2.22}$$

for independent and identically distributed normal variables $X_1,\ldots,X_n$.

Equation (2.22) is the only known exact formula for $\mathcal{E}[R_n^{**}]$ for general n. Formulas for small n have been derived, for example, a tedious direct calculation yields

$$\mathcal{E}[R_3^{**}] = \frac{3\sqrt{1-\rho_2}}{\sqrt{2}\,\pi} \left\{ \frac{1}{\sqrt{\rho_1-\rho_2}} \arctan \left[\sqrt{\frac{3(\rho_1-\rho_2)}{3-4\rho_1+\rho_2}} \right] + \left[\left(\frac{1}{6}\right) \sqrt{\frac{3-4\rho_1+\rho_2}{(\rho_1-\rho_2)(1-\rho_2)}} \right] \log \left[\frac{3+2\rho_1-5\rho_2+3\sqrt{3(\rho_1-\rho_2)(1-\rho_2)}}{3+2\rho_1-5\rho_2-3\sqrt{3(\rho_1-\rho_2)(1-\rho_2)}} \right] \right\} \tag{2.23}$$

for X_1, X_2, X_3 stationary normal with $\mu = \mathcal{E}[X_i]$, $\sigma^2 = \mathrm{var}[X_i]$, $\rho_1 = \rho(X_1,X_2) = \rho(X_2,X_3)$, $\rho_2 = \rho(X_1,X_3)$ and $\rho_1 > \rho_2$. Similar formulas for small n for the raw range and adjusted range are given in Troutman (1974) for normally distributed summands X_1, X_2, X_3, and X_4.

In summary, natural phenomena suggest that the range grows as n^H for $H > 1/2$, whereas nice traditional models exhibit $n^{1/2}$ growth. Further, some exact results for arbitrary n are known for the expected value of the three ranges.

3. Heavy-tailed Marginal Distributions

Since the ranges are functions of the partial sums of the net inputs $\{X_t\}$ it was surmised early on that heavy-tailed marginal distributions for X_t could effect n^H, $H > 1/2$, growth. After all, the essential assumption in the $n^{1/2}$ results listed in the last section was that the $\{X_t\}$ have a Brownian domain of attraction which is compatible with light-tailed marginal distributions.

Moran (1964) showed that for $X_1, X_2, \ldots$ independent random variables with common characteristic function $\psi(t) = \exp(-|t|^\alpha)$ where $1 < \alpha \le 2$,

$$\mathcal{E}[R_n] = \mathcal{E}[|X_1|] \sum_{i=1}^{n} i^{(1/\alpha)-1}. \tag{3.1}$$

Here the X_t's have symmetric stable distribution with characteristic exponent α and finite first moment. It is readily seen from Equation (3.1) that $\mathcal{E}[R_n]$ is asymptotically proportional to $n^{1/\alpha}$ and as α ranges over the interval (1,2), n^H growth for any $1/2 < H < 1$ is obtainable and thus a simple model that preserves Hurst behavior in the mean for the raw range does exist. Similarly, Boes and Salas (1973), for the same independent symmetric stable distributions, derived

$$\mathcal{E}[R_n^{*}] = \mathcal{E}[|X_1|] \sum_{i=1}^{n} \{i[(1/i) - (1/n)]^\alpha + (n-i)(1/n)^\alpha\}^{1/\alpha} \tag{3.2}$$

which again leads to

$$\mathcal{E}[R_n^*] \sim \text{constant times } n^{1/\alpha} \tag{3.3}$$

giving Hurst behavior in the mean for the adjusted range. That similar behavior does not carry over to the rescaled adjusted range is given in Mandelbrot and Taqqu (1979) or Leiva and Boes (1984), where it is shown, for $\{X_t\}_{t=1}^{\infty}$ independent and identically distributed belonging to the domain of attraction of a stable random variable with characteristic exponent α, $1 < \alpha < 2$, that $R_n^{**}/n^{1/2}$ converges in law, giving $n^{1/2}$ growth. So, although the numerator of the right hand side of $R_n^{**} = R_n^*/D_n$ grows like $n^{1/\alpha}$, the denominator also grows with n at a rate that yields $n^{1/2}$ growth for the ratio.

It will be seen in Section 6 that heavy-tailed distributions in the construction of a shifting level model can be used to preserve the Hurst effect for the rescaled adjusted range.

The assumption of independence of the stably distributed inputs can be relaxed. Davis (1983) considers stable limits for partial sums of dependent random variables. The weak convergence results given in the previous section for the Brownian domain of attraction can be replaced by similar results for the stable domain of attraction but such will not be pursued here.

We have seen that heavy-tailed marginals, with or without serial dependence, give Hurst behavior for raw and adjusted range but not the rescaled adjusted range.

4. Nonstationary Processes

Bhattacharya, Gupta, Waymire (1983) constructed a nonstationary process capable of exhibiting a Hurst effect by adding to a weakly dependent stationary process a deterministic, slowly damping sequence of the form $f(t) = c(m + t)^{\beta}$. When $-1/2 < \beta < 0$, the Hurst exponent H equalled $1 + \beta$. In this case, weakly dependent means that the process is ergodic with summable correlations such that the process is attracted to ordinary Brownian motion. The authors in fact give necessary and sufficient conditions for preservation of a Hurst effect in the case of weakly dependent stationary process perturbed by a trend. Their investigation was motivated by the observed presence of monotonic trends and weak dependence in certain geophysical records.

An argument for the inclusion of a monotonic trend term can be advanced from a physical viewpoint. Consider, for example, mud varves measured at the bottom of a lake in a enclosed basin. The easily eroded silt may be deposited in the lower layers and then as time proceeds the silt gets progressively harder to erode and consequently the layers tend to get narrower as time advances. Similarly, certain climatic conditions may lead to a warming trend which in turn would produce a concomitant increase in temperature readings.

Specifically, Bhattacharya et.al. (1983) considered the model for the input process $\{X_t\}$ to be

$$X_t = Y_t + f(t), \quad t = 1,2,\ldots \tag{4.1}$$

where $\{Y_t\}$ was a stationary sequence of weakly dependent random variables having zero mean and $f(\cdot)$ was an arbitrary real-valued function. Necessary and sufficient conditions on $f(\cdot)$ so that the ranges of the partial sums of the $\{X_t\}$ process would possess a Hurst effect were given. Trend of the form $f(t) = c(m+t)^{\beta}$, $t = 1,2,\ldots$ where m and c were positive constants and $-1/2 < \beta < 0$ yielded a Hurst exponent of $H = 1 + \beta$ for the rescaled adjusted range in the sense that R_n^{**}/n^H converged in law as $n \to \infty$ to a limit random variable that was not almost surely zero. One intuitively suspects that the deterministic trend could be replaced with an appropriate stochastic trend term. Whence, a simple mathematical model, incorporating nonstationarity, that preserves Hurst behavior does exist.

5. Pre-asymptotic Interpretation

Known exact formulas for the expected value of the range, for the various ranges, are given in Equations (2.3), (2.18), (2.19), (2.20), (2.21), (2.22), (3.1), and (3.2). None of these exact formulas in n are proportional to their asymptotic n^H counterparts. So an obvious query is: how fast do the exact formulas approach their asymptotic values? And might not the slope of the exact formula in a log-log scale be greater than 1/2, and possibly near the slopes Hurst observed in natural data for the range of sample sizes he considered? And if so the so-called Hurst behavior could be interpreted as pre-asymptotic or transient. Several investigators have considered such a possibility, and a popular attack has been to assume serial dependence of the input process $\{X_t\}$ in such a way so as to delay the ultimate $n^{1/2}$ behavior. Another attack that has been explored is the skewness of the marginal distribution of the X_t's but that approach will not be pursued here.

For the case where the $\{X_t\}$ are independent standard normal random variables the exact and asymptotic formulas for the expected values of the ranges are given by, respectively,

$$\mathcal{E}[R_n] = \sqrt{\frac{2}{\pi}} \sum_{i=1}^{n} i^{-1/2} \sim 2\sqrt{\frac{2}{\pi}}\, n^{1/2}, \tag{5.1}$$

$$\mathcal{E}[R_n^*] = \sqrt{\frac{n}{2\pi}} \sum_{i=1}^{n-1} [i(n-i)]^{-1/2} \sim \sqrt{\frac{\pi}{2}}\, n^{1/2}, \tag{5.2}$$

$$\mathcal{E}[R_n^{**}] = \frac{\Gamma[(n-1)/2]}{\Gamma[n/2]\pi^{1/2}} \sum_{i=1}^{n-1} [(n-i)/i]^{1/2} \sim \sqrt{\frac{\pi}{2}}\, n^{1/2} \tag{5.3}$$

These formulas are plotted in Figures 5.1 a, b and c on a log - log scale. Using the exact formulas, a local slope, say H_n, defined as

$$H_n = \frac{\log(\mathcal{E}[R_{n+1}]) - \log(\mathcal{E}[R_{n-1}])}{\log(n+1) - \log(n-1)}, \tag{5.4}$$

was computed. Similarly, H_n^* and H_n^{**} were obtained by replacing R_n by R_n^* and R_n^{**} in Equation (5.4). In the case of the rescaled adjusted range, the Hurst slope, say K_n defined as

$$K_n = \log(\mathcal{E}[R_n^{**}])/\log(n/2), \tag{5.5}$$

was also computed. (See Equation (2.6).) These slopes are plotted in Figures 5.2 a, b and c. One can see that there is a definite pre-asymptotic or transient region where the slopes exceed 1/2, however, these preasymptotic slopes are not as large as Hurst had observed.

Using formulas of Equations (3.1) and (3.2) and the local slope formula of Equations (5.5), Figure 5.2d exhibits the transient behavior of the expected value of the raw range and adjusted range when sampling from symmetric stable distributions. It can be noted that the transient effect is less severe for $H > 1/2$ than for $H = 1/2$.

The above formulas, tabled values and figures are based on the assumption that the $\{X_t\}$ are independent and identically distributed. The effect of removing the independence assumption and replacing it with serial dependence will be explored through two simple examples. That dependence structure may contribute to prolonging the transient region and thereby provide a possible explanation of Hurst's findings was suggested early on. In fact Feller (1951) first suggested autocorrelation as a possible explanation. Moran (1959) also suggested that the serial correlation of the underlying process could extend the transient region. Gomide (1978) computed the slopes of a conditional range, different from any of the ranges considered here, for a first-order autoregressive model with different ρ, using $0 \leq \rho \leq .90$. He found that the computed slopes for n up to 1000 were similar to those found by Hurst in his data. The mean of Gomide's computed slopes was around 0.75 as compared to Hurst's 0.73.

O'Connell (1971) suggested the ARMA (1,1) model as a candidate for explaining the Hurst phenomenon. This model has autocorrelation function given by

$$\rho_1 = \frac{(1-\phi\theta)(\phi-\theta)}{1 + \theta^2 - 2\,\phi\,\theta}$$

and (5.6)

$$\rho_k = \phi\,\rho_{k-1} \quad \text{for } k = 2, 3, \ldots ,$$

which decays geometrically for $k \geq 2$. Such decay is slow for ϕ near one. Boes and Salas (1978) (see also Salas et al. (1979)). selected two sets of values of (ϕ, θ) for a simulation study designed to check the effect of dependence on the transient region. These values are

$$\text{parameter set \#1:} \quad (\phi, \theta) = (.99,\ 0.8676) \tag{5.7}$$

and

$$\text{" \quad " \#2:} \quad (\phi, \theta) = (.90,\ 0.6268). \tag{5.8}$$

Figure 5.3 plots the estimated (via the simulation) expected adjusted range and estimated expected rescaled adjusted range for parameter set #1 along with the asymptotic curve for the expected rescaled adjusted range. Figure 5.4 does the same for parameter set #2. These curves definitely demonstrate the transience, and seem to show three regions; an initial region with slopes somewhat greater than 1/2, a pre-asymptotic transient region with slopes much greater than 1/2, and the asymptotic region with slopes converging to 1/2. The slope of the S-curve in Figure 5.3 is greater than unity for n between about 100 and 500. Figure 5.5 gives a scatter plot of the computed rescaled adjusted range for three groups of Hurst's data with n ranging between 35 and 1100, together with the S-curves of the estimated rescaled adjusted range corresponding to parameter sets #1 and 2. Figure 5.6 shows a scatter plot of other of Hurst's data, this time with n ranging from 37 to 237, together with the S-curve corresponding to parameter set #1. Since these parameter sets were not selected to fit Hurst's data it is somewhat surprising how closely the data fit the curves.

In any case, it is quite clear that Hurst's empirical findings are compatible with the pre-asymptotic/transient behavior of relatively simple models within the Brownian domain of attraction, so that a pre-asymptotic interpretation is indeed viable.

6. Long-Range Dependence

It was noted in Section 2 that for the class of stationary processes belonging to the Brownian domain of attraction, any of the three ranges has $n^{1/2}$ growth. Reference to Troutman (1976) and (1978) and Siddiqui(1976) was made. There are, of course, domains of attraction other than the Brownian domain of attraction, and two of them become the subject of this section. The first is the domain of attraction of fractional Brownian motion and the second is the stable domain of attraction.

In our consideration of the possibility of explaining Hurst-like behavior via pre-asymptotics in the previous section, a model that had positive correlations which fell off slowly as the time lag increased seemed to suffice. There the decay in the correlations was geometric, which is still relatively fast, fast enough that the correlation function is summable. What if the decay was of the form lag raised to a negative exponent instead of geometric, then the decay could be slow enough that the correlation function is not summable. It is precisely such models that have the capability of preserving Hurst-type growth asymptotically. And the most successful of such models is that of fractional Brownian noise, advanced in the work of Mandelbrot and coauthors. (See Mandelbrot, 1965; Mandelbrot and Van Ness, 1968; Mandelbrot and Wallis, 1968; and Mandelbrot and Wallis, 1969a, 1969b, 1969c.)

Fractional Brownian motion and fractional Gaussian noise can be defined several ways. One way is by use of the concept of self-similarity. Self-similar processes are distribution invariant under appropriate scaling. They play an important role as limiting processes. (See Davydov 1970, Mandelbrot 1975, Taqqu 1975 and Mandelbrot and Taqqu 1979.)

Definition 6.1 A real-valued process $\{Z(t): t \geq 0\}$ is defined to be *self-similar* with *index* H if for all $a > 0$ the finite dimensional distributions of $Z(a\cdot)$ and $a^H Z(\cdot)$ are identical and if $Z(0) \equiv 0$.

Definition 6.2 A real-valued process $\{Z(t): t \geq 0\}$ has *stationary increments* if the finite dimensional distributions of $Z(\cdot + s) - Z(s)$ and $Z(\cdot) - Z(0)$ are the same for all $s \geq 0$.

Remark If $\{Z(t): t \geq 0\}$ is self-similar with stationary increments satisfying $E[Z(t)] \equiv 0$ and $\mathrm{var}[Z(1)] = \sigma^2$ then $\mathrm{var}[Z(t)] = t^{2H}\sigma^2$ and

$$\mathrm{cov}[Z(t), Z(t + \tau)] = (1/2)\{t^{2H} + (t + \tau)^{2H} - \tau^{2H}\}\sigma^2 \qquad (6.1)$$

for $t \geq 0$, $\tau \geq 0$.

Proof $\mathrm{var}[Z(t)] = \mathrm{var}[t^H Z(1)] = t^{2H}\sigma^2$.

$$\mathrm{cov}[Z(t), Z(t + \tau)] = (1/2)\ \{\mathcal{E}[Z^2(t)] + \mathcal{E}[Z^2(t + \tau)] - \mathcal{E}[(Z(t + \tau) - Z(t))^2]\}$$
$$= (1/2)\{t^{2H} + (t+\tau)^{2H} - \tau^{2H}\}\sigma^2.$$

Definition 6.3 A real-valued process, say $\{B_H(t): t \geq 0\}$, that is self-similar with index H, $0 < H < 1$, and has stationary increments and is Gaussian with $\mathcal{E}[B_H(t)] = 0$ and $\mathrm{var}\ [B_H(1)] = \sigma^2$ is called *fractional Brownian motion* (FBM) with index H. When $H = 1/2$ it becomes *Brownian motion*.

Mandelbrot and Van Ness (1968) write the increments of fractional Brownian motion as an integral of weighted past increments of ordinary Brownian motion, but such representation will not be pursued here. Actually fractional Brownian motion can be thought of as a Gaussian process with mean 0 and covariance given by equation (6.1).

Fractional Brownian motion will be used to define fractional Brownian noise and it is fractional Brownian noise that will be used as the $\{X_t\}$ process in our modeling. FBM is continuous-time indexed, whereas fractional Brownian noise will be discrete-time indexed as $\{X_t\}$ has been throughout.

Definition 6.4 Let $\{B_H(t): t \geq 0\}$ be fractional Brownian motion with index H. Define $X_t^{(H)} = \mu + B_H(t) - B_H(t - 1)$ for $t = 1,2,\ldots$. $\{X_t^{(H)}\}_{t=1}^{\infty}$ is called *fractional Gaussian* (or Brownian) *noise* with index H, mean μ, and variance σ^2.

Properties of fractional Gaussian noise with index H

(i) $\{X_t^{(H)}\}_{t=1}^{\infty}$ is strictly stationary since $\{B_H(t): t \geq 0\}$ has stationary increments.

(ii) $S_n = \sum_{t=1}^{n} X_t^{(H)} = n\mu + \sum_{t=1}^{n} [B_H(t) - B_H(t-1)] = n\mu + B_H(n)$ which implies

$\mathrm{var}[S_n] = \mathrm{var}[B_H(n)] = n^{2H}\sigma^2$.

(iii)
$$\mathrm{cov}[X_1^{(H)}, X_{k+1}^{(H)}] = \mathcal{E}[B_H(1)(B_H(k+1) - B_H(k))]$$
$$= \mathrm{cov}[B_H(1), B_H(k+1)] - \mathrm{cov}[B_H(1), B_H(k)]$$
$$= (1/2)\{1 + (k+1)^{2H} - k^{2H} - 1 - k^{2H} + (k-1)^{2H}\}\sigma^2$$
$$= (1/2)\{(k+1)^{2H} - 2k^{2H} + (k-1)^{2H}\}\sigma^2 \qquad (6.2)$$
$$\sim \sigma^2 H(2H-1)k^{2H-2} \text{ for large } k \text{ and } H \neq 1/2.$$

Note that for $1/2 < H < 1$ the correlations, which are all positive, fall off so slowly that they are not summable. Also, for $H = 1/2$, these correlations are zero and fractional Gaussian noise reduces to regular Gaussian white noise as it should.

(iv) For $\mu = 0$, $S_{mn} = \sum_{t=1}^{mn} X_t^{(H)}$ has the same distribution as $m^H S_n = m^H \sum_{t=1}^{n} X_t^{(H)}$ for all $m, n \geq 1$, and in fact fractional Gaussian noise with index H is the unique Gaussian process with this property.

In order to illustrate the range behavior for fractional Gaussian noise, let us concentrate on the adjusted range and define a continuous-time indexed version of the adjusted range. Since $S_j = \sum_{t=1}^{j} X_t^{(H)} = B_H(j)$ for $j = 1,2,\ldots$ the partial sums in the definition of the adjusted range are to be replaced by B_H's. In R_n^*, replace the discrete time index n by a continuous-time index, say T. Now the continuous-time analogue of R_n^* becomes

$$R_T^* = \sup_{0 \leq x \leq T} [B_H(x) - (x/T)B_H(T)] - \inf_{0 \leq x \leq T} [B_H(x) - (x/T)B_H(T)]. \qquad (6.3)$$

Using that $B_H(T\cdot)$ has the same distribution as $T^H B_H(\cdot)$ and continuity of sample paths, R_T^* has the same distributions as $T^H R_1^*$ and it follows that

$$\mathcal{E}[R_T^*] = T^H \mathcal{E}[R_1^*],$$

where $\mathcal{E}[R_1^*]$ does not depend on T, and hence, in continuous-time, the *Hurst behavior* is *exact* for all T, and one has an exact Hurst law. A similar argument suffices for the other ranges. Thus fractional Brownian noise has rightly become the prototype model for preservation of the Hurst phenomenon.

Starting with $\{X_t\}_{t=1}^{\infty}$ fractional Gaussian noise with index H and $\mu = 0$, and defining the partial sum process in the usual way (see Billingsley (1968)), one sees that the partial sum process converges weakly to fractional Brownian motion and by the continuous mapping theorem, one obtains Hurst behavior for all three ranges. Further, any process in the domain of attraction of fractional Brownian motion will also possess a Hurst effect. One important class of processes in the domain of attraction of fractional Brownian motion is a particular class of infinite moving averages defined as follows. Let $\{Z_n\}$ be independent and identically distributed random variables with zero mean and finite variance, and let $\{C_j\}$ be a sequence of constants satisfying $\sum_{j=-\infty}^{\infty} C_j^2 < \infty$, then $\{X_n\}$ defined by

$X_n = \sum_{j=-\infty}^{\infty} C_j Z_{n-j}$ is an infinite moving average. Now, according to Davylov (1970), $\{X_n\}$ is in the domain of attraction of fractional Brownian motion with index H if $\mathcal{E}[|Z_j|^{2K}] < \infty$ for some $K \geq 2$ and $\mathrm{var}[S_n] = n^{2H}h(n)$ where $1/(K+2) < H \leq 1$ for some slowly varying function $h(\cdot)$. A special case of such infinite moving averages is that of fractionally differenced autoregressive integrated moving average (ARIMA) models considered by Granger and Joyeux (1980) and Hoskings (1981). Such models are conveniently defined via the backward shift operator B where $BX_t = X_{t-1}$, $B^jX_t = B(B^{j-1}X_t)$ and $(1-B)X_t = X_t - X_{t-1}$.

Definition 6.5 $\{X_t\}_{t=-\infty}^{\infty}$ is called an *ARIMA (p,d,q) process* if $\{X_t\}$ satisfies

$$\phi(B)(1-B)^dX_t = \theta(B)Z_t \qquad (6.4)$$

where $\{Z_t\}_{t=-\infty}^{\infty}$ is an independent and identically distributed zero mean and finite variance sequence of random variables,

$$(1-B)^dX_t = \sum_{k=0}^{\infty}\binom{d}{k}(-B)^kX_t, \qquad (6.5)$$

p and q are non-negative integers,

$$\phi(B) = 1 - \phi_1B - \dots - \phi_pB^p,$$

$$\theta(B) = 1 - \theta_1B - \dots - \theta_qB^q,$$

and d is real satisfying $d < 1/2$.

It can be shown that for $p = q = 0$ and $-1/2 < d < 1/2$ that the correlation function of $\{X_t\}$ is given by

$$\rho_k = \frac{d(1+d)\cdot\ldots\cdot(k-1+d)}{(1-d)(2-d)\cdot\ldots\cdot(k-d)} \qquad (6.6)$$

$$= \frac{\Gamma(1-d)\Gamma(k+d)}{\Gamma(d)\Gamma(k+1-d)} \sim \frac{\Gamma(1-d)}{\Gamma(d)}k^{2d-1} \text{ for large k.}$$

Comparing (6.2) and (6.6) one notes similar asymptotic behavior of the correlation functions of fractional Gaussian noise and ARIMA (0,d,0) if $2H - 2 = 2d - 1$, that is

$$H = d + 1/2; \qquad (6.7)$$

and, so for $0 < d < 1/2$, the Hurst effect will be preserved for the ARIMA (0, d, 0). In essence, it is the autoregressive and moving average components in ARIMA (p, d, q) that model short term components and the fractional difference d that models long term dependence. Various estimators of d have been proposed, see e.g. Geweke & Porter-Hudak (1983) and Yajima (1985), and after d is estimated the autoregressive & moving-average parameters $\phi_1,\dots,\phi_p$, $\theta_1,\dots,\theta_q$ can be estimated as for ARMA processes; so, ARIMA's provide a class of practical models for dealing with the Hurst phenomenon.

A final class of models to be considered is that of shifting level processes. Such processes were introduced by Boes and Salas (1978), as generalizations of models of Hurst (1957), Klemes (1974), and Potter (1975). The motivating concept behind the notion of shifting level is that the observed process consists of a sample path of a stochastic process at a given level for a random period of time and then shifts to a sample path of a stochastic process at a different level for different random period of time, and continues in this way.

Definition 6.6 Consider the doubly indexed family of stochastic processes $\{\{X_t^{(\ell)}: t = 1,2,\dots\}\ \ell \in L\}$, where L is an arbitrary index set. Let $\{L_k\}_{k=1}^{\infty}$ be any process taking values in L and let $\{E_k\}_{k=1}^{\infty}$ be a strictly increasing sequence of positive integer valued random variables. Define $\{X_t\}_{t=1}^{\infty}$ by

$$X_t = \sum_{j=1}^{\infty} X_t^{(L_j)} I_{\{E_{j-1}+1,\dots,E_j\}}(t) \tag{6.8}$$

where $E_0 \equiv 0$ and $I_{\{E_{j-1}+1,\dots,E_j\}}(t) = 1$ if $t = E_{j-1}+1,\dots,E_j$ and zero otherwise. $\{X_t\}_{t=1}^{\infty}$ is called a *shifting level process*.

Note that the process X_t follows the process $X_t^{(L_j)}$, randomly selected from the family of stochastic processes $\{\{X_t^{(\ell)}: t = 1,2,..\}\ \ell \in L\}$ for the random time span $\{E_{j-1}+1,\dots,E_j\}$ of length $E_j - E_{j-1}$. The processs $\{L_k\}$ is the *level process* and the process $\{E_k\}$ is the *shift-epoch* process. The shift-epoch process is often modeled by modeling the intershifts, i.e. the $(E_k - E_{k-1})$'s.

It will suffice for our purposes to consider a special case the shifting level process, namely a strictly stationary version known as a *level plus noise* process.

Definition 6.7 Let $\{N_j\}_{j=1}^{\infty}$ be a discrete, stationary, delayed-renewal sequence taking values on the positive integers. Let $\{L_k\}_{k=1}^{\infty}$ and $\{Y_t\}_{t=1}^{\infty}$ be strictly stationary processes, independent of each other and of $\{N_j\}_{j=1}^{\infty}$. $\{X_t\}_{t=1}^{\infty}$, defined by

$$X_t = \sum_{j=1}^{t} L_j I_{\{E_{j-1}+1,\dots,E_j\}}(t) + Y_t, \tag{6.9}$$

where $E_j = \sum_{i=1}^{j} N_i$, is called a *level plus noise* process.

In Definition 6.7, $\{Y_t\}$ is the *noise* process and $\{L_k\}$ is the *level* process; $X_t = L_j + Y_t$ over time span $\{N_1 + N_2 + \dots + N_{j-1} + 1,\dots,N_1 + \dots + N_j\}$ of length N_j. Recall that for the discrete, stationary, delayed-renewal process $\{N_j\}_{j=1}^{\infty}$ that the $\{N_j\}_{j=2}^{\infty}$ are iid with cdf, say $F_N(\cdot)$, and finite mean and N_1 is independent of $\{N_j\}_{j=2}^{\infty}$ and $F_{N_1}(n) = \sum_{j=0}^{n-1}(1-F_N(j))/\sum_{j=0}^{\infty}(1-F_N(j))$ for $n = 1, 2,\dots$. Although N_2 must have finite mean, it is possible that N_1 have infinite mean.

It was indicated at the outset of this section that two types of domains of attraction would be considered, that of fractional Brownian motion, covered earlier, and the stable domain of attraction, to be covered now.

Definition 6.8. A sequence $\{Z_n\}_{n=1}^{\infty}$ is said to be in the *domain of attraction of a stable* distribution, say $G_\alpha(\cdot)$, of *index* α if there exists a sequence of centering constants $\{a_n\}_{n=1}^{\infty}$ and of positive scaling constants $\{b_n\}_{n=1}^{\infty}$ with $b_n = n^{1/\alpha}h(n)$, where $h(\cdot)$ is slowly varying at infinity, such that $\{b_n^{-1}\sum_{j=1}^{n} Z_j - a_n\}$ converges in distribution to a random variable with distribution $G_\alpha(\cdot)$.

The crucial result to be employed here is that if $\{Z_n\}_{n=1}^{\infty}$ is any strictly stationary process with finite second moment in the domain of attraction of a stable distribution with index $\alpha \in (1,2)$ then the process $\{Z_n\}$ exhibits a Hurst effect for all three ranges with Hurst exponent $H = 1/\alpha$. On the other hand, if $\{Z_n\}_{n=1}^{\infty}$ is strictly stationary with infinite second moment in the domain of attraction of a stable with index α then the process $\{Z_n\}$ exhibits a Hurst effect for the raw range and adjusted range but not the rescaled adjusted range. The details of these results appear in Ballerini and Boes (1985), Leiva (1983), and Leiva and Boes (1984). Level plus noise processes form a convenient framework for constructing stationary processes, having either finite or infinite second moment, belonging to the stable domain of attraction. For instance, if both the noise process $\{Y_t\}$ and the level process $\{L_k\}$ have finite second moment, but $\{N_j\}_{j=2}^{\infty}$ belongs to the domain of attraction of a stable distribution with index $\alpha \in (0, 1)$ then it can be shown that the level plus noise process $\{X_t\}$ is stationary, belongs to the stable domain of attraction, and has finite second moment; and, hence such level plus noise process does preserve the Hurst effect in all three ranges. On the other hand, if both the noise process $\{Y_t\}$ and the gap or span process $\{N_j\}$ have finite second moment and the level process $\{L_k\}$ belongs to the stable domain of attraction with index $\alpha \in (0, 1)$, then again the level plus noise process $\{X_t\}$ is stationary, belongs to the stable domain of attraction with index α, and preserves the Hurst phenomenon for the raw range and adjusted range but not for the rescaled adjusted range. A somewhat curious result here is that one can have asymptotic behavior of the rescaled adjusted range identical to that of fractional Brownian noise without belonging to the domain of attraction of fractional Brownian motion.

Acknowledgement. This work was supported by the National Science Foundation under Grant CEE-8110782.

References

Anis, A. A., and E. H. Lloyd, 1953. "On the range of partial sums of a finite number of independent normal variates," Biometrika, 40, pp. 35-42.

Anis, A. A., and E. H. Lloyd, 1976. "The expected value of the adjusted rescaled Hurst range of independent normal summands," Biometrika, Vol. 63, pp. 111-116.

Ballerini, R. and D. C. Boes, 1985. "Hurst behavior of shifting level processes," Water Resources Research, Vol. 21, No. 11, pp. 1642-1648.

Bhattacharya, R. N., V. K. Gupta, and E. Waymire, 1983. "The Hurst effect under trends", J. Appl. Prob., 20, 649-662.

Billingsley, P., 1968. "Convergence of probability measures," John Wiley, New York.

Boes, D. C., and J. D. Salas-La Cruz, 1973. "On the expected range and expected adjusted range of partial sums of exchangeable random variables," Journal of Applied Probability, 10, pp. 671-677.

Boes, D. C., and J. D. Salas, "Nonstationarity of the mean and the Hurst phenomenon," Water Resour. Res., 14(1), 135-143.

Davis, R., 1983 "Stable limits for partial sums of dependent random variables," Annals of Probability, Vol. 11, No. 2 pp. 262-269.

Davydov, Y. A., 1970. "The invariance principle for stationary processes," Theor. Prob. Appl., 15, 487-498.

Feller, W., 1951. "The asymptotic distribution of the range of sums of independent random variables," Ann. Math. Stat., 22, 427-432.

Geweke, J. and S. Porter-Hudak, 1983. "The estimation and application of long-memory time series models." Journal of Time Series Analysis, 4, 221-238.

Gomide, F. L. S., 1978. "Markovian inputs and the Hurst phenomenon," J. Hydrol., 37, pp.23-45.

Granger, C. W. J., and R. Joyeux, 1980 "An introduction to long memory time series models and fractional differencing," J. Time Ser. Anal., 1(1), 15-29.

Hosking, J. R. M., 1981. "Fractional differencing," Biometrika, 68, 165-176.

Hurst, H. E., 1951. "Long-term storage capacity of reservoirs," Trans. Am. Soc. Civ. Eng., 116, 770-808.

Hurst, H. E., R. P. Black, and Y. M. Simaika, 1965. "Long term Storage: An Experimental Study," Constable press, London.

Hurst, H. E., R. P. Black, and Y. M. Simaika, 1966. The Nile Basin Vol. X, The Major Nile Projects, Ministry of Irrigation, Cairo.

Klemes, V., 1974. "The Hurst phenomenon: A puzzle?," Water Resour. Res., 10(4), 675-688.

Leiva, R., D. C. Boes, 1984. "Asymptotic behavior of partial sums of level plus noise processes", Dept. of Statistics, C.S.U., Technical Report #87.

Leiva, R., 1983. "Properties, convergence and range behavior of shifting level processes," Ph.D. dissertation, Dept. of Statistics, C.S.U., Fort Collins.

Mandelbrot, B., 1965. "Une classe de processus homothetiques a soi; application a loi climatologique de H. E. Hurst," C. R. Acad. Sci. Paris 260, 3274-3277.

Mandelbrot, B., 1975. "Limit theorems on the self-normalized range for weakly and strongly dependent processes," Z. Wahrschein. Gebiete, 31, 271-285.

Mandelbrot, B., and J. W. Van Ness, 1968. "Fractional Brownian motion, fractional noises and applications," SIAM Rev., 10, 422-437.

Mandelbrot, B. B., and J. R. Wallis, 1968. "Noah, Joseph and operational hydrology," Water Resour. Res., 4(5), pp. 909-918.

Mandelbrot, B. B., and J. R. Wallis, 1969a. "Computer experiments with fractional Gaussian noises, parts 1, 2 and 3," Water Resour. Res., 5(1), pp. 228-267.

Mandelbrot, B. B., and J. R. Wallis, 1969b. "Some long run properties of geophysical records," Water Resour. Res., 5(2), pp. 321-340.

Mandelbrot, B. B., and J. R. Wallis, 1969c. "Robustness of the rescaled range R/S in the measurement of non-cyclic long-run statistical dependence," Water Resour. Res., 5(5), pp. 967-988.

Moran, P. A. P., 1959. "The theory of storage," Methuen and Co. Ltd., London.

Moran, P. A. P., 1964. "On the range of cumulative sums," Ann. Inst. Stat. Math., 16, pp. 109-112.

O'Connell, P. E., 1971. "A simple stochastic modeling of Hurst's law," Int. Symp. on Mathematical Models in Hydrology, Warsaw, pp. 327-358.

Potter, K. W., 1975. Comment on "The Hurst phenomenon: A puzzle?" be V. Klemes, Water Resour. Res., 11(2), 373-374.

Salas, J.D., D. C. Boes, V. Yevjevich, and G. G. S. Pegram, 1979. "Hurst phenomenon as a preasymptotic behavior," Jour. of Hydr., 44, pp. 1-15.

Siddiqui, M. M., 1976. "The asymptotic distribution of the range and other functions of partial sums of stationary processes," Water Resour. Res., 12(6), pp. 1271-1275.

Solari, M. E., and A. A. Anis, 1957. "The mean and variance of the maximum of the adjusted partial sums of a finite number of independent normal variates," Ann. Math. Stat., 28, pp. 705-716.

Spitzer, F., 1956. "A combinational lemma and its application to probability theory," Amer. Math. Soc. Trans., Vol. 82. pp. 323-339.

Taqqu, M., 1975. "Weak convergence to fractional Brownian motion and to the Rosenblatt process," Z. Wahrschein Gebiete, 31, 287-302.

Troutman, B. M., 1974. "Expected range of partial sums," unpublished M.S. thesis, Colorado State University, Fort Collins.

Troutman, B. M., 1976. "Limiting distributions in storage theory," Ph.D. dissertation, Colorado State University, Fort Collins.

Troutman, B. M., 1978. "Reservoir storage with dependent periodic net inputs," Water Resour. Res., 14(3), pp. 395-401.

Yajima, Y., 1985. "On estimation of long-memory time series models." Australian Journal of Statistics, 27(3), 303-320.

Original paper received: 29.07.86
Final paper received: 24.04.87

Paper recommended by M.M. Siddiqui

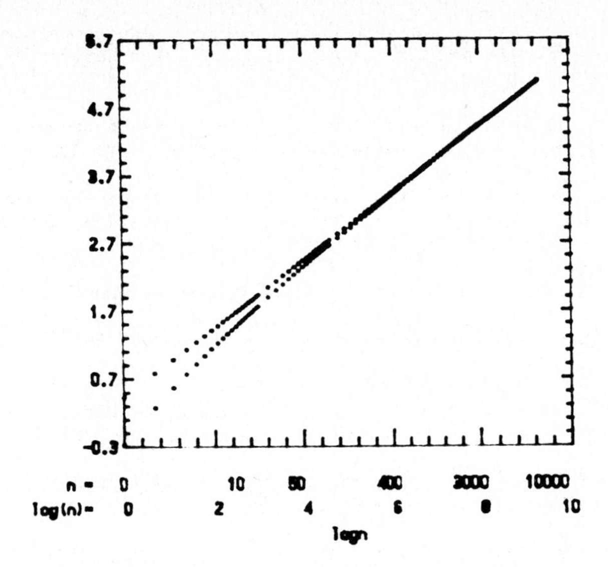

Figure 5.1a log $\mathcal{E}[R_n]$ versus log n

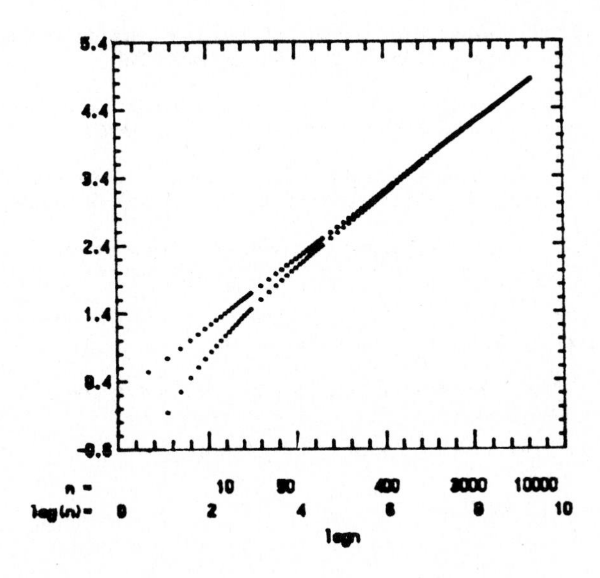

Figure 5.1b log $\mathcal{E}[R_n^*]$ versus log n

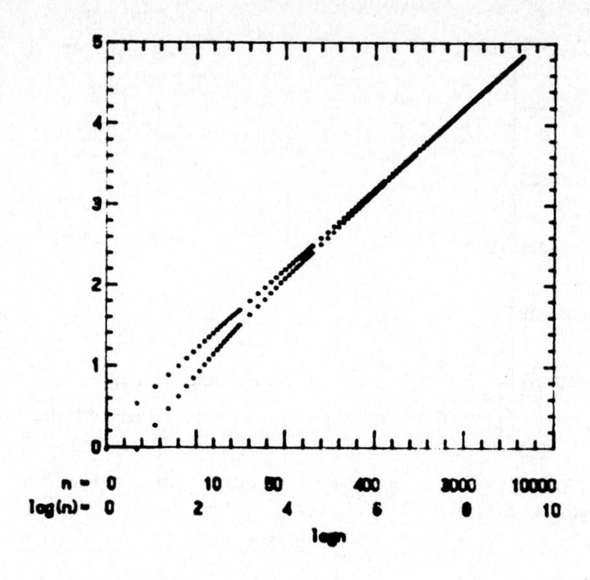

Figure 5.1c log $\mathcal{E}[R_n^{**}]$ versus log n

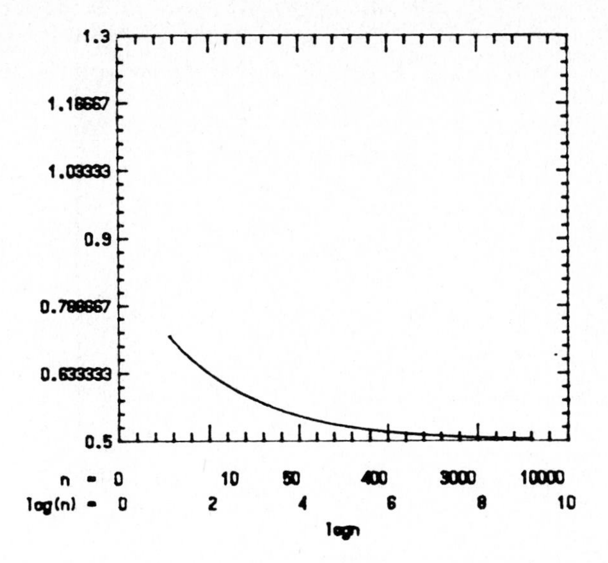

Figure 5.2a H_n (local slope of log $\mathcal{E}[R_n]$ versus log n) versus n

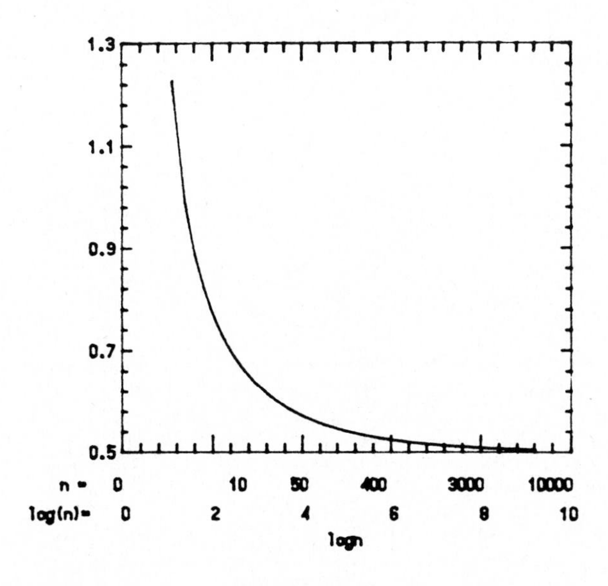

Figure 5.2b H_n^* versus n

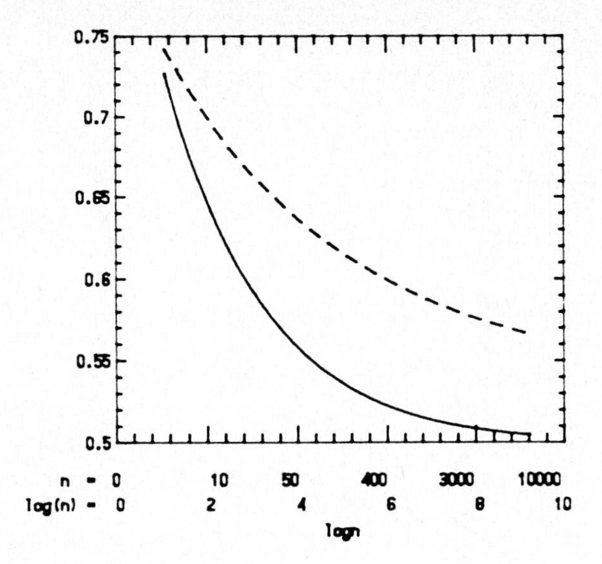

Figure 5.2c H_n^{**} and K_n versus n

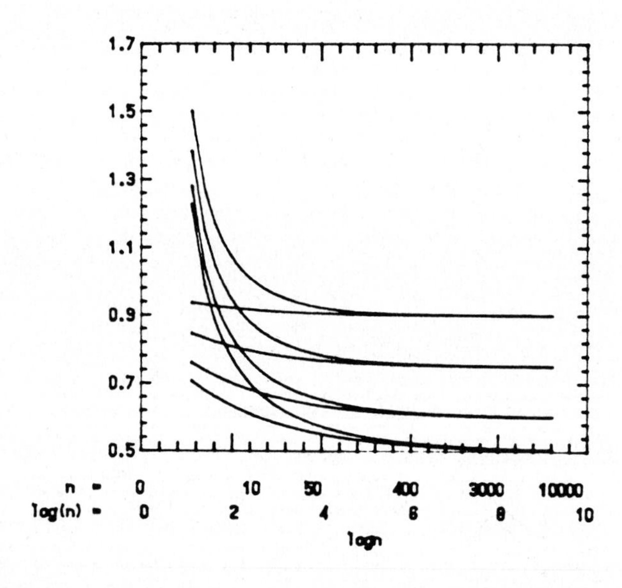

Figure 5.2d:

Local slopes H_n and H_n^* versus n for independent symmetric stable random variables having characteristic function $\exp(-|t|^\alpha)$ for $H = 1/\alpha = .5, .6, .75$ and .9.

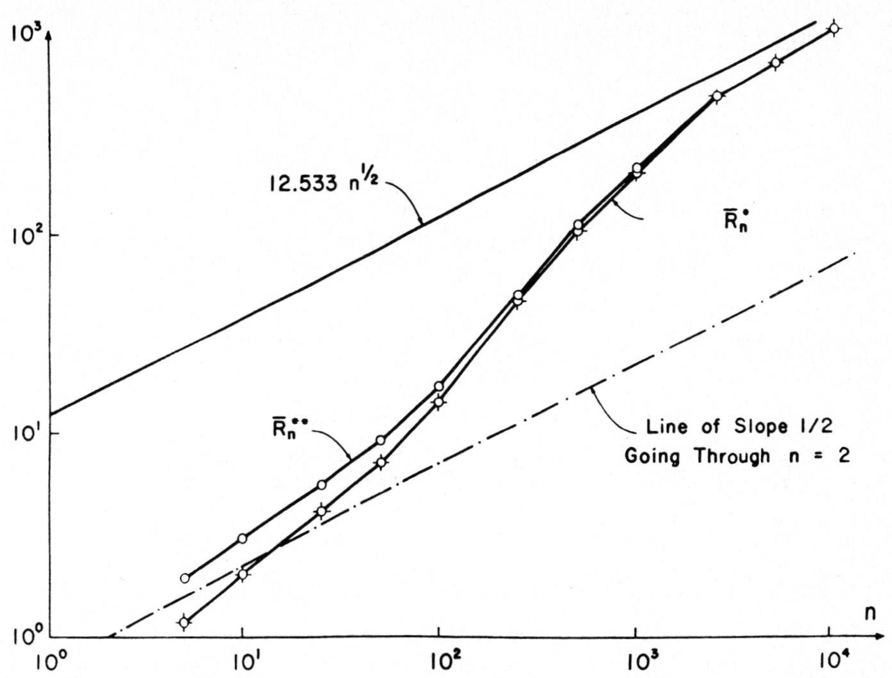

Figure 5.3 Simulation estimates of the expected adjusted range and expected rescaled adjusted range for Parameter Set #1.

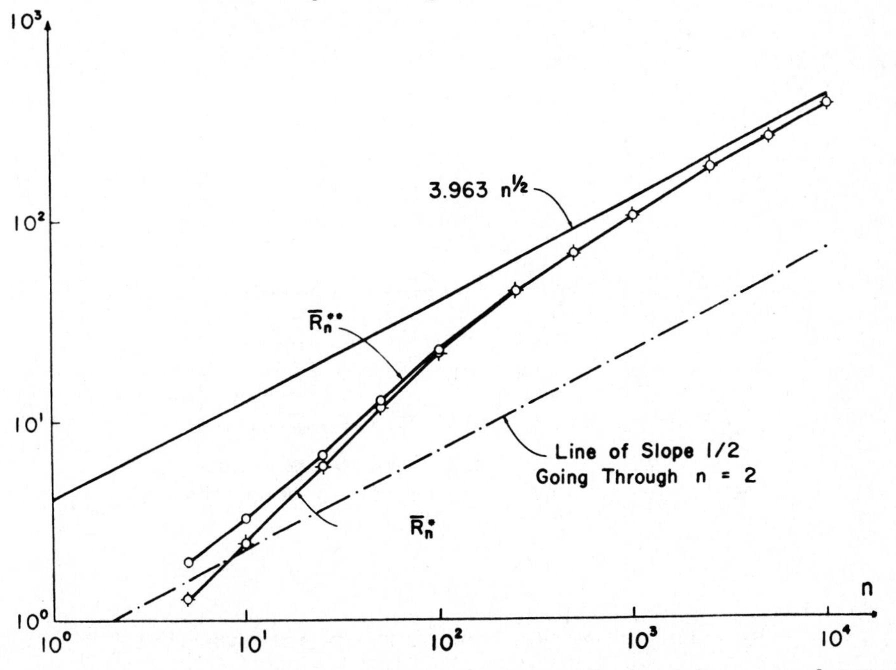

Figure 5.4 Simulation estimates of the expected adjusted range and expected rescaled adjusted range for Parameter Set #2.

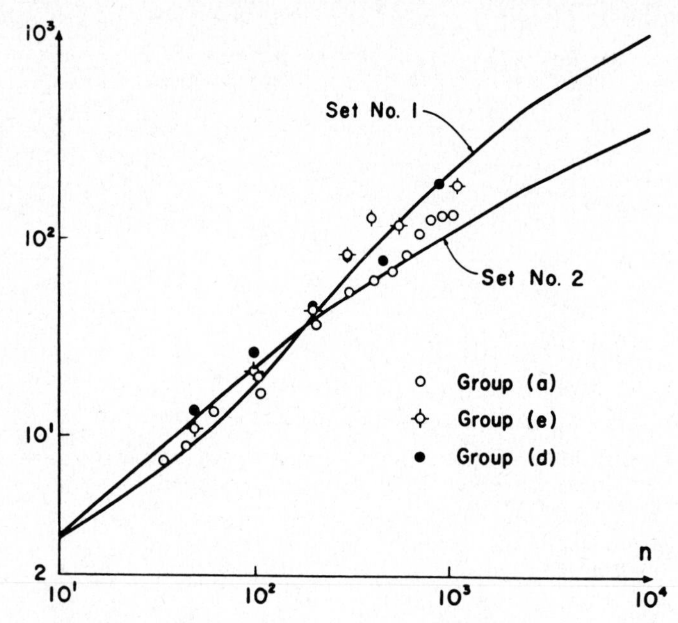

Figure 5.5 Scatter plot of computed rescaled adjusted range for three groups of Hurst's data and estimated rescaled adjusted range for Parameter Sets #1 and #2.

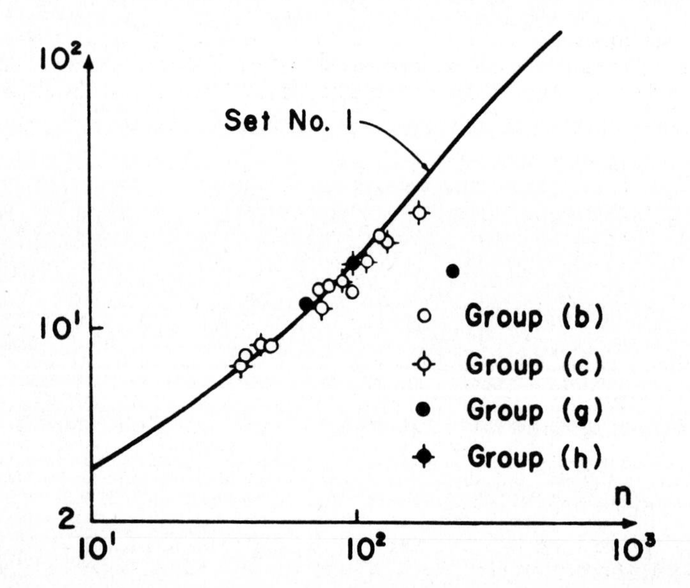

Figure 5.6 Scatter plot of computed rescaled adjusted range for four groups of Hurst's data and estimated rescaled adjusted range for parameter Set #1.

PROBABILITY AND STATISTICS
Essays in Honor of Franklin A. Graybill
J.N. Srivastava (Editor)

THE EXTENSION OF GRUBBS' ESTIMATORS FOR CLASSES OF INCOMPLETE BLOCK DESIGNS

Ralph A. Bradley
Department of Statistics
University of Georgia
Athens, Georgia, U.S.A.

and

Dwayne A. Rollier
Department of Industrial Engineering
Arizona State University
Tempe, Arizona, U.S.A.

The paper begins with the usual model for the general connected block design, except that error components for the model have variances σ_j^2, j indexing b blocks. Grubbs' estimators, G_j of σ_j^2, are derived and shown to exist when the inverse of a specified matrix exists. Specifically, they are shown to exist for PBIB(2) designs that are SRGD, $\lambda_1 = 0$ or $\lambda_2 \neq 0$, $r \leq 10$, and SGD. Explicit forms for the estimators are given for some very special designs. The Russell-Bradley test of H_0: $\sigma_1^2 = \sigma^2$, against H_a: $\sigma_1^2 \neq \sigma^2$, given $\sigma_j^2 = \sigma^2$, $j = 2,\ldots,b$, is extended to all BIB designs and to certain PBIB(m) designs. The test is based on a central F-statistic, as is its power function.

INTRODUCTION

Consider the general block design as discussed, for example, by John (1971, Chapter 11). The design parameters are v, the number of treatments, b, the number of blocks, r_i, the number of replications of treatment i, $i = 1,\ldots,v$, k_j, the number of experimental units in block j, $j = 1,\ldots,b$, r_i, $k_j > 0$. The general model for the intrablock analysis with fixed effects has

$$y_{ijm} = \mu + \tau_i + \beta_j + \varepsilon_{ijm},\ m = 1,\ldots,n_{ij}, \quad i = 1,\ldots,v,\ j = 1,\ldots,b, \tag{1}$$

where y_{ijm} is the m-th observation on treatment i in block j, μ, τ_i, and β_j are parameters representing respectively mean, treatment and block effects, and the ε_{ijm} are independent, zero-mean, random error components, taken to be normally distributed in distribution theory aspects of this article. The parameters n_{ij}, elements of the incidence matrix $\underset{\sim}{N}$, represent the number of occurrences of treatment i in block j. The design is taken to be connected. Matrices $\underset{\sim}{R}$ and $\underset{\sim}{K}$, diagonal matrices with typical elements r_i and k_j of sizes v-square and b-square respectively, are needed. Grubbs' estimators, in the context of the basic randomized block design or two-way classification, were developed to estimate σ_j^2 on the

---

AMS 1980 subject classifications. Primary 62K10, secondary 62K99.

Key Words and Phrases. Connected designs, heterogeneous block variances, residual analysis, PBIB(m) designs, BIB designs, F-tests.

assumption that $E(\varepsilon^2_{ijm}) = \sigma^2_j$, $j = 1,\ldots,b$, a departure from the usual assumption of homogeneity of variances. The purpose of this article is to extend definition of Grubbs' estimators, first to the general block design and then to particular classes of incomplete block designs, and to consider some distributional problems.

Maloney (1973) summarized the fairly extensive bibliography on the variance estimators stimulated by Grubbs' (1948) initial paper on the subject. Russell and Bradley (1958) presented two tests of variance homogeneity, an exact test on the assumption that the first (b-1) variances are homogeneous and an approximate test when b = 3. Ellenberg (1977) obtained the joint distribution of $S_1,\ldots,S_t$, $t \leq b$, S_j being the sum of squared residuals for block j, under the homogeneity assumption, but regarded the complicated series expressions obtained as intractable. Brindley and Bradley (1985) obtained the small-sample joint distribution of S_1, S_2 and S_3 when b = 3 and the distribution of a variance-homogeneity test statistic based on them in remarkably simple forms, both under the assumption of homogeneous variances and the general alternative to the assumption. They gave an approximate procedure for b > 3 and obtained also the power function of the exact test of Russell and Bradley. More details on the approximate procedure, together with some Monte Carlo results, are given by Bradley and Brindley (1986). While progress has been made, use of Grubbs' estimators has been hampered by difficulties encountered in developing appropriate inference procedures and by lack of extensions to more general block designs.

This paper has two objectives. The first is to obtain extensions of Grubbs' estimators for classes of incomplete block designs and the second is to consider inference procedures based on them to the extent possible.

ESTIMATORS FOR THE GENERAL BLOCK DESIGN

Let

$$S_j = \sum_{i=1}^{v} \sum_{m=1}^{n_{ij}} (y_{ijm} - \hat{\mu} - \hat{\tau}_i - \hat{\beta}_j)^2, \quad j = 1,\ldots,b, \tag{2}$$

where $\hat{\mu}$, $\hat{\tau}_i$ and $\hat{\beta}_j$ are the usual least squares estimators of μ, τ_i and β_j, and improper sums are zero. The procedure followed is to find the expectation of S_j and to adjust it for bias.

Define $\underset{\sim}{T}$ and $\underset{\sim}{B}$ to be the vectors of treatment and block totals and $\hat{\underset{\sim}{\tau}}$ and $\hat{\underset{\sim}{\beta}}$ to be the vectors of estimators. We let

$$\underset{\sim}{Q} = \underset{\sim}{T} - \underset{\sim}{N}\underset{\sim}{K}^{-1}\underset{\sim}{B} \quad \text{and} \quad \underset{\sim}{C} = \underset{\sim}{R} - \underset{\sim}{N}\underset{\sim}{K}^{-1}\underset{\sim}{N}'. \tag{3}$$

Then

$$\underset{\sim}{Q} = \underset{\sim}{C}\hat{\underset{\sim}{\tau}} \quad \text{and} \quad \underset{\sim}{1}_b\hat{\mu} + \hat{\underset{\sim}{\beta}} = \underset{\sim}{K}^{-1}(\underset{\sim}{B} - \underset{\sim}{N}'\hat{\underset{\sim}{\tau}}) \tag{4}$$

while

$$\hat{\underset{\sim}{\tau}} = \underset{\sim}{C}^{-}\underset{\sim}{Q}\ , \tag{5}$$

where $\underset{\sim}{C}^{-}$ is a generalized inverse of $\underset{\sim}{C}$ and $\underset{\sim}{1}_b$ is the column vector of b unit elements. The estimators are substituted in (2) with the result that

$$(y_{ijm} - \hat{\mu} - \hat{\tau}_i - \hat{\beta}_j) = \varepsilon_{ijm} - \sum_s \sum_t \sum_m c^{-}_{is}\varepsilon_{stm} + \sum_h \sum_s \sum_t \sum_m \frac{n_{hj}}{k_j} c^{-}_{hs}\varepsilon_{stm}$$

$$+ \sum_h \sum_s \sum_t \sum_m \frac{n_{ht}}{k_t} c^-_{ih} \varepsilon_{stm} - \sum_h \sum_s \sum_t \sum_u \sum_m \frac{n_{hj} n_{ut}}{k_j k_t} c^-_{hu} \varepsilon_{stm} - \sum_s \sum_m \frac{\varepsilon_{sjm}}{k_j},$$

where c^-_{hs} is the (h,s)-element of $\underset{\sim}{C}^-$. We may now obtain $E(S_j)$ using the fact that $E(\varepsilon^2_{ijm}) = \sigma^2_j$. The result is that

$$E(S_j) = \sigma^2_j \left(k_j - 1 - 2 \sum_s n_{sj} c^-_{ss} + 2 \sum_h \sum_s \frac{n_{sj} n_{hj}}{k_j} c^-_{hs} \right)$$

$$+ \sum_t \sigma^2_t \left[\sum_h \sum_s n_{ht} n_{sj} (c^-_{hs})^2 + \left(\frac{1}{k_j k_t} \right) \left(\sum_h \sum_s n_{hj} n_{st} c^-_{hs} \right)^2 \right. \tag{6}$$

$$\left. - \sum_h \frac{n_{ht}}{k_j} \left(\sum_s n_{sj} c^-_{hs} \right)^2 - \sum_h \frac{n_{hj}}{k_t} \left(\sum_s n_{st} c^-_{hs} \right)^2 \right].$$

In the special case of a randomized complete block design, $\underset{\sim}{N} = \underset{\sim}{J}_{vb}$, $\underset{\sim}{K} = v\underset{\sim}{I}_b$, $\underset{\sim}{C} = b(\underset{\sim}{I}_v - \frac{1}{v}\underset{\sim}{J}_{vv})$ and $\underset{\sim}{C}^- = \frac{1}{b}(\underset{\sim}{I}_v - \frac{1}{v}\underset{\sim}{J}_{vv})$, where $\underset{\sim}{I}_v$ is the v-square identity matrix and $\underset{\sim}{J}_{vb}$ is the v by b matrix of unit elements. Now (6) reduces to

$$E(S_j) = \frac{(v - 1)(b - 2)}{b} \sigma^2_j + \frac{(v - 1)}{b^2} \sum_t \sigma^2_t. \tag{7}$$

Let $E^* = \sum_j S_j$ be the error sum of squares for the design. It follows from (7) that $E(E^*) = [(v - 1)(b - 1)/b] \sum_t \sigma^2_t$ and that

$$G_j = \frac{b}{(v - 1)(b - 2)} S_j - \frac{1}{(v - 1)(b - 1)(b - 2)} E^* \tag{8}$$

is an unbiased estimator of σ^2_j. Grubb's estimator is G_j and we have summarized the procedure to be generalized.

For the general block design, if we write $\underset{\sim}{S}$ as the column vector with elements S_j and $\underset{\sim}{\sigma}^2$ and $\underset{\sim}{G}$ similarly in terms of σ^2_j and G_j, (6) may be rewritten

$$E(\underset{\sim}{S}) = \underset{\sim}{D}\, \underset{\sim}{\sigma}^2 \tag{9}$$

and

$$\underset{\sim}{G} = \underset{\sim}{D}^{-1} \underset{\sim}{S}, \tag{10}$$

thus defining the desired extension of Grubb's estimators. Elements of $\underset{\sim}{D}$ in (9) are determined by (6) and we assume that $\underset{\sim}{D}$ is non-singular for the moment. The estimators in $\underset{\sim}{D}$ are unbiased, but may in some cases be negative.

We examine next well known classes of incomplete block designs to obtain special forms of (6) and of $\underset{\sim}{D}$ in (9).

ESTIMATORS FOR CLASSES OF INCOMPLETE BLOCK DESIGNS

(i) PBIB Designs with m Associate Classes

Consider m-associate class PBIB designs. We follow the notation of Shah (1959). The design parameters are v, b, $r_i = r$, $i = 1,\ldots,v$, $k_j = k$, $j = 1,\ldots,b$, n_u, λ_u, and p^u_{vw}, u, v, w = 1,...,m, where, given a specific treatment, n_u is the number of its u-th associates and λ_u is the number of times a u-th associate treatment appears with it in blocks, while, given two treatments that are u-th associates, p^u_{vw} is the number of treatments common to the v-th associates of the first and the w-th associates of the second. It is convenient to take $\delta_{vw} = 1$ or 0 as $v = w$ or $v \neq w$ and to define $n_0 = 1$, $\lambda_0 = r$, $p^0_{vw} = \delta_{vw}n_v$, $p^u_{0w} = p^u_{w0} = \delta_{uw}$. It is helpful also to define the v-square association matrices, $\underset{\sim}{B}_u = (b^u_{hs})$, where $b^u_{hs} = 1$ or 0 as treatments h and s are or are not u-th associates, $u = 1,\ldots,m$, $\underset{\sim}{B}_0 = \underset{\sim}{I}_v$, $\sum_s b^u_{hs} = n_u$. In addition, let

$$c^*_u = \lambda_u(\delta_{u0}k - 1)/k, \quad u = 0,1,\ldots,m. \tag{11}$$

Given the notation defined,

$$\underset{\sim}{C} = \sum_{u=0}^{m} c^*_u\underset{\sim}{B}_u \tag{12}$$

and

$$\underset{\sim}{C}^- = \sum_{u=0}^{m} a_u\underset{\sim}{B}_u, \tag{13}$$

where the a_u are the solution of the equations,

$$\sum_{w=0}^{m}\sum_{x=0}^{m} p^u_{wx}\, c^*_w a_x = \delta_{u0} - (1/v), \quad u = 0,1,\ldots,m, \tag{14}$$

from whence a unique solution results through use of the additional restriction of convenience suggested by Shah,

$$\sum_{u=0}^{m} n_u a_u = 0. \tag{15}$$

Substitution of the elements of $\underset{\sim}{C}^-$ in (13) into (6) provides the new expression,

$$\begin{aligned} E(S_j) = {} & \sigma^2_j\Big(k - 1 - 2ka_0 + \frac{2}{k}\sum_h\sum_s n_{hj}n_{sj}\sum_{u=0}^{m} a_u b^u_{hs}\Big) \\ & + \sum_t \sigma^2_t\Bigg[\sum_h\sum_s n_{hj}n_{st}\sum_{u=0}^{m} a^2_u b^u_{hs} + \frac{1}{k^2}\Big(\sum_h\sum_s n_{ht}n_{sj}\sum_{u=0}^{m} a_u b^u_{hs}\Big)^2 \\ & - \frac{1}{k}\sum_h\sum_s\sum_i n_{hj}n_{sj}n_{it}\sum_{u=0}^{m}\sum_{v=0}^{m} a_u a_v b^u_{ih} b^v_{is} \\ & - \frac{1}{k}\sum_h\sum_s\sum_i n_{ht}n_{st}n_{ij}\sum_{u=0}^{m}\sum_{v=0}^{m} a_u a_v b^u_{ih} b^v_{is}\Bigg]. \end{aligned} \tag{16}$$

The matrix $\underset{\sim}{D}$ in (9) may be specified. Let $\underset{\sim}{A}$ and $\underset{\sim}{B}$ be two commensurate matrices and let $\underset{\sim}{A} * \underset{\sim}{B}$ be their Hadamard product; the (i,j)-element of $\underset{\sim}{A} * \underset{\sim}{B}$ is $a_{ij}b_{ij}$, the product of the corresponding elements of $\underset{\sim}{A}$ and $\underset{\sim}{B}$. Thus, from (16),

$$\begin{aligned}
\underset{\sim}{D} = \{&(k-1) - 2ka_0 + (2/k)\sum_{u=0}^{m} a_u\,[(\underset{\sim}{N}'\underset{\sim}{B}_u\underset{\sim}{N}) * \underset{\sim}{I}_b]\}\underset{\sim}{I}_b \\
&- (1/k)\sum_{u=0}^{m}\sum_{v=0}^{m} a_u a_v\,[(\underset{\sim}{N}'\underset{\sim}{B}_u) * (\underset{\sim}{N}'\underset{\sim}{B}_v)]\underset{\sim}{N} \qquad (17)\\
&- (1/k)\sum_{u=0}^{m}\sum_{v=0}^{m} a_u a_v\,\underset{\sim}{N}'[(\underset{\sim}{N}'\underset{\sim}{B}_u) * (\underset{\sim}{N}'\underset{\sim}{B}_v)]' + \sum_{u=0}^{m} a_u^2(\underset{\sim}{N}'\underset{\sim}{B}_u\underset{\sim}{N}) \\
&+ (1/k^2)\left[\sum_{u=0}^{m} a_u(\underset{\sim}{N}'\underset{\sim}{B}_u\underset{\sim}{N})\right] * \left[\sum_{v=0}^{m} a_v(\underset{\sim}{N}'\underset{\sim}{B}_v\underset{\sim}{N})\right] .
\end{aligned}$$

The most common PBIB designs in use have two associate classes and Bose, Clatworthy and Shrikhande (1954) have provided an extensive catalog of such designs. When $m = 2$, there are no major simplifications in (16) or (17), but values of a_0, a_1 and a_2 are available. The most convenient form has

$$rv(k-1)a_0 = (v-1)(k-c_2) - n_1(c_1-c_2),$$

$$rv(k-1)a_1 = (n_2+1)(c_1-c_2) - (k-c_2) \qquad (18)$$

and

$$rv(k-1)a_2 = -n_1(c_1-c_2) - (k-c_2),$$

values of c_1 and c_2 being given with each design in the referenced catalog. Definitions of c_1 and c_2 follow:

$$\begin{aligned}
k\Delta c_1 &= \lambda_1[r(k-1)+\lambda_2] + (\lambda_1-\lambda_2)(p_{12}^1\lambda_2 - p_{12}^2\lambda_1) \\
k\Delta c_2 &= \lambda_2[r(k-1)+\lambda_1] + (\lambda_1-\lambda_2)(p_{12}^1\lambda_2 - p_{12}^2\lambda_1) \qquad (19)\\
k^2\Delta &= [r(k-1)+\lambda_1][r(k-1)+\lambda_2] + (\lambda_1-\lambda_2)[r(k-1)(p_{12}^1 - p_{12}^2) \\
&\quad + p_{12}^1\lambda_2 - p_{12}^2\lambda_1].
\end{aligned}$$

Rao (1947) gives similar information for $m = 3$.

The inverse of $\underset{\sim}{D}$ does not always exist, although exceptions noted below are rather trivial. If $\underset{\sim}{D}$ is a positive, generalized, doubly stochastic matrix, Brauer's Theorem can assist in showing the existence of $\underset{\sim}{D}^{-1}$ (See Definition 2 and Theorem 1 below). Consideration of (17) shows that it is necessary that

$$\sum_{u=0}^{m}\sum_{w=0}^{m}\sum_{x=0}^{m}\sum_{s=1}^{v}\sum_{t=1}^{v} a_u a_w n_{sj} n_{tj} p_{uw}^x b_{st}^x$$

not be dependent on j if $\underset{\sim}{D}$ is to have the desired stochastic form. This is not generally true for PBIB(m) designs. Similar difficulties arise with other approaches. In subsection (ii), the existence of $\underset{\sim}{D}^{-1}$ is established for two broad classes of PBIB(2) designs and, for more specialized designs, forms of $\underset{\sim}{D}^{-1}$ are given in (iii).

(ii) <u>Special Cases of PBIB(2) Designs</u>

Consider PBIB(2) designs, PBIB designs with two associate classes. We begin with group divisible (GD) designs. The existence of $\tilde{D}^{-1}$ in (10) will be considered and computation of $\tilde{D}^{-1}$ given in feasible special cases.

A GD PBIB(2) design has $v = mn$, the treatments being divided into m groups of n in such a way that any two treatments in a group are first associates and any two treatments from different groups are second associates; $n_1 = n - 1$, $n_2 = n(m - 1)$. (Note the new use of m.) It is clear that

$$(n - 1)\lambda_1 + n(m - 1)\lambda_2 = r(k - 1)$$

and

$$\tilde{P}_1 = \begin{bmatrix} (n-2) & 0 \\ 0 & n(m-1) \end{bmatrix} \quad \text{and} \quad \tilde{P}_2 = \begin{bmatrix} 0 & (n-1) \\ (n-1) & n(m-2) \end{bmatrix},$$

where p^i_{uv} is the typical element of $\tilde{P}_i$, i, u, v = 1,2. Treatments may be exhibited in an m by n association array with column j having treatments $(j - 1)m + 1, \ldots, jm$, $j = 1,\ldots,n$. Treatments with treatment numbers identical modulo m are in groups. It is now easy to determine that

$$\tilde{B}_0 = \tilde{I}_{mn},\ \tilde{B}_1 = (\tilde{J}_{nn} - \tilde{I}_n) \otimes \tilde{I}_m,\ \text{and}\ \tilde{B}_2 = \tilde{J}_{nn} \otimes (\tilde{J}_{mm} - \tilde{I}_m), \tag{20}$$

where $\otimes$ denotes the Kronecker product of the indicated matrices. We may find $\tilde{D}$ because needed values of a_0, a_1, a_2 follow from substitution in and simplification of (18) and (19). GD designs have $r \geq \lambda_1$ and $rk - \lambda_2 v \geq 0$; they are Singular (S) if $r = \lambda_1$, Semi-Regular (SR) if $r > \lambda_1$, $rk - \lambda_2 v = 0$, and Regular (R) if $r > \lambda_1$ and $rk - \lambda_2 v > 0$. Sufficient simplifications in $\tilde{D}$ occur for SRGD and SGD designs to demonstrate the existence of $\tilde{D}^{-1}$ for designs of practical use in these classes.

(a) SRGD Designs: The equality defining an SRGD design and the fact that every block of such a design contains k/m treatments from each of the m rows of the association array as shown by Bose and Conner (1952) permit further simplification of the elements of $\tilde{D}$ in (17). Those elements are

$$d_{jj} = (m - 1)\left(\frac{\lambda_1 - \lambda_2}{m\Delta} - 1\right)^2 + (k - 1)\left(\frac{v\lambda_2}{\Delta k} - 1\right)^2 + (m - 1)\left[\frac{2n\lambda_2(\lambda_1 - \lambda_2)}{\Delta^2 k} - 1\right],\quad j = 1,\ldots,b,$$

and (21)

$$d_{jj'} = \frac{(m - 1)(\lambda_1 - \lambda_2)^2}{m^2\Delta^2} + \left(\frac{v\lambda_2}{\Delta k^2}\right)\left[\mu^2_{jj'} + k(k - 2)\mu_{jj'}\right] + \left[\frac{2n(m - 1)\lambda_2(\lambda_1 - \lambda_2)}{\Delta^2 k^2}\right]\mu_{jj'},\quad j \neq j',\ j, j' = 1,\ldots,b.$$

In (21), $\mu_{jj'}$ is the typical element of $\tilde{N}'\tilde{N}$, the number of treatments common to blocks j and j'.

The existence of $\underset{\sim}{D}^{-1}$ for SRGD designs with $\lambda_1 = 0$ or $\lambda_1 \neq 0$, $r \leq 10$ is now considered. Two definitions and a theorem from matrix algebra are needed:

Definition 1. A $b \times b$ matrix $\underset{\sim}{M} = [m_{jj'}]$ is a positive stochastic matrix of order b if each element $m_{jj'} > 0$ and $\sum_{j'=1}^{b} m_{jj'} = 1$, $j = 1,\ldots,b$.

Definition 2. The matrix $\underset{\sim}{M}$ is a generalized doubly stochastic matrix of order b if $m_{jj'} \geq 0$ and $\sum_{j=1}^{b} m_{jj'} = \sum_{j'=1}^{b} m_{jj'} = c > 0$.

Theorem 1. [Brauer (1952)]: If $\underset{\sim}{M}$ is a positive stochastic matrix of order b, s_M is the smallest element of $\underset{\sim}{M}$, and

$$m_{jj} + m_{j'j'} > 1 - (b - 2)s_M, \; j \neq j', \; j, j' = 1,\ldots,b,$$

then each characteristic root of $\underset{\sim}{M}$ has a positive real part and $|\underset{\sim}{M}| > 0$.

Theorem 1 applies at once to a positive, generalized, doubly stochastic matrix $\underset{\sim}{D}$ of the same order if the inequality is replaced by

$$d_{jj} + d_{j'j'} > \sum_{j=1}^{b} d_{jj'} - (b - 2)s_D, \; j \neq j', \; j, j' = 1,\ldots,b, \tag{22}$$

where s_D is the smallest element of $\underset{\sim}{D}$. For SRGD designs,

$$\sum_{j,j \neq j'} \mu_{jj'} = k(r - 1) \text{ and } \sum_{j,j \neq j'} \mu_{jj'}^2 = k(r - k) + [k(k - m)\lambda_1 + k^2(m - 1)\lambda_2]/m, \; j' = 1,\ldots,b. \tag{23}$$

When $\lambda_1 = 0$, $k = m$, the design parameters for the SRGD design reduce to $v = mn$, $b = n^2\lambda_2$, $r = n\lambda_2$, $k = m$, $n_1 = (m - 1)$, $n_2 = n(m - 1)$, $\lambda_1 = 0$, λ_2, while $\Delta k = n^2(m - 1)\lambda_2^2$, and, from (21),

$$d_{jj} = [n^2(m - 1)\lambda_2 - (v - 1)]^2/n^4(m - 1)\lambda_2^2, \; j = 1,\ldots,b,$$

and

$$d_{jj'} = \{(m - 1) + n^2[\mu_{jj'}^2 + m(m - 2)\mu_{jj'}] - 2n(m - 1)\mu_{jj'}\}/n^4(m - 1)^2\lambda_2^2, \quad j \neq j', \; j, j' = 1,\ldots,b. \tag{24}$$

It is clear that $d_{jj} > 0$, $j = 1,\ldots,b$. Let $d_{jj'} = [(m - 1) + a_{jj'}]/n^4(m - 1)^2\lambda_2^2$. Since $m,n \geq 2$ and $\mu_{jj'} \geq 0$, replace $\mu_{jj'}^2$ in $a_{jj'}$ by $\mu_{jj'}$ and it follows that

$$a_{jj'} \geq n[n(m - 1)^2 - 2(m - 1)]\mu_{jj'} \geq 0, \; d_{jj'} > 0, \; j \neq j', \; j, j' = 1,\ldots,b,$$

and $s_D > 1/n^4(m - 1)\lambda_2^2$. Thus $\underset{\sim}{D}$ is a positive, generalized doubly stochastic matrix.

To show that $\underset{\sim}{D}$ satisfies (22), consider $d_{jj} - \sum_{j',j'\neq j} d_{jj'}$. The results of (23) are used:

$$\sum_{j',j'\neq j} d_{jj'} = (v - 1)[n^2(m - 1)\lambda_2 - (v - 1)]/n^4(m - 1)\lambda_2^2, \; j = 1,\ldots,b,$$

and (25)

$$d_{jj} - \sum_{j',j'\neq j} d_{jj'} = [n^2(m - 1)\lambda_2 - (v - 1)][n^2(m - 1)\lambda_2 - 2(v - 1)]/n^4(m - 1)\lambda_2^2, \; j = 1,\ldots,b.$$

If $\lambda_2 \geq 2$, all factors on the right-hand sides of (25) are positive and (22) follows. If $\lambda_2 = 1$, we first consider $d_{jj} - \sum_{j',j'\neq j} d_{jj'} + (b - 2)/n^4(m - 1)$ with $b = n^2$; it is easy to show that the specified expression is positive if and only if $n \geq 3$ and (22) follows. When $\lambda_1 = 0$, SRGD designs are such that the integer $m \leq (n^2\lambda_2 - 1)/(n - 1)$ -- see Plackett and Burman (1946). Thus, when $n = 2$ and $\lambda_2 = 1$, $m \leq 3$ and only two rather simple designs exist. In each case it is easy to evaluate d_{jj} and $d_{jj'}$ in (24) and verify that (22) holds. We have proved that $\underset{\sim}{D}^{-1}$ exists for any SRGD design with $\lambda_1 = 0$ and next consider SRGD designs with $\lambda_1 \neq 0$, $r \leq 10$.

Bose, Shrikhande and Bhattacharya (1953) showed that the only SRGD designs with $\lambda_1 \neq 0$ and $r \leq 10$ are those listed below in Table 1. For each series, it can be shown that $\underset{\sim}{D}$ is

Table 1. Parameters of SRGD Designs with $\lambda_1 \neq 0$, $r \leq 10$

Series	v	b	r	k	m	n	λ_1	λ_2	Maximum m
(1)	4m	12	6	2m	m	4	2	3	3
(2)	3m	9	6	2m	m	3	3	4	4
(3)	6m	20	10	3m	m	6	4	5	3

a positive, generalized, doubly stochastic matrix satisfying (22) and hence that $\underset{\sim}{D}^{-1}$ exists. Every SRGD design listed by Bose, Clatworthy and Shrikhande (1954, pp. 139-140) either has $\lambda_1 = 0$ or $\lambda_1 \neq 0$, $r \leq 10$, and estimators G_j of σ_j^2, $j = 1, \ldots, b$, exist.

When $\mu_{jj'} = \lambda \geq 1$, $j \neq j'$, $j,j' = 1,\ldots,b$, the SRGD design is λ times linked and $b = v - m + 1$. Each $d_{jj'}$ in (21) has the same value, say d_1, and each d_{jj} is the same, say d. When $\underset{\sim}{D}^{-1}$ exists, its elements are

$$D_{jj} = [d + (b - 2)d_1]/(d - d_1)[d + (b - 1)d_1], \; j = 1,\ldots,b,$$

and

$$D_{jj'} = -d_1/(d - d_1)[d + (b - 1)d_1], \; j \neq j', \; j,j' = 1,\ldots,b.$$

In this case, an explicit form G_j of the estimator of σ_j^2 is

$$G_j = \{[d + (b - 1)d_1]S_j - d_1 \sum_{j'=1}^{b} S_{j'}\}/(d - d_1)[d + (b - 1)d_1], \quad j = 1,\ldots,b, \tag{26}$$

where S_j is defined in (2).

(b) <u>SGD Designs</u>: We turn to consideration of SGD designs. Every SGD design may be generated from a balanced incomplete block (BIB) design through replacement of each treatment in the BIB design by a group of n treatments. Given a BIB design with $v^* = m$, $b^* = b$, $r^* = r$, $k^* = k/n$, $\lambda^* = \lambda_2$, a treatment p of the BIB design is replaced by treatments p, m + p, 2m + p,...,(n - 1)m + p for the SGD design, p = 1, ...,m, this group of treatments constituting the pth row of the m × n association scheme of the SGD design. The matrix $\underset{\sim}{D}$ in (17) basically depends on $\underset{\sim}{N}'\underset{\sim}{B}_1$ and $\underset{\sim}{N}'\underset{\sim}{B}_2$. It is obvious from the association scheme and the design construction that the (p,j)-element of $\underset{\sim}{N}'\underset{\sim}{B}_1$ is (n - 1) if treatment p is in block j and zero otherwise. Thus $\underset{\sim}{N}'\underset{\sim}{B}_1 = (n - 1)\underset{\sim}{N}'$ and $\underset{\sim}{N}'\underset{\sim}{B}_2$ may be deduced from this. Then from (17)-(19) and $\underset{\sim}{P}_1$ and $\underset{\sim}{P}_2$ for GD designs,

$$d_{jj} = \frac{k(n - 1)(\lambda_1 - \lambda_2)^2}{\Delta^2 k^2} + k\left[\frac{k\lambda_1 - (n - 1)(\lambda_1 - \lambda_2)}{\Delta k} - 1\right]^2 - \left(\frac{\lambda_1}{\Delta} - 1\right)^2, \quad j = 1,\ldots,b.$$

and (27)

$$d_{jj'} = (\mu_{jj'}/\Delta^2 k^2)\{\mu_{jj'}\lambda_1^2 + k(k - 2)\lambda_1^2 + n(n - 1)(\lambda_1 - \lambda_2)^2 - 2k(n - 1)\lambda_1(\lambda_1 - \lambda_2)\}, \quad j \neq j',\ j, j' = 1,\ldots,b,$$

where $\Delta k = \lambda_1[(k - n)\lambda_1 + n\lambda_2]$.

It can be demonstrated that $d_{jj} > 0$ and $d_{jj'} \geq 0$, the equality arising if and only if $\mu_{jj'} = 0$. Since $k^* = k/n \geq 2$ is an integer, $k \geq 2n$. Also, for an SGD design, $r = \lambda_1 > \lambda_2 \geq 1$. It follows that $\Delta > \lambda_1 > 0$, $k\lambda_1 - (n - 1)(\lambda_1 - \lambda_2) > 0$ and the second positive term exceeds the negative term in the right-hand side of d_{jj} in (27) so that $d_{jj} > 0$, j = 1,...,b. Since $k(k - 2) > 2k(n - 1)$ and $\lambda_1 > \lambda_2 > 1$, the second factor of $d_{jj'}$ in (27) is positive and $d_{jj'} > 0$, $j \neq j'$, j,j' = 1,...,b, the equality holding if and only if $\mu_{jj'} = 0$. To show that $\underset{\sim}{D}^{-1}$ exists, a new approach is required.

Definition 3. A b × b complex matrix $\underset{\sim}{M}$ is diagonally dominant if

$$|m_{jj}| \leq \sum_{j',j'\neq j} |m_{jj'}|$$

for all j = 1,...,b, $\underset{\sim}{M}$ being strictly diagonally dominant if the inequality is strict for all j.

Theorem 2. (See Taussky 1949). If $\underset{\sim}{M}$ is strictly diagonally dominant, then $\underset{\sim}{M}$ is nonsingular. If all of the diagonal elements of $\underset{\sim}{M}$ are positive real numbers, the characteristic roots of $\underset{\sim}{M}$ have positive real parts.

For SGD designs, $\sum_{j',j'\neq j} \mu_{jj'} = k(r - 1)$ and

$\sum_{j',j'\neq j} \mu_{jj'}^2 = k[n\lambda_1 - k + (k - n)\lambda_2]$, $j = 1,\ldots,b$. It follows that $\underset{\sim}{D}$ is a generalized doubly stochastic matrix. Since $rk - \lambda_2 v = 0$ and $(n - 1)\lambda_1 + n(m - 1)\lambda_2 = r(k - 1)$, it follows also from (27) that

$$\begin{aligned} d_{jj} - \sum_{j',j'\neq j} d_{jj'} = [&(k - 1)(k - n)^2\lambda_1^4 - 3k(k - n)^2\lambda_1^3 + 2k(k - n)(k - n + 1)\lambda_1^2 \\ &- 3k(k - n)(2n - 1)\lambda_1^2\lambda_2 + 2n(k - n)(k - 1)\lambda_1^3\lambda_2 \\ &+ n^2(k - 1)\lambda_1^2\lambda_2^2 - 3nk(n - 1)\lambda_1\lambda_2^2 + 2nk(n - 1)\lambda_2^2 \\ &+ 4k(n - 1)(k - n)\lambda_1\lambda_2]/\Delta^2k^2. \end{aligned} \tag{28}$$

Groupings of the first three terms and the next two pairs of terms in (28) are nonnegative since $k \geq 2n$ when $\lambda_1 \geq 3$ and then $\underset{\sim}{D}$ is strictly diagonally dominant with positive diagonal elements. From Theorem 2, $\underset{\sim}{D}$ is positive definite and $\underset{\sim}{D}^{-1}$ exists when $\lambda_1 \geq 3$.

When $\lambda_2 = 1$, inequalities and design parameter relationships for an SGD design specify all of the design parameters except n : $v = 3n$, $b = m = 3$, $r = 2$, $k = 2n$, $n_1 = n - 1$, $n_2 = 2n$, $\lambda_1 = 2$, $\lambda_2 = 1$. For such designs the expression in (28) is negative; however, Roy and Laha (1957) prove that an SGD design is λ times linked if and only if $b = m$, implying that $\lambda = n\lambda_2$. Thus the case with $\lambda_2 = 1$ is covered as $\underset{\sim}{D}^{-1}$ is obtained below for SGD designs that are $n\lambda_2$ times linked and demonstration of the existence of $\underset{\sim}{D}^{-1}$ for SGD designs is complete.

If an SGD design is $n\lambda_2$ times linked, $\mu_{jj'} = n\lambda_2$, $j \neq j'$, $j, j' = 1,\ldots,b$. We can set each diagonal element of $\underset{\sim}{D}$ equal to d and each off-diagonal element equal to d_1. Then, if $d \neq d_1$, we have the situation discussed above for linked SRGD designs, although d and d_1 are now different. The estimator G_j of σ_j^2 in (26) applies as do the expressions for the elements of $\underset{\sim}{D}^{-1}$. It is easy to calculate $d - d_1 = (n - 1)/4$ from (27) for the special SGD design parameters resulting when $\lambda_1 = 2$ and $\underset{\sim}{D}^{-1}$ exists. SGD designs do not arise with $\lambda_1 = 1, 0$ and the existence of the estimators G_j of σ_j^2 has been established for all SGD designs.

The existence of $\underset{\sim}{D}^{-1}$ and G_j for regular, group divisible designs as a class has not been established. For a specific design, elements of $\underset{\sim}{D}$ in (17) may be calculated and the existence or computation of $\underset{\sim}{D}^{-1}$ considered. There seems to be no reason to expect $\underset{\sim}{D}$ to be singular except perhaps in very special cases.

(iii) Other Special Block Designs

Bradley and Rollier (1985) considered some special block designs in detail. Results are summarized here and the reference may be obtained if derivations are needed.

(a) Two Replicate, Singly Linked, Triangular PBIB(2) Designs: Bose (1951) and Nair (1950) studied two-replicate PBIB(2) designs. Nair studied particularly such designs that are singly linked with triangular association schemes. For such designs, $p \geq 5$, p an integer,

$$v = \tfrac{1}{2}p(p-1),\ b = p,\ r = 2,\ k = p-1,\ n_1 = 2(p-2),$$
$$n_2 = \tfrac{1}{2}(p-2)(p-3),\ \lambda_1 = 1,\ \lambda_2 = 0, \tag{29}$$

and

$$\underset{\sim}{P}_1 = \begin{bmatrix} (p-2) & (p-3) \\ (p-3) & (p-3)(p-4)/2 \end{bmatrix} \text{ and } \underset{\sim}{P}_2 = \begin{bmatrix} 4 & 2(p-4) \\ 2(p-4) & (p-4)(p-5)/2 \end{bmatrix}. \tag{30}$$

It can be shown that

$$4p^2\underset{\sim}{D} = (p-2)(p-1)^2\underset{\sim}{I}_b + (p+1)(p-2)(\underset{\sim}{J}_{bb} - \underset{\sim}{I}_b),$$
$$[(p-1)(p-2)(p-3)/2]\underset{\sim}{D}^{-1} = (2p^2 - 3p - 1)\underset{\sim}{I}_b - (p+1)(\underset{\sim}{J}_{bb} - \underset{\sim}{I}_b),$$

and the explicit estimator of σ_j^2 is

$$G_j = [2/(p-1)(p-2)(p-3)][2p(p-1)S_j - (p+1)\sum_{j'=1}^{b} S_{j'}], \tag{31}$$
$$j = 1,\ldots,b.$$

(b) BIB Designs, Existence Results: For BIB designs, $\underset{\sim}{D}$ is obtained from (6) with use of a particular $\underset{\sim}{C}^-$ and $\lambda(v-1) = r(k-1)$. Then

$$d_{jj} = (k-1)(\lambda v - k)^2/\lambda^2 v^2,\ j = 1,\ldots,b,$$
$$d_{jj'} = [k(k-2)\mu_{jj'} + \mu_{jj'}^2]/\lambda^2 v^2,\ j \neq j',\ j,j' = 1,\ldots,b, \tag{32}$$

and

$$\sum_{j'} \mu_{jj'} = rk \text{ and } \sum_{j'} \mu_{jj'}^2 = (r-\lambda)k + \lambda k^2. \tag{33}$$

It can be demonstrated that $\underset{\sim}{D}$ is a positive, generalized, doubly stochastic matrix and, if $\lambda v > 2k$, $\underset{\sim}{D}$ is strictly diagonally dominant. Theorem 2 then establishes the existence of $\underset{\sim}{D}^{-1}$. If $\lambda v = 2k$, $v = k$, $\lambda = b = r = 2$ and $\underset{\sim}{D}$ is singular. If $\lambda v = 2k$, $v > k$, Plan 11.1 of Cochran and Cox (1957) is the only design and $\underset{\sim}{D}^{-1}$ exists. If $\lambda v < 2k$, $v = k$, no BIB design exists. If $\lambda v < 2k$, $v > k$, then $v = b = 3$, $r = k = 2$, $\lambda = 1$ and $\underset{\sim}{D}$ is singular.

(c) BIB Designs with BIB Duals: Only symmetric BIB designs have BIB duals. Now $b = v$ and $\mu_{jj'} = \lambda$, $j \neq j'$. From (32), $\underset{\sim}{D} = d\,\underset{\sim}{I}_b + d_1\,(\underset{\sim}{J}_{bb} - \underset{\sim}{I}_b)$ with

$$d = (k-1)(\lambda v - k)^2/\lambda^2 v^2 \text{ and } d_1 = [\lambda + k(k-2)]/\lambda v^2,$$

and G_j may be obtained from (26).

(d) BIB Designs with PBIB Duals: Consider a BIB design with parameters, $v = b^*$, $b = v^*$, $r = k^*$, $k = r^*$, λ, where the PBIB dual design has parameters, v^*, b^*, r^*, k^*, n_u, λ_u, p^u_{vw}, $u, v, w = 1,\ldots,m$, the dual design having m associate classes. From (32),

$$\underset{\sim}{D} = \sum_{u=0}^{m} d_{1u}\underset{\sim}{B}^*_u,$$

where

$$d_{10} = (k - 1)(\lambda v - k)^2/\lambda^2 v^2,$$
$$d_{1u} = [k(k - 2)\lambda_u + \lambda_u^2]/\lambda^2 v^2, \quad u = 1,\ldots,m, \tag{34}$$

and $\underset{\sim}{B}^*_u$, $u = 0,\ldots,m$, are the association matrices of the dual design. Use of Theorem 3.2 of Shah (1959) assures us that

$$\underset{\sim}{D}^{-1} = \sum_{u=0}^{m} s_u\underset{\sim}{B}^*_u, \tag{35}$$

where the s_u are solutions of the equations,

$$\sum_{u=0}^{m} n_u d_{1u} s_u = 1 \text{ and } \sum_{v=0}^{m}\sum_{w=0}^{m} p^u_{vw} d_{1v} s_w = 0, \quad u = 1,\ldots,m.$$

Explicit variance estimators are

$$G_j = s_0 S_j + \sum_{u=1}^{m} s_u \phi_u(S_j), \quad j = 1,\ldots,b, \tag{36}$$

where $\phi_u(S_j)$ is the sum of the S_j for those blocks having λ_u treatments in common with block j.

(e) Balanced Lattice Designs: Bradley and Rollier (1985) obtained some special results for BIB designs with GD PBIB(2) duals. One result of interest is for balanced lattice designs. Suppose that the BIB design is affine resolvable with $b = v + r - 1$ and let the design parameters be $v = nk$, $b = mn$, $r = m$, k, $\lambda = (k - 1)/(n - 1)$, $mn = nk + m - 1$, k/n and $(k - 1)/(n - 1)$ integers. The dual design is an SRGD PBIB(2) design. When $m = k + 1$, $n = k$, a balanced lattice design results. Then $d_{10} = (k - 1)^3/k^2$, $d_{11} = 0$, $d_{12} = (k - 1)^2/k^4$,

$$s_0 = (k^3 - k - 1)/(k - 1)^2(k^2 - k - 1), \quad s_1 = 1/(k - 1)^3(k^2 - k - 1)$$

and

$$s_2 = -1/(k - 1)^2(k^2 - k - 1).$$

The estimators G_j result from substitution in (36).

BIB designs with $\lambda = 1$, $b \neq v$ and BIB designs with triangular, singly linked block design duals were also considered in Bradley and Rollier (1985).

GENERALIZED TEST PROCEDURES

The first test of Russell and Bradley (1958) may be generalized to certain incomplete block designs. We limit consideration to connected general binary block designs, $n_{ij} = 0$ or 1, then consider a special class of designs, and provide examples. Without loss of generality, attention is focused on block 1. We wish to test the null hypothesis, $H_0:\sigma_1^2 = \sigma^2$, against the alternative hypothesis, $H_a:\sigma_1^2 \neq \sigma^2$, given that $\sigma_j^2 = \sigma^2$, $j = 2,\ldots,b$.

Some general results are needed under H_a. Let r_{ij} be the observed residual for treatment i in block j if $n_{ij} = 1$. Then

$$S_j = \sum_{i=1}^{v} n_{ij} r_{ij}^2 \tag{37}$$

and

$$n_{ij}r_{ij} = n_{ij}(\varepsilon_{ij} - \sum_s n_{sj}\varepsilon_{sj}/k_j - \sum_s\sum_t n_{st}c_{is}^-\varepsilon_{st}$$
$$+ \sum_s\sum_t n_{st}b_{it}\varepsilon_{st} + \sum_s\sum_t n_{st}b_{sj}\varepsilon_{st} - \sum_s\sum_t n_{st}e_{jt}\varepsilon_{st}), \tag{38}$$

where

$$b_{st} = \sum_h n_{ht}c_{hs}^-/k_t,$$

and (39)

$$e_{tt'} = \sum_h\sum_s n_{ht}n_{st'}c_{hs}^-/k_t k_{t'},\quad s = 1,\ldots,v,\ t, t' = 1,\ldots,b.$$

Note that (38) follows from the expression below (5), may be written with the ε_{st}'s replaced by y_{st}'s, and is a linear function of the usual design error contrasts. In addition, $\sum_i n_{ij}r_{ij} = 0$.

The variance-covariance matrix of $n_{ij}r_{ij}$, $i = 1,\ldots,v$, is provided for each j under H_a. We obtain

$$\begin{aligned}\operatorname{cov}(n_{ij}r_{ij}, n_{i'j}r_{i'j}) = n_{ij}n_{i'j}\{&[\delta_{1j}\sigma_1^2 + (1 - \delta_{1j})\sigma^2](\delta_{ii'} - c_{ii'}^- - \frac{1}{k_j} + b_{ij} \\ &+ b_{i'j} - e_{jj}) + (\sigma_1^2 - \sigma^2)[\sum_s n_{s1}c_{is}^-c_{i's}^- - k_1 b_{i1}b_{i'1} \\ &- \sum_s n_{s1}b_{sj}c_{is}^- - \sum_s n_{s1}b_{sj}c_{i's}^- + k_1 b_{i1}e_{1j} + k_1 b_{i'1}e_{1j} \\ &+ \sum_s n_{s1}b_{sj}^2 - k_1 e_{1j}^2 - \delta_{1j}(c_{ii'}^- - b_{i1} - b_{i'1} + e_{11})]\}.\end{aligned} \tag{40}$$

Since $\underset{\sim}{C}^-\underset{\sim}{C}\underset{\sim}{C}^- = \underset{\sim}{C}^-$ and $\underset{\sim}{C} = \underset{\sim}{R} - \underset{\sim}{N}\underset{\sim}{K}^{-1}\underset{\sim}{N}'$,

$$\sum_i r_i c_{hi}^- c_{si}^- = c_{hs}^- + \sum_j k_j b_{hj} b_{sj}. \tag{41}$$

Repeated use of (41) is required to obtain (40). If we set

$$g_{ii'} = \sum_h n_{h1}(c_{ii'}^- - c_{ih}^- - c_{i'h}^- + b_{h1}),$$

then (40) may be rewritten for $j = 1$,

$$\mathrm{Cov}(n_{i1}r_{i1}, n_{i'1}r_{i'1}) = n_{i1}n_{i'1}[\sigma_1^2(\delta_{ii'} - \frac{1}{k} - \frac{2}{k} g_{ii'} + \frac{1}{k^2} \sum_s n_{s1}g_{is}g_{i's}) + \sigma^2(\frac{1}{k} g_{ii'} - \frac{1}{k^2} \sum_s n_{s1}g_{is}g_{i's})]. \tag{42}$$

We restrict attention now to proper, binary, equireplicate incomplete block designs. Without loss of generality, let the treatments in block 1 be 1,...,k. Let

$$\underset{\sim}{H} = \underset{\sim}{I}_k - \frac{1}{k} \underset{\sim}{J}_{kk}$$

and $\underset{\sim}{H}$ is idempotent. From (42) and the definition of $g_{ii'}$, it follows that $\underset{\sim}{\Sigma}_1$, the dispersion matrix of $r_{11},\ldots,r_{k1}$, is

$$\underset{\sim}{\Sigma}_1 = \sigma_1^2(\underset{\sim}{H} - 2\underset{\sim}{H}\underset{\sim}{C}_1^-\underset{\sim}{H} + \underset{\sim}{H}\underset{\sim}{C}_1^-\underset{\sim}{H}\underset{\sim}{C}_1^-\underset{\sim}{H}) + \sigma^2(\underset{\sim}{H}\underset{\sim}{C}_1^-\underset{\sim}{H} - \underset{\sim}{H}\underset{\sim}{C}_1^-\underset{\sim}{H}\underset{\sim}{C}_1^-\underset{\sim}{H}), \tag{43}$$

where $\underset{\sim}{C}_1^-$ is the k-square principal minor of $\underset{\sim}{C}^-$.

It is clear that the distribution of $S_1 = \sum_{i=1}^{k} r_{i1}^2$ depends on the nature of $\underset{\sim}{C}_1^-$ and its relationship with $\underset{\sim}{H}$.

Substantial simplification of (43) seems to result only when

$$\underset{\sim}{C}_1^- = a\underset{\sim}{I}_k + a^*\underset{\sim}{J}_{kk}, \tag{44}$$

a, a* constants, $0 < a < 1$. Then $\underset{\sim}{C}_1^-\underset{\sim}{H} = a\underset{\sim}{H}$ and

$$\underset{\sim}{\Sigma}_1 = (1 - a)[\sigma_1^2 + a(\sigma^2 - \sigma_1^2)]\, \underset{\sim}{H} = (1 - a)(1 + \gamma)\sigma^2\underset{\sim}{H}, \tag{45}$$

$$\gamma = (1 - a)(\sigma_1^2 - \sigma^2)/\sigma^2. \tag{46}$$

Let $\underset{\sim}{r}_1$ be the k-element vector with typical element r_{i1}, $\underset{\sim}{r}_1 = (1 - a)^{\frac{1}{2}}\underset{\sim}{u}_1$, $\underset{\sim}{q} = \underset{\sim}{\Gamma}^*\underset{\sim}{u}_1$, and $\underset{\sim}{q}_1 = \underset{\sim}{\Gamma}\underset{\sim}{u}_1$, where $\underset{\sim}{\Gamma}^*$ is the k-square Helmert matrix, $\underset{\sim}{\Gamma}$ consists of the first (k - 1) zero-sum rows of $\underset{\sim}{\Gamma}^*$, and $\underset{\sim}{q}_1$ consists of the first (k - 1) elements of $\underset{\sim}{q}$. Note that $q_k = 0$ since $\sum_i r_{i1} = 0$. Now $(1 + \gamma)\sigma^2\underset{\sim}{H}$ is the dispersion matrix of the elements of $\underset{\sim}{u}_1$ and Theorem 9.2.1 of Rao and Mitra (1971) is enough to establish that $\underset{\sim}{u}_1'\underset{\sim}{u}_1/(1 + \gamma)\sigma^2 = \underset{\sim}{q}_1'\underset{\sim}{q}_1/(1 + \gamma)\sigma^2$ has the central chi-square distribution with (k - 1) degrees of freedom. Finally, $\underset{\sim}{q}_1'\underset{\sim}{q}_1 = S_1/(1 - a)$.

Let $\underset{\sim}{y}_1$ be the column vector of the k observations in block 1, let $\underset{\sim}{y}_2$ be the vector of the remaining k(b - 1) observations, and let $\underset{\sim}{r}_2$ be the vector of the corresponding remaining residuals. We may write

$$\begin{bmatrix} \underset{\sim}{r}_1 \\ \underset{\sim}{r}_2 \end{bmatrix} = \begin{bmatrix} \underset{\sim}{A}_{11} & \underset{\sim}{A}_{12} \\ \underset{\sim}{A}_{21} & \underset{\sim}{A}_{22} \end{bmatrix} \begin{bmatrix} \underset{\sim}{y}_1 \\ \underset{\sim}{y}_2 \end{bmatrix}$$

and define a set of orthogonal error constrasts as

$$\begin{bmatrix} q_1 \\ q_2 \end{bmatrix} = \begin{bmatrix} \Gamma & 0 \\ 0 & X \end{bmatrix} \begin{bmatrix} r_1 \\ r_2 \end{bmatrix},$$

where q_1 and q_2 have respectively $(k - 1)$ and $(k - 1)(b - 1)-(v - 1)$ elements, Γ is defined above, and X completes the transformation. Now

$$\begin{bmatrix} q_1 \\ q_2 \end{bmatrix} = \begin{bmatrix} \Gamma A_{11} & \Gamma A_{12} \\ 0 & X A_{22} \end{bmatrix} \begin{bmatrix} y_1 \\ y_2 \end{bmatrix}. \tag{47}$$

By choice of X, the matrix in (47) is orthogonal, $X A_{21} = 0$, and

$$\Gamma A_{12} A_{22}' X' = 0. \tag{48}$$

Our interest centers on two quadratic forms, $Q_1 = q_1' q_1 = y' B_1 y$ and $Q_2 = q_2' q_2 = y' B_2 y$, $y' = [y_1', y_2']$, the dispersion matrix of y being Σ_y, where

$$B_1 = \begin{bmatrix} 0 & 0 \\ 0 & A_{22}' X' X A_{22} \end{bmatrix}, \quad B_2 = \begin{bmatrix} A_{11}' \Gamma' \Gamma A_{11} & A_{11}' \Gamma' \Gamma A_{12} \\ A_{12}' \Gamma' \Gamma A_{11} & A_{22}' \Gamma' \Gamma A_{22} \end{bmatrix}, \quad \Sigma_y = \begin{bmatrix} \sigma_1^2 I_k & 0 \\ 0 & \sigma^2 I_{k(b-1)} \end{bmatrix}.$$

Now $E^* = Q_1 + Q_2$ and $Q_2 = E^* - [S_1/(1 - a)]$. It has been demonstrated that $Q_1/(1 + \gamma)\sigma^2$ has the central chi-square distribution with $(k - 1)$ degrees of freedom. Since q_2 depends only on y_2, Q_2 has the applicable distribution under the standard design model; Q_2/σ^2 has the central chi-square distribution with $[(k - 1)(b - 1) - (v - 1)]$ degrees of freedom. Theorem 9.4.2 of Rao and Mitra is enough to establish the independence of the two distributions since (48) confirms that $B_1 \Sigma_y B_2 = 0$. The desired result follows: Let

$$F = [(k - 1)(b - 1) - (v - 1)]S_1/(k - 1)[1 - a)E^* - S_1]. \tag{49}$$

Under H_0, $\gamma = 0$, and F has the central F-distribution with ν_1 and ν_2 degrees of freedom,

$$\nu_1 = (k - 1), \quad \nu_2 = [(k - 1)(b - 1) - (v - 1)]; \tag{50}$$

under H_a, $F/(1 + \gamma)$ has the same F-distribution. The test statistic for a generalization of the first test of Russell and Bradley and its distributions under both H_0 and H_a are now known.

Three examples follow.

(a) Randomized Complete Block Designs: For an RCB design, a form of C_1^- is (44) with $a = 1/b$, $a^* = -1/bv$, $k = v$. then

$$F = b(b - 2)S_1/[(b - 1)E^* - bS_1]$$

with $\gamma = (b - 1)(\sigma_1^2 - \sigma^2)/b\sigma^2$. The degrees of freedom for the F-distribution associated with F under H_0 and $F/(1 + \gamma)$ under H_a are $(v - 1)$ and $(v - 1)(b - 2)$.

This result is equivalent to that stated by Brindley and Bradley (1985).

(b) BIB Designs: The form of $\underset{\sim}{C}_1^-$ for BIB designs is again (44) where now $a = k/\lambda v$ and $a^* = -k/\lambda v^2$. Thus

$$F = \lambda v[(k - 1)(b - 1) - (v - 1)]S_1/(k - 1)[(\lambda v - k)E^* - \lambda v S_1]$$

with $\gamma = (\lambda v - k)(\sigma_1^2 - \sigma^2)/\lambda v\sigma^2$. The degrees of freedom are given in (50).

(c) Designs with One-Associate Class Blocks: Consider any PBIB(m) design such that any pair of treatments in block 1 are u-th associates, u = 1,2,...,or m. Then $\underset{\sim}{C}_1^-$ has the form (44) with a and a* determined by $\underset{\sim}{C}^-$ and (49) and (50) apply. Numerous examples exist, for example, PBIB(2) designs for which either λ_1 or λ_2 is zero.

REMARKS

The existence of Grubbs' estimators for broad classes of designs has been established and specific forms given for special classes of designs. Test 1 of Russell and Bradley has been extended to a substantial set of designs and other extensions may be possible. The general homogeneity test of block variances, Test 2 of Russell and Bradley, has not been considered. Test 2 is very difficult; Brindley and Bradley were able to obtain the exact distribution of the test statistic only for the RCB design with b = 3.

No numerical example has been given. The computation of S_j, j = 1,...,b, is trivial, as is application of the test procedure given. The test proposed has potential for important applications. If blocks are associated with material producers and treatments with processing units, the F-test provides a means of comparing variability of a selected producer with other producers. If blocks are associated with analysts, the inexperienced analyst may be compared with the experienced. In each suggested application, a "control chart" may be developed to monitor the producer or analyst. Indeed, each producer or analyst may be so monitored as a control procedure with little distortion likely due to lack of independence.

FELICITATIONS

The authors are pleased to have this opportunity to add their felicitations to Franklin A. Graybill to those of his many other friends. The subject of the paper seems particularly appropriate because of Frank's early interest, Graybill (1954).

Dwayne Rollier was Frank's student in graduate study at Oklahoma State University. Ralph Bradley participated with him at Southern Regional Education Board sponsored regional graduate summer sessions and in American Statistical Association activities, most recently being appreciative of Frank's support on the Coordinating Committee for the ASA Building and Development Fund.

ACKNOWLEDGEMENTS

This research was initiated at the Florida State University and supported in part at that University by the Army, Navy and Air Force through ONR Contract, NONR 988(08), Task Order NR 042-004, and a National Institute of General Medical Sciences Training Grant, 5-T01-GM00913. We appreciate also the support of the University of Georgia and Arizona State University as the research was completed.

REFERENCES

[1] Bose, R.C., Partially balanced incomplete block designs with two associate classes involving only two replications, Calcutta Statist. Assoc. Bulletin 3 (1951) 120-125.

[2] Bose, R.C., Clatworthy, W.H. and Shrikhande, S.S., Tables of partially balanced designs with two associate classes, Institute of Statistics, Univ. of North Carolina (1954).

[3] Bose, R.C. and Conner, W.S., Combinatorial properties of group divisible incomplete block designs, Ann. Math. Statist. 23 (1952) 367-383.

[4] Bose, R.C., Shrikhande, S.S. and Bhattacharya, K.N., On the construction of group divisible block designs, Ann. Math. Statist. 24 (1953) 167-195.

[5] Bradley, R.A. and Brindley, D.A., An approximate test of variance homogeneity based on Grubbs' estimators, Commun. Statist.-Simula. 15 (1986) 27-34.

[6] Bradley, R.A. and Rollier, D.A., The extension of Grubbs' estimators for classes of incomplete block designs, Tech. Rpt. No. 14, Dept. of Statist., Univ. of Georgia (April 1985).

[7] Brauer, A., Limits for the characteristic roots of a matrix. IV: Applications to stochastic matrices, Duke Math. Journal 19 (1952) 75-91.

[8] Brindley, D.A. and Bradley, R.A., Some new results on Grubbs' estimators, J. Amer. Statist. Assoc. 80 (1985) 711-714.

[9] Cochran, W.G. and Cox, G.M., Experimental Designs, 2nd ed. (John Wiley and Sons, New York, 1957).

[10] Ellenberg, J.H., The joint distribution of the standardized row sums of squares from a balanced two-way layout, J. Amer. Statist. Assoc. 72 (1977) 407-411.

[11] Graybill, F.A., Variance heterogeneity in a randomized block design, Biometrics 10 (1954) 516-520.

[12] Grubbs, F.E., On estimating precision of measuring instruments and product variability, J. Amer. Statist. Assoc. 43 (1948) 243-264.

[13] John, P.W.M., Statistical Design and Analysis of Experiments (John Wiley and Sons, New York, 1971).

[14] Maloney, C.J., Grubbs' estimators to date, Proc. 18th Conf. on Design of Experiments in Army Res. Devel. and Testing, Part 2 (1973) 523-547.

[15] Nair, K.R., Partially balanced incomplete block designs involving only two replications, Calcutta Statist. Assoc. Bulletin 3 (1950) 83-86.

[16] Plackett, R.L. and Burman, J.P., The design of optimum multifactorial experiments, Biometrika 33 (1946) 305-325.

[17] Rao, C.R., General methods of analysis for incomplete block designs, J. Amer. Statist. Assoc. 42 (1947) 541-561.

[18] Rao, C.R. and Mitra, S.K., Generalized inverse of matrices and its applications (John Wiley and Sons, New York, 1971).

[19] Roy, J. and Laha, R.G., On partially balanced linked block designs, Ann. Math. Statist. 28 (1957) 488-493.

[20] Russell, T.S. and Bradley, R.A., One-way variances in a two-way classification, Biometrika 45 (1958) 111-129.

[21] Shah, B.V., A generalization of partially balanced incomplete block designs, Ann. Math. Statist. 30 (1959) 1041-1050.

[22] Taussky, O., A recurring theorem on determinants, Amer. Math. Monthly 56 (1949) 672-676.

Original paper received: 14.05.85
Final paper received: 10.06.86

Paper recommended by J.S. Williams

PROBABILITY AND STATISTICS
Essays in Honor of Franklin A. Graybill
J.N. Srivastava (Editor)

APPLICATIONS OF INNOVATION REPRESENTATIONS IN TIME SERIES ANALYSIS

P.J. Brockwell and R.A. Davis

Department of Statistics
Colorado State University

ABSTRACT

If $\{X_t\}$ is a zero-mean second order process, it is known that the best linear mean-square predictor $\hat{X}_{n+1}$ of X_{n+1} based on $X_1, \cdots, X_n$ is expressible in terms of the innovations $(X_j - \hat{X}_j)$, $j = 1, \cdots, n$ as $\sum_{j=1}^{n} \theta_{nj}(X_{n+1-j} - \hat{X}_{n+1-j})$, where the coefficients θ_{nj} and the mean squared errors $v_n = E(X_{n+1} - \hat{X}_{n+1})^2$ can be found recursively from the covariance function of $\{X_t\}$. If $\{X_t\}$ is an ARMA(p,q) process defined by the equations, $\phi(B)X_t = \theta(B)Z_t$, then application of the recursions to the process $\{\phi(B)X_t\}$ gives the more compact representation,

$$\hat{X}_{n+1} = \phi_1 X_n + \cdots + \phi_p X_{n-p} + \sum_{j=1}^{q} \theta_{nj}(X_{n+1-j} - \hat{X}_{n+1-j}), n \geq max(p,q).$$

Applications of these representations to inference problems for time series are investigated.

1. INTRODUCTION

The aim of this paper is to collect together and extend some of the many applications of innovation representations in time series analysis. If $\{X_t, t = 1, 2, \cdots\}$ is an arbitrary zero-mean second order process, we define the innovations to be the orthogonal random variables $\{X_t - \hat{X}_t, t = 1, 2, \cdots\}$, where $\hat{X}_1 = 0$ and $\hat{X}_t, t \geq 2$, is the best linear predictor of X_t based on $\{X_1, \cdots, X_{t-1}\}$, i.e. the projection in L^2 of X_t on span$\{X_1, \cdots, X_{t-1}\}$. At the heart of the applications of innovations-based techniques is the representation,

$$\hat{X}_{n+1} = \sum_{j=1}^{n} \theta_{nj}(X_{n+1-j} - \hat{X}_{n+1-j}), \tag{1.1}$$

which is clearly valid for any zero-mean second order process since

$$\hat{X}_{n+1} \in \text{span}\{X_1, \cdots, X_n\} = \text{span}\{X_1 - \hat{X}_1, \cdots, X_n - \hat{X}_n\}.$$

The coefficients $\theta_{n1}, \cdots, \theta_{nn}, n = 1, 2, \cdots$, and the mean squared errors $v_n = E(X_{n+1} - \hat{X}_{n+1})^2$ can be found recursively from the autocovariances $\text{Cov}(X_i, X_j), i, j = 1, 2, \cdots$, as described in Proposition 2.1 below (or equivalently from the Cholesky factorization of the covariance matrix; see Rissanen and Barbosa (1969)).

In the special case when $\{X_t\}$ is the stationary ARMA process satisfying the equation,

$$X_t - \phi_1 X_{t-1} - \cdots - \phi_p X_{t-p} = Z_t + \theta_1 Z_{t-1} + \cdots + \theta_q Z_{t-q}, \tag{1.2}$$

Research supported by NSF Grant No. MCS 8501763

AMS 1980 subject classification: 62M20, 60G15, 60G10.

Key words and phrases: Best predictor, Gaussian likelihood, ARMA process, linear process, spectral density estimation, simulation, identification.

where $\{Z_t\} \sim WN(0,\sigma^2)$ and $(1-\phi_1 z - \cdots - \phi_p z^p) \neq 0$ for $|z| \leq 1$, we have the simpler representation,

$$\hat{X}_{n+1} = \sum_{j=1}^{p} \phi_j X_{n+1-j} + \sum_{j=1}^{q} \theta_{nj}(X_{n+1-j} - \hat{X}_{n+1-j}),\ n \geq max(p,q), \tag{1.3}$$

where θ_{nj} and $v_n = E(X_{n+1} - \hat{X}_{n+1})^2$ are found recursively by applying Proposition 2.1 to the transformed process,

$$Y_t = \begin{cases} X_t, & \text{if } t \leq max(p,q), \\ X_t - \sum_{j=1}^{p} \phi_j X_{t-j}, & \text{if } t > max(p,q), \end{cases}$$

used by Ansley (1979). If $(1+\theta_1 z + \cdots + \theta_q z^q) \neq 0$ for $|z| < 1$ then the coefficients θ_{nj} and mean squared errors v_n in the representation (1.3) have the properties $\theta_{nj} \to \theta_j$ and $v_n \to \sigma^2$ as $n \to \infty$.

The value of the representations (1.1) and (1.3) has been recognized for many years. The exact likelihood of the observations $\{X_1, \cdots, X_n\}$ of an arbitrary Gaussian process was expressed in terms of the one-step predictors $\hat{X}_{j+1}$ and their mean squared errors $v_j, j = 1, \cdots, n$, by Schweppe (1965). The one-step predictors were expressed in terms of the innovations (as in (1.1) and (1.3)) by Rissanen and Barbosa (1969) and Rissanen (1973). Ansley (1979) used the innovations representation of the Gaussian likelihood and Cholesky factorization to compute the exact likelihood of the observations $\{X_1, \cdots, X_n\}$ of an ARMA process. This is equivalent to the method of likelihood calculation described in Section 4 below. The innovations representation of the likelihood of observations of an ARMA process can also be used in conjunction with the Kalman filter, as in Jones(1980) and Gardner, Harvey and Phillips(1980), to compute exact Gaussian likelihoods. Kailath and co-workers, in a series of papers (see for example Kailath (1970)), have fruitfully exploited innovation representations of both discrete and continuous time processes in estimation and signal-processing problems.

In spite of the widespread use of techniques based on innovations, there appears to be no unified account of their application to the simulation, modelling and prediction of time series data. This paper attempts to fill the gap. In Section 2 we derive the basic recursions for the coefficients θ_{nj} and mean squared errors v_n in the representation (1.1). These are then used in Section 3 for simulation and exact likelihood calculation for zero-mean Gaussian processes which are not necessarily stationary. The only assumption made is that the covariance matrices $\Gamma_n = [E(X_i X_j)]_{i,j=1,\cdots,n}$ are all non-singular. In Section 4 we specialize to zero-mean Gaussian ARMA processes, and for these we discuss (a) prediction, (b) likelihood calculation, (c) simulation, (d) preliminary estimation, (e) goodness of fit tests and (f) spectral estimation. We conclude with an application of the modelling, prediction and estimation techniques to the International Airline Passenger data of Box and Jenkins (1976).

The techniques proposed for preliminary estimation (important for accelerating likelihood maximization) and spectral estimation are new. The underlying idea for these techniques is to estimate the coefficients $\psi_1, \psi_2, \cdots$, in the MA(∞) representation,

$$X_t = Z_t + \sum_{j=1}^{\infty} \psi_j Z_{t-j},$$

of the process (1.2). To do this we apply Proposition 2.1 to the sample covariance matrix $\hat{\Gamma}_{m+1}$ giving estimates $\hat{\theta}_{m1}, \cdots, \hat{\theta}_{mm}$ of the corresponding quantities $\theta_{m1}, \cdots, \theta_{mm}$ obtained when Proposition 2.1 is applied to the true covariance matrix Γ_{m+1}. Provided m depends on the sample size n in such a way that $m \to \infty$ and $m = o(n^{1/3})$ as $n \to \infty$, then $(\hat{\theta}_{m1}, \cdots, \hat{\theta}_{mm}, 0, 0, \cdots)$ is consistent for $(\psi_1, \psi_2, \cdots)$ and its asymptotic distribution has a simple form (see Theorem 4.2). This permits the specification of large-sample confidence intervals for the MA(∞) coefficients, the corresponding autoregressive and moving average coefficients, and the spectral density (see Section 4).

The use of $\hat{\theta}_m = (\hat{\theta}_{m1}, \cdots, \hat{\theta}_{mm})$ to estimate $\psi_m = (\psi_1, \cdots, \psi_m)$ is completely analogous to the use of the Yule-Walker estimators $\hat{\phi}_m = (\hat{\phi}_{m1}, \cdots, \hat{\phi}_{mm})' = \hat{\Gamma}_m^{-1}(\hat{\gamma}(1), \cdots, \hat{\gamma}(m))'$ to estimate the coefficient vector $-\pi_m = -(\pi_1, \cdots, \pi_m)'$ in the AR(∞) representation,

$$X_t = Z_t - \sum_{j=1}^{\infty} \pi_j X_{t-j},$$

of the process (1.2). The asymptotic behaviour of $\hat{\phi}_m$ as $n \to \infty$ where $m(n) \to \infty$ and $m(n) = o(n^{1/3})$ is given in Theorem 4.1, the proof of which follows the ideas of Berk (1974). The estimators $\hat{\phi}_m$ and $\hat{\theta}_m$ are connected by the relation,

$$\begin{bmatrix} 1 & 0 & \cdots & 0 & 0 \\ -\hat{\phi}_{11} & 1 & \cdots & 0 & 0 \\ \vdots & \vdots & \ddots & \vdots & \vdots \\ -\hat{\phi}_{m-1,m-1} & -\hat{\phi}_{m-1,m-2} & \cdots & 1 & 0 \\ -\hat{\phi}_{mm} & -\hat{\phi}_{m,m-1} & \cdots & -\hat{\phi}_{m1} & 1 \end{bmatrix} = \begin{bmatrix} 1 & 0 & \cdots & 0 & 0 \\ \hat{\theta}_{11} & 1 & \cdots & 0 & 0 \\ \vdots & \vdots & \ddots & \vdots & \vdots \\ \hat{\theta}_{m-1,m-1} & \hat{\theta}_{m-1,m-2} & \cdots & 1 & 0 \\ \hat{\theta}_{mm} & \hat{\theta}_{m,m-1} & \cdots & \hat{\theta}_{m1} & 1 \end{bmatrix}^{-1}.$$

Since the purpose of this paper is to describe and illustrate the practical use of innovation-based techniques, we have omitted the derivations of the asymptotic results. These can be found in Brockwell and Davis (1986).

2. THE INNOVATIONS ALGORITHM

Let $\{X_n, n = 0, \pm 1, \pm 2, \cdots\}$ be any zero-mean, finite-variance process defined on some probability space (Ω, F, P). We shall assume that the covariance function

$$\kappa(i,j) = E(X_i X_j)$$

is such that the matrices $\Gamma_n = [\kappa(i,j)]_{i,j=1,\cdots,n}$ are all non-singular. Let H be the Hilbert space of real-valued square-integrable random variables on (Ω, F, P) with inner product $\langle X, Y \rangle = E(XY)$, and let $M_n, n \geq 0$, be the subspaces $M_0 = \text{span}\{1\}$ and $M_n = \text{span}\{1, X_1, \cdots, X_n\}, n \geq 1$. Define $\hat{X}_n, n = 1, 2, \cdots$, to be the projections (or one-step linear predictors),

$$\hat{X}_n = P_{M_{n-1}} X_n, \quad n = 1, 2, \cdots, \tag{2.1}$$

and v_{n-1}, $n = 1, 2, \cdots$, to be the corresponding mean squared errors,

$$v_{n-1} = E(X_n - \hat{X}_n)^2. \tag{2.2}$$

Note that if $\{X_n\}$ is a Gaussian process then $\hat{X}_{n+1} = E(X_{n+1}|X_1, \cdots, X_n)$, $n \geq 0$.

It is easy to see that

$$M_n = \text{span}\{1, X_1 - \hat{X}_1, X_2 - \hat{X}_2, \cdots, X_n - \hat{X}_n\},$$

and since the process has zero mean, $\hat{X}_{n+1}$ can be expressed as a linear combination of the *innovations* $X_k - \hat{X}_k$, $k = 1, \cdots, n$, i.e.

$$\hat{X}_{n+1} = \sum_{j=0}^{n-1} \theta_{n,n-j}(X_{j+1} - \hat{X}_{j+1}).$$

A recursive scheme for computing $\{\theta_{nj}, j = 1, \cdots, n; v_n\}, n = 1, 2, \cdots$, is given in the following proposition.

Proposition 2.1 The Innovations Algorithm If $\{X_t\}$ has zero mean and $E(X_iX_j) = \kappa(i,j)$, then the one-step predictors $\hat{X}_n, n \geq 1$, and their mean-squared errors $v_{n-1}, n \geq 1$, are given by

$$(2.3) \qquad \hat{X}_{n+1} = \begin{cases} 0 & \text{if } n = 0, \\ \sum_{j=0}^{n-1} \theta_{n,n-j}(X_{j+1} - \hat{X}_{j+1}) & \text{if } n \geq 1. \end{cases}$$

and

$$(2.4) \qquad \begin{cases} v_0 = \kappa(1,1), \\ \theta_{n,n-k} = v_k^{-1}(\kappa(n+1,k+1) - \sum_{j=0}^{k-1} \theta_{k,k-j}\theta_{n,n-j}v_j), \ k = 0,1,\cdots,n-1, \\ v_n = \kappa(n+1,n+1) - \sum_{j=0}^{n-1} \theta_{n,n-j}^2 v_j. \end{cases}$$

(It is trivial to solve (2.4) recursively in the order $v_0; \theta_{11}, v_1; \theta_{22}, \theta_{21}, v_2; \theta_{33}, \theta_{32}, \theta_{31}, v_3; \cdots$.)

Proof The set $\{X_1 - \hat{X}_1, X_2 - \hat{X}_2, \cdots, X_n - \hat{X}_n\}$ is orthogonal since $(X_i - \hat{X}_i) \in M_{j-1}$ for $i < j$ and $(X_j - \hat{X}_j) \perp M_{j-1}$ by definition of $\hat{X}_j$. Taking the inner product on both sides of (2.3) with $X_{k+1} - \hat{X}_{k+1}, 0 \leq k < n$, we have

$$\langle \hat{X}_{n+1}, X_{k+1} - \hat{X}_{k+1} \rangle = \theta_{n,n-k} v_k.$$

Since $(X_{n+1} - \hat{X}_{n+1}) \perp (X_{k+1} - \hat{X}_{k+1})$, the coefficients $\theta_{n,n-k}, k = 0, \cdots, n-1$ are given by

$$(2.5) \qquad \theta_{n,n-k} = v_k^{-1} \langle X_{n+1}, X_{k+1} - \hat{X}_{k+1} \rangle.$$

Making use of the representation (2.3) with n replaced by k, we obtain

$$\theta_{n,n-k} = v_k^{-1}(\kappa(n+1,k+1) - \sum_{j=0}^{k-1} \theta_{k,k-j} \langle X_{n+1}, X_{j+1} - \hat{X}_{j+1} \rangle),$$

and since $\langle X_{n+1}, X_{j+1} - \hat{X}_{j+1} \rangle = v_j \theta_{n,n-j}$, $0 \leq j < n$, this may be rewritten in the form

$$\theta_{n,n-k} = v_k^{-1}(\kappa(n+1,k+1) - \sum_{j=0}^{k-1} \theta_{k,k-j}\theta_{n,n-j}v_j),$$

as required. Since $\hat{X}_{n+1} \perp (X_{n+1} - \hat{X}_{n+1})$, we can write $v_n = EX_{n+1}^2 - E\hat{X}_{n+1}^2$, and from (2.3) and the orthogonality of the innovations we then obtain,

$$v_n = \kappa(n+1,n+1) - \sum_{k=0}^{n-1} \theta_{n,n-k}^2 v_k. \qquad \square$$

The above algorithm can also be used to find the h-step predictors $P_{M_n} X_{n+h}$ for $h \geq 1$. In particular it is easy to show that

$$P_{M_n} X_{n+h} = \sum^{n-1+h} \theta_{n+h-1,j}(X_{n+h-j} - \hat{X}_{n+h-j}),$$

where the coefficients $\theta_{n,n-j}$ are determined as before by (2.4). Moreover the mean squared error can be expressed as

$$E(X_{n+h} - P_{M_n}X_{n+h})^2 = \kappa(n+h, n+h) - \sum_{j=h}^{n-1+h} \theta^2_{n+h-1,j} v_{n+h-j-1}.$$

3. APPLICATIONS TO GENERAL GAUSSIAN PROCESSES

In this section we consider some applications of Proposition 2.1. Throughout this section, $\{X_n\}$ is assumed to be Gaussian with zero mean and covariance function $\kappa(i,j) = EX_iX_j$. Let $\underset{\sim}{X}_N = (X_1, \cdots, X_N)'$, $\hat{\underset{\sim}{X}}_N = (\hat{X}_1, \cdots, \hat{X}_N)'$, $\Gamma_N = E(\underset{\sim}{X}_N \underset{\sim}{X}_N')$, and assume that Γ_N is nonsingular.

(a) **Factorization of covariance matrices** Let C be the $N \times N$ matrix

$$\begin{bmatrix} 1 & 0 & \cdots & 0 & 0 \\ \theta_{11} & 1 & \cdots & 0 & 0 \\ \vdots & \vdots & \ddots & \vdots & \vdots \\ \theta_{N-2,N-2} & \theta_{N-2,N-3} & \cdots & 1 & 0 \\ \theta_{N-1,N-1} & \theta_{N-1,N-2} & \cdots & \theta_{N-1,1} & 1 \end{bmatrix},$$

and let V be the $N \times N$ diagonal matrix, $V = \mathrm{diag}(v_0, v_1, \cdots, v_{N-1})$. Then we have from Proposition 2.1,

$$\hat{\underset{\sim}{X}}_N = (C - I_N)(\underset{\sim}{X}_N - \hat{\underset{\sim}{X}}_N),$$

where I_N is the $N \times N$ identity matrix. This implies that

$$\underset{\sim}{X}_N = \underset{\sim}{X}_N - \hat{\underset{\sim}{X}}_N + \hat{\underset{\sim}{X}}_N = C(\underset{\sim}{X}_N - \hat{\underset{\sim}{X}}_N), \tag{3.1}$$

and hence that

$$\Gamma_N = CVC'. \tag{3.2}$$

(b) **Likelihood calculation** From (3.1) and (3.2), we immediately obtain

$$\underset{\sim}{X}_N' \Gamma_N^{-1} \underset{\sim}{X}_N = (\underset{\sim}{X}_N - \hat{\underset{\sim}{X}}_N)' V^{-1} (\underset{\sim}{X}_N - \hat{\underset{\sim}{X}}_N) = \sum_{j=1}^{N} v_{j-1}^{-1} (X_j - \hat{X}_j)^2 \tag{3.3}$$

and

$$|\Gamma_N| = |C|^2 |V| = v_0 v_1 \cdots v_{N-1}. \tag{3.4}$$

The likelihood of the vector $\underset{\sim}{X}_N$ is therefore

$$L(\Gamma_N) = (2\pi)^{-N/2} (v_0 \cdots v_{N-1})^{-1/2} \exp\left\{ -\frac{1}{2} \sum_{j=0}^{N-1} v_j^{-1} (X_{j+1} - \hat{X}_{j+1})^2 \right\}. \tag{3.5}$$

(c) **Simulation** In view of (3.1) it is a simple matter to generate realizations of a zero-mean Gaussian process with specified autocovariance function, κ, provided

$$\Gamma_N = [\kappa(i,j)]_{i,j=1,\cdots,N}$$

is non-singular for every N. If $\underset{\sim}{Z}_N = (Z_1, \cdots, Z_N)'$ has independent standard normal components, then by the factorization (3.2),

$$\underset{\sim}{X}_N = CV^{1/2}\underset{\sim}{Z}_N$$

is a zero-mean Gaussian random vector with covariance matrix Γ_N for every N.This implies that

$$X_n = v_{n-1}^{1/2} Z_n + \sum_{j=1}^{n-1} \theta_{n-1,j} v_{n-j-1}^{1/2} Z_{n-j}, n = 1, 2, \cdots, \tag{3.6}$$

has the required joint distributions. Simulation is performed by generating the sequence $Z_1, Z_2, \cdots$, and then substituting in (3.6) with θ_{nj} and v_n determined recursively by (2.4).

Example 3.1 Consider the stationary Gaussian process $\{X_n\}$ with zero mean, spectral density

$$f_X(\omega) = \sigma^2(\pi - |\omega|)/\pi^2, \quad -\pi \leq \omega \leq \pi,$$

and corresponding autocovariance function,

$$\kappa(i,j) = \gamma(i-j) = \begin{cases} \sigma^2, & |i-j| = 0, \\ 4\pi^{-2}\sigma^2|i-j|^{-2}, & |i-j| = 1, 3, 5, \cdots, \\ 0 & \text{otherwise.} \end{cases}$$

It is a trivial matter to compute the coefficients θ_{ij}, the variances v_j, the predictors $\hat{X}_j$, and hence the likelihood of $\underset{\sim}{X}_N = (X_1, \cdots, X_N)'$ from (2.3), (2.4) and (3.5). We find that

$$L(\sigma^2) = (2\pi\sigma^2)^{-N/2}(r_0 \cdots r_{N-1})^{-1/2} \exp\left\{-\frac{1}{2}\sigma^{-2}\sum_{j=1}^{N} r_{j-1}^{-1}(X_j - \hat{X}_j)^2\right\}, \tag{3.7}$$

where $r_i = v_i/\sigma^2$ and θ_{ij}, $j \leq i$, are shown for $i \leq 9$ in Table 1 and $\hat{\underset{\sim}{X}}_N$ is found from $\underset{\sim}{X}_N$ using (2.3). Maximizing (3.7) with respect to σ^2 gives the maximum likelihood estimator,

$$\hat{\sigma}^2 = N^{-1}\sum_{j=1}^{N} r_{j-1}^{-1}(X_j - \hat{X}_j)^2. \tag{3.8}$$

Having found θ_{ij}, $j \leq i$, and r_i, $i = 0, \cdots, N-1$, we can use (3.6) to simulate $\{X_1, \cdots, X_N\}$ for any specified value of σ^2. For any realization it is equally simple to compute the maximum likelihood estimate of σ^2 from (3.8). It should be noted that Kalman filtering techniques for computing the likelihood are not applicable in this case since the process $\{X_n\}$ has no finite-dimensional state-space representation.

TABLE 1
The coefficients θ_{ij} and $r_i = v_i/\sigma^2, i = 0, \cdots, 9$, for Example 3.1

$i \backslash j$	1	2	3	4	5	6	7	8	9	r_i
0										1.0000
1	.4053									.8357
2	.4849	0								.8035
3	.5154	-.0218	.0450							.7841
4	.5325	-.0272	.0539	0						.7747
5	.5430	-.0328	.0600	-.0079	.0162					.7675
6	.5503	-.0354	.0630	-.0098	.0194	0				.7631
7	.5556	-.0379	.0656	-.0122	.0222	-.0040	.0083			.7594
8	.5596	-.0395	.0672	-.0134	.0235	-.0050	.0099	0		.7568
9	.5628	-.0410	.0686	-.0147	.0249	-.0061	.0115	-.0024	.0050	.7546

4. APPLICATIONS TO ARMA PROCESSES

Prediction, likelihood calculation and simulation for ARMA processes can be carried out simply as a special case of Section 3. However, as we shall see in this section, the linear structure of ARMA processes can be exploited to simplify the general results. Consider the stationary Gaussian ARMA process

$$X_n = \phi_1 X_{n-1} + \cdots + \phi_p X_{n-p} + \theta_0 Z_n + \cdots + \theta_q Z_{n-p}, \tag{4.1}$$

where $1 - \phi_1 z - \cdots - \phi_p z^p \neq 0$ for $|z| \leq 1, \{Z_n\}$ is an i.i.d. sequence of $N(0, \sigma^2)$ random variables, and $\theta_0 = 1$. Without loss of generality we shall assume that the coefficients θ_i and white noise variance σ^2 have been adjusted (without affecting the joint distributions of $\{X_n\}$) to ensure that $1 + \theta_1 z + \cdots + \theta_q z^q \neq 0$ for $|z| < 1$. Since we allow any of the coefficients to be zero, we may also assume that $p \geq 1$ and $q \geq 1$.

(a) **Prediction** Instead of applying the innovations algorithm (Proposition 2.1) directly to $\{X_n\}$, we apply it to the transformed process (cf. Ansley (1979)),

$$Y_n = \begin{cases} X_n, & n \leq m = \max(p, q), \\ X_n - \phi_1 X_{n-1} - \cdots - \phi_p X_{n-p}, & n > m. \end{cases} \tag{4.2}$$

The covariance function $\gamma_Y(i, j)$ of $\{Y_n\}$ is easily expressed in terms of the covariance function $\gamma(h) = E(X_{n+h} X_n)$ as follows:

$$\gamma_Y(i, j) = \begin{cases} \gamma(i - j), & 1 \leq i, j \leq m, \\ \gamma(i - j) - \sum_{r=1}^{p} \phi_r \gamma(r - |i - j|), & \min(i, j) \leq m < \max(i, j) \leq 2m, \\ \sigma^2 \sum_{r=0}^{q} \theta_r \theta_{r+|i-j|}, & \min(i, j) > m, \\ 0, & \text{otherwise}, \end{cases} \tag{4.3}$$

where we have adopted the convention $\theta_j = 0$ for $j > q$.

Notice that span$\{1, X_1, \cdots, X_n\}$ and span$\{1, Y_1, \cdots, Y_n\}$ are the same, so we shall denote them both by M_n and write $\hat{X}_n$ and $\hat{Y}_n$ for the projections on M_{n-1} of X_n and Y_n respectively ($M_0 = \text{span}\{1\}$).

The crucial property of $\{Y_n\}$ which facilitates the application of Proposition 2.1 is the zero value of $\gamma_Y(i,j)$ for $i,j > m$ and $|i-j| > q$. As a result, when we substitute γ_Y for κ in the proposition, we obtain

$$\theta_{nj} = 0 \text{ for } n \geq m \text{ and } j > q, \tag{4.4}$$

whence

$$\hat{Y}_{n+1} = \begin{cases} \sum_{j=1}^{n} \theta_{nj}(Y_{n+1-j} - \hat{Y}_{n+1-j}), & 1 \leq n < m, \\ \sum_{j=1}^{q} \theta_{nj}(Y_{n+1-j} - \hat{Y}_{n+1-j}), & n \geq m. \end{cases} \tag{4.5}$$

To determine $\hat{X}_n$ from $\hat{Y}_n$ we project each side of (4.2) onto M_{n-1} to obtain

$$\hat{Y}_n = \begin{cases} \hat{X}_n, & 1 \leq n \leq m, \\ \hat{X}_n - \phi_1 X_{n-1} - \cdots - \phi_p X_{n-p}, & n > m, \end{cases} \tag{4.6}$$

which, with (4.2), shows that

$$Y_n - \hat{Y}_n = X_n - \hat{X}_n \text{ for all } n \geq 1. \tag{4.7}$$

From (4.5), (4.6) and (4.7) we finally obtain $\hat{X}_1 = 0$,

$$\hat{X}_{n+1} = \begin{cases} \sum_{j=1}^{n} \theta_{nj}(X_{n+1-j} - \hat{X}_{n+1-j}), & 1 \leq n < m, \\ \phi_1 X_n + \cdots + \phi_p X_{n+1-p} + \sum_{j=1}^{q} \theta_{nj}(X_{n+1-j} - \hat{X}_{n+1-j}), & n \geq m, \end{cases} \tag{4.8}$$

and

$$E(X_{n+1} - \hat{X}_{n+1})^2 = v_n, \tag{4.9}$$

where θ_{nj} and v_n are obtained from Proposition 2.1 with $\kappa(i,j) = \gamma_Y(i,j)$.

Remarks 1. If Proposition 2.1 is applied to the covariance function

$$\kappa^*(i,j) = \gamma_Y(i,j)/\sigma^2,$$

(or equivalently to $\gamma_Y(i,j)$ with σ^2 set equal to one), the resulting coefficients and mean squared errors are $\theta^*_{nj} = \theta_{nj}$ and $v^*_n = v_n/\sigma^2$. We shall use the notation

$$r_n = v_n/\sigma^2, \tag{4.10}$$

observing that r_n depends on $\underset{\sim}{\phi} = (\phi_1, \cdots, \phi_p)'$ and $\underset{\sim}{\theta} = (\theta_1, \cdots, \theta_q)'$ but not on σ^2.

2. The coefficients θ_{nj} and mean squared errors v_n in (4.8) and (4.9) have the properties

$$\theta_{nj} \to \theta_j \text{ as } n \to \infty, \tag{4.11}$$

$$v_n \downarrow \sigma^2 \text{ as } n \to \infty. \tag{4.12}$$

(Recall that $\theta_1, \cdots, \theta_q$ and σ^2 are the parameter values satisfying $1 + \theta_1 z + \cdots + \theta_q z^q \neq 0$ for $|z| < 1$.)

3. Since we know that $\hat{X}_1 = 0$, equations (4.8) allow the calculation of $\hat{X}_2, \hat{X}_3, \cdots$. For $n > m$ the recursive calculation of $\hat{X}_n$ requires the storage of only p observations and q innovations. Moreover for large n the computations can be greatly simplified by using the asymptotic values from (4.11) and (4.12).

4. For higher order ARMA processes the covariance function $\gamma_x(i,j)$ and the corresponding coefficients θ_{nj} and mean squared errors v_n must be computed numerically. For the case $p = q = 1$ however, the recursions can be written in an elementary algebraic form. Details are given in the following example.

Example 4.1 The ARMA(1,1) Process Suppose that

$$X_n - \phi X_{n-1} = Z_n + \theta Z_{n-1}, \text{ with } |\phi| < 1 \text{ and } Z_n \sim N(0,\sigma^2). \tag{4.13}$$

Then the recursion relations for $\hat{X}_n$ and $r_n (= v_n/\sigma^2)$ reduce to

$$\hat{X}_{n+1} = \phi X_n + \theta r_{n-1}^{-1}(X_n - \hat{X}_n), \quad n \geq 1, \tag{4.14}$$

and

$$r_n = 1 + \theta^2 - \theta^2 / r_{n-1}, \tag{4.15}$$

with initial conditions $\hat{X}_1 = 0$ and $r_0 = (1 + 2\theta\phi + \phi^2)/(1 - \phi^2)$. The substitution $y_n = r_n/(r_n - 1)$ reduces (4.15) to a linear difference equation which is easily solved to give

$$r_n = \frac{(1+\phi\theta)^2 - (\phi+\theta)^2\theta^{2n+2}}{(1+\phi\theta)^2 - \theta^{-2}(\phi+\theta)^2\theta^{2n+2}} \quad \text{if } \theta^2 \neq 1, \tag{4.16}$$

and

$$r_n = 1 + (1+\phi)/[(n+1)(1-\phi)] \quad \text{if } \theta^2 = 1. \tag{4.17}$$

We observe from these results that $v_n (= r_n\sigma^2) \to \sigma^2$ and $\theta_{n1} \to \theta$ if $|\theta| \leq 1$ in accordance with (4.11) and (4.12). If however $|\theta| > 1$, then $v_n \to \theta^2\sigma^2$ and $\theta_{n1} \to \theta^{-1}$. In the latter case it is necessary to reparametrize the process before applying (4.11) and (4.12). Thus if $|\theta| > 1$, the process $\{X_n\}$ defined by

$$X_n - \phi X_{n-1} = \tilde{Z}_n + \tilde{\theta}\tilde{Z}_{n-1}, \qquad \tilde{Z}_n \sim N(0,\tilde{\sigma}^2),$$

with $\tilde{\theta} = \theta^{-1}$ and $\tilde{\sigma}^2 = \theta^2\sigma^2$, has the same distribution as the original process and satisfies the condition $1 + \tilde{\theta}z \neq 0$ for $|z| < 1$. We can therefore apply (4.11) and (4.12) to deduce that $v_n \to \theta^2\sigma^2$ and $\theta_{n1} \to \theta^{-1}$ as observed directly from (4.16) and (4.17).

Recursive calculation of $\hat{X}_1, \hat{X}_2, \cdots$ from (4.14) using the observations $X_1, X_2, \cdots$ and the initial value $\hat{X}_1 = 0$ is quite elementary. In the next example we illustrate the general numerical procedure for a higher order model.

Example 4.2 An ARMA(2,3) Process Simulated values, $X_1, \cdots, X_{10}$, of the ARMA(2,3) process,

$$X_t - X_{t-1} + 0.24X_{t-2} = Z_t + 0.4Z_{t-1} + 0.2Z_{t-2} + 0.1Z_{t-3}, \qquad Z_t \sim N(0,1), \tag{4.18}$$

are shown in Table 2.

In order to find the one-step predictors $\hat{X}_{n+1}$, $n = 1, \cdots, 10$, we first need the covariances $\gamma(h)$, $h = 0,1,2$ of $\{X_t\}$, which are easily found on multiplying (4.18) by X_{t-h}, $h = 0,1,2$ and taking expectations to get three equations for $\gamma(0), \gamma(1)$ and $\gamma(2)$. Thus we find

$$\gamma(0) = 7.17133, \quad \gamma(1) = 6.44139 \quad \gamma(2) = 5.06027.$$

Substituting in (4.3) we find that the symmetric matrix $\Gamma_Y = [\gamma_Y(i,j)]_{i,j=1,2,\cdots}$ is given by,

$$(4.19) \qquad \Gamma_Y = \begin{bmatrix} 7.17133 & \cdot & \cdot & \cdot & \cdot & \cdot & \cdot & \cdot \\ 6.44139 & 7.17133 & \cdot & \cdot & \cdot & \cdot & \cdot & \cdot \\ 5.06027 & 6.44139 & 7.17133 & \cdot & \cdot & \cdot & \cdot & \cdot \\ 0.10 & 0.34 & 0.816 & 1.21 & \cdot & \cdot & \cdot & \cdot \\ 0 & 0.10 & 0.34 & 0.50 & 1.21 & \cdot & \cdot & \cdot \\ 0 & 0 & 0.10 & 0.24 & 0.50 & 1.21 & \cdot & \cdot \\ \cdot & 0 & 0 & 0.10 & 0.24 & 0.50 & 1.21 & \cdot \\ \cdot & \cdot & 0 & 0 & \cdot & \cdot & \cdot & \cdot \\ \cdot & \cdot & \cdot & \cdot & \cdot & \cdot & \cdot & \cdot \\ \cdot & \cdot & \cdot & \cdot & \cdot & \cdot & \cdot & \cdot \end{bmatrix}.$$

The next step is to apply Proposition 2.1 with $\kappa(i,j) = \gamma_Y(i,j)$, giving θ_{nj} and v_{n-1}, $j = 1,\cdots,n$; $n = 1,\cdots,10$. Then

$$\hat{X}_{n+1} = \sum_{j=1}^{n} \theta_{nj}(X_{n+1-j} - \hat{X}_{n+1-j}), \qquad n = 1,2,$$

$$\hat{X}_{n+1} = X_n - 0.24X_{n-1} + \sum_{j=1}^{3} \theta_{nj}(X_{n+1-j} - \hat{X}_{n+1-j}), \qquad n = 3,4,\cdots,$$

and

$$E(X_{n+1} - \hat{X}_{n+1})^2 = v_n.$$

TABLE 2

Calculation of X_{n+1} for data from the ARMA(2,3) process of (4.18)

n	X_{n+1}	v_n	θ_{n1}	θ_{n2}	θ_{n3}	$\hat{X}_{n+1}$
0	1.704	7.1713				0
1	0.527	1.3856	0.8982			1.5305
2	1.041	1.0057	1.3685	0.7056		-0.1710
3	0.942	1.0019	0.4008	0.1806	0.0139	1.2428
4	0.555	1.0016	0.3998	0.2020	0.0722	0.7443
5	-1.002	1.0005	0.3992	0.1995	0.0994	0.3138
6	-0.585	1.0000	0.4000	0.1997	0.0998	-1.7293
7	0.010	1.0000	0.4000	0.2000	0.0998	-0.1688
8	-0.638	1.0000	0.4000	0.2000	0.0999	0.3193
9	0.525	1.0000	0.4000	0.2000	0.1000	-0.8731
10		1.0000	0.4000	0.2000	0.1000	1.0638

h-step prediction If $P_n X_{n+h}$ denotes the projection of X_{n+h} onto $M_n = \text{span}\{1, X_1, \cdots, X_n\}$, i.e. the best m.s. predictor of X_{n+h} in M_n, then the same arguments used to derive (4.8) lead to the equations,

$$(4.20) \qquad P_n X_{n+h} = \begin{cases} \sum_{j=h}^{n+h-1} \theta_{n+h-1,j}(X_{n+h-j} - \hat{X}_{n+h-j}), & n+h \le m, \\ \phi_1 P_n X_{n+h-1} + \cdots + \phi_p P_n X_{n+h-p} \\ \qquad + \sum^{q} \theta_{n+h-1,j}(X_{n+h-j} - \hat{X}_{n+h-j}), & n+h > m. \end{cases}$$

The mean squared error of $P_n X_{n+h}$ (assuming $n > m$, as is usually the case in practice) is given by,

$$\sigma_n^2(h) = E(X_{n+h} - P_n X_{n+h})^2 = \sum_{j=1}^{h} \Big(\sum_{r=1}^{j} \chi_{r-1}\theta_{n+h-r,j-r}\Big)^2 v_{n+h-j}, \tag{4.21}$$

where θ_{n0} is defined to be 1 and

$$\chi(z) = \sum_{r=0}^{\infty} \chi_r z^r = (1 - \phi_1 z - \cdots - \phi_p z^p)^{-1}, \quad |z| < 1.$$

Given the observations $X_1, \cdots, X_n$, it is a trivial matter to compute $P_n X_{n+1}, P_n X_{n+2}, P_n X_{n+3}, \cdots$ recursively from (4.20) and the corresponding mean squared errors from (4.21).

Using the asymptotic behaviour of v_n and θ_{nj} for large n, we can approximate (4.20) and (4.21) for large n by

$$P_n X_{n+h} = \phi_1 P_n X_{n+h-1} + \cdots + \phi_p P_n X_{n+h-p} + \sum_{j=h}^{q} \theta_j (X_{n+h-j} - \hat{X}_{n+h-j}) \tag{4.22}$$

and

$$\sigma_n^2(h) = \sum_{j=0}^{h-1} \psi_j^2 \sigma^2, \tag{4.23}$$

where

$$\psi(z) = \sum_{r=0}^{\infty} \psi_r z^r = (1 + \theta_1 z + \cdots + \theta_q z^q)(1 - \phi_1 z - \cdots - \phi_p z^p)^{-1}, \quad |z| < 1.$$

The following table shows the h-step predicted values $P_4 X_{4+h}$, $h = 1, \cdots, 7$, based on the first four observations in Table 2, together with the corresponding mean squared errors $\sigma_4^2(h)$. The last two columns are the asymptotic approximations based on (4.22) and (4.23).

TABLE 3
h-step predictors based on the first four observations in Table 2

h	X_{4+h}	$P_4 X_{4+h}$	$\sigma_4^2(h)$	$P_4 X_{4+h}$ (approx.)	$\sigma_4^2(h)$ (approx.)
1	0.555	0.7443	1.0016	0.7771	1.0000
2	-1.002	0.5787	2.9583	0.6121	2.9600
3	-0.585	0.3700	4.8127	0.3955	4.8096
4	0.010	0.2311	6.0755	0.2486	6.0730
5	-0.638	0.1423	6.7104	0.1537	6.7091
6	0.525	0.0869	6.9885	0.0940	6.9878
7		0.0527	7.1013	0.0571	7.1009

In this particular case the asymptotic and exact forms of $P_n X_{n+h}$ and $\sigma^2(h)$ are in good agreement.

(b) **Likelihood calculation** From (3.5) we can write the likelihood of the observations $X_1, \cdots, X_n$ in the form,

$$L(\underset{\sim}{\phi},\underset{\sim}{\theta},\sigma^2) = (2\pi\sigma^2)^{-n/2}(r_0 \cdots r_{n-1})^{-1/2} \exp\left\{-\frac{1}{2}\sigma^{-2}\sum_{j=1}^{n} r_{j-1}^{-1}(X_j - \hat{X}_j)^2\right\}, \tag{4.24}$$

where

$$\psi(z) = \sum_{r=0}^{\infty} \psi_r z^r = (1 + \theta_1 z + \cdots + \theta_q z^q)(1 - \phi_1 z - \cdots - \phi_p z^p)^{-1}, \quad |z| < 1.$$

The following table shows the h-step predicted values $P_4 X_{4+h}$, $h = 1, \cdots, 7$, based on the first four observations in Table 2, together with the corresponding mean squared errors $\sigma_4^2(h)$. The last two columns are the asymptotic approximations based on (4.22) and (4.23).

TABLE 3
h-step predictors based on the first four observations in Table 2

h	X_{4+h}	$P_4 X_{4+h}$	$\sigma_4^2(h)$	$P_4 X_{4+h}$ (approx.)	$\sigma_4^2(h)$ (approx.)
1	0.555	0.7443	1.0016	0.7771	1.0000
2	-1.002	0.5787	2.9583	0.6121	2.9600
3	-0.585	0.3700	4.8127	0.3955	4.8096
4	0.010	0.2311	6.0755	0.2486	6.0730
5	-0.638	0.1423	6.7104	0.1537	6.7091
6	0.525	0.0869	6.9885	0.0940	6.9878
7		0.0527	7.1013	0.0571	7.1009

In this particular case the asymptotic and exact forms of $P_n X_{n+h}$ and $\sigma_n^2(h)$ are in good agreement.

(b) **Likelihood calculation** From (3.5) we can write the likelihood of the observations $X_1, \cdots, X_n$ in the form,

$$L(\underset{\sim}{\phi},\underset{\sim}{\theta},\sigma^2) = (2\pi\sigma^2)^{-n/2}(r_0 \cdots r_{n-1})^{-1/2} \exp\left\{-\frac{1}{2}\sigma^{-2}\sum_{j=1}^{n} r_{j-1}^{-1}(X_j - \hat{X}_j)^2\right\}, \tag{4.24}$$

where $\hat{X}_j$ is defined by (4.8) and the coefficients θ_{nj} and $r_n (= v_n/\sigma^2)$ are computed recursively from Proposition 2.1 with $\kappa(i,j) = \gamma_Y(i,j)$ defined as in (4.3) and $\sigma^2 = 1$. Noting that $\hat{X}_j$ and r_j are independent of σ^2, we deduce from (4.24) that the maximum likelihood estimators, $\hat{\underset{\sim}{\phi}}$, $\hat{\underset{\sim}{\theta}}$, and $\hat{\sigma}^2$ satisfy

$$\hat{\sigma}^2 = n^{-1} S(\hat{\underset{\sim}{\phi}}, \hat{\underset{\sim}{\theta}}), \tag{4.25}$$

where

$$S(\hat{\underset{\sim}{\phi}}, \hat{\underset{\sim}{\theta}}) = \sum_{j=1}^{n} (X_j - \hat{X}_j)^2 / r_{j-1}, \tag{4.26}$$

and $\hat{\phi}, \hat{\theta}$ are the values of ϕ, θ which minimize

(4.27)
$$\ell(\underset{\sim}{\phi},\underset{\sim}{\theta}) = \ln S(\underset{\sim}{\phi},\underset{\sim}{\theta}) + n^{-1}\sum_{j=1}^{n} \ln r_{j-1}.$$

Maximum likelihood estimation of $\underset{\sim}{\phi}$ and $\underset{\sim}{\theta}$ is carried out using the algorithm in conjunction with a non-linear minimization program, and the maximum likelihood estimate of σ^2 is then found from (4.25). Least squares estimation is carried out by minimizing $S(\underset{\sim}{\phi},\underset{\sim}{\theta})$ instead of $\ell(\underset{\sim}{\phi},\underset{\sim}{\theta})$.

(c) Simulation Simulation of the process $\{Y_n\}$ defined by (4.2) can be carried out exactly as described in (3.6) starting with an i.i.d. $N(0,1)$ sequence $\{Z_n\}$. Since $\{Y_n, n > m = \max(p,q)\}$ is an MA(q) process, equation (3.6) takes the simple form,

$$Y_n = \begin{cases} v_{n-1}^{1/2} Z_n + \sum_{j=1}^{n-1} \theta_{n-1,j} v_{n-1-j}^{1/2} Z_{n-j}, & n \le m, \\ v_{n-1}^{1/2} Z_n + \sum_{j=1}^{q} \theta_{n-1,j} v_{n-1-j}^{1/2} Z_{n-j}, & n > m, \end{cases}$$

where θ_{nj} and v_n are found from Proposition 2.1 with $\kappa(i,j) = \gamma_Y(i,j)$. From (4.2), the corresponding simulated values of the ARMA process $\{X_n\}$ are then found recursively from

$$X_n = \begin{cases} Y_n, & n \le m, \\ Y_n + \phi_1 X_{n-1} + \cdots + \phi_p X_{n-p}, & n > m. \end{cases}$$

(d) Preliminary estimation The algorithm of Section 2 may be used to give preliminary estimates of the parameters in an ARMA(p,q) process. Let $\{X_t\}$ be the causal invertible ARMA(p,q) process,

(4.28)
$$X_t - \phi_1 X_{t-1} - \cdots - \phi_p X_{t-p} = Z_t + \theta_1 Z_{t-1} + \cdots + \theta_q Z_{t-q},$$

where $\{Z_t\}$ is an i.i.d. sequence of random variables with $EZ_t = 0, EZ_t^2 = \sigma^2$ and $EZ_t^4 < \infty$. The causality and invertibility assumptions ensure that $X_t = \sum_{j=0}^{\infty} \psi_j Z_{t-j}$ and $Z_t = \sum_{j=0}^{\infty} \pi_j X_{t-j}$, where the sequences $\{\psi_j\}$ and $\{\pi_j\}$ converge to zero at a geometric rate.

For $m < n$, define $\Gamma_m = [\gamma(i-j)]_{i,j=1}^{m}$, $\hat{\Gamma}_m = [\hat{\gamma}(i-j)]_{i,j=1}^{m}$ where $\gamma(h) = EX_t X_{t+h}$ is the autocovariance function of the process and

$$\hat{\gamma}(h) = \frac{1}{h}\sum_{t=1}^{n-|h|} X_t X_{t+|h|}$$

is the sample autocovariance function based on the observations $X_1, X_2, \cdots, X_n$. If we let $\underset{\sim}{\phi}_m = (\phi_{m1}, \cdots, \phi_{mm})'$ be the solution of the Yule-Walker equations,

(4.29)
$$\underset{\sim}{\phi}_m = \Gamma_m^{-1}\underset{\sim}{\gamma}_m, \quad \underset{\sim}{\gamma}_m = [\gamma(1), \cdots, \gamma(m)]',$$

then $\underset{\sim}{\phi}_m$ minimizes $E(X_t - \phi_{m1}X_{t-1} - \cdots - \phi_{mm}X_{t-m})^2$ and

(4.30)
$$\hat{X}_{m+1} = \sum_{j=1}^{m} \phi_{mj} X_{m+1-j}.$$

An obvious estimate of $\underset{\sim}{\phi}_m$ is given by

(4.31)
$$\hat{\underset{\sim}{\phi}}_m = \hat{\Gamma}_m^{-1}\hat{\underset{\sim}{\gamma}}_m, \quad \hat{\underset{\sim}{\gamma}}_m = [\hat{\gamma}(1), \cdots, \hat{\gamma}(m)]'.$$

The vector $\hat{\underset{\sim}{\phi}}_m$ may be computed recursively using the Durbin-Levinson algorithm. Now $-\underset{\sim}{\phi}_m \to \underset{\sim}{\pi} = (\pi_1, \pi_2, \cdots)'$ as $m \to \infty$, which suggests estimating $\underset{\sim}{\pi}_m = (\pi_1, \cdots, \pi_m)'$ by $-\hat{\underset{\sim}{\phi}}_m$. The asymptotic distribution of $\hat{\underset{\sim}{\phi}}_m$ (which demonstrates that $-\hat{\underset{\sim}{\phi}}_m$ is a consistent estimator of $\underset{\sim}{\pi}_m$) is given in the following theorem. In the statement of the theorem, Γ_∞ denotes the covariance matrix $[\gamma(i-j)]_{i,j=1}^{\infty}$ of $\{X_t\}$, which is easily seen to have the form

$$\Gamma_\infty = \sigma^2 TT', \tag{4.32}$$

where

$$T = \begin{bmatrix} 1 & \psi_1 & \psi_2 & \psi_3 & \cdot & \cdot & \cdot \\ 0 & 1 & \psi_1 & \psi_2 & \cdot & \cdot & \cdot \\ 0 & 0 & 1 & \psi_1 & \cdot & \cdot & \cdot \\ 0 & 0 & 0 & 1 & \cdot & \cdot & \cdot \\ \cdot & \cdot & \cdot & \cdot & \cdot & \cdot & \cdot \\ \cdot & \cdot & \cdot & \cdot & \cdot & \cdot & \cdot \end{bmatrix}. \tag{4.33}$$

The inverse Γ_∞^{-1} has the form

$$\Gamma_\infty^{-1} = \sigma^{-2}(T')^{-1}T^{-1}, \tag{4.34}$$

and since $(1 + \pi_1 z + \pi_2 z^2 + \cdots)^{-1}(1 + \psi_1 z + \psi_2 z^2 + \cdots) \equiv 1$ for $|z| \le 1$, we have

$$T^{-1} = \begin{bmatrix} 1 & \pi_1 & \pi_2 & \pi_3 & \cdot & \cdot & \cdot \\ 0 & 1 & \pi_1 & \pi_2 & \cdot & \cdot & \cdot \\ 0 & 0 & 1 & \pi_1 & \cdot & \cdot & \cdot \\ 0 & 0 & 0 & 1 & \cdot & \cdot & \cdot \\ \cdot & \cdot & \cdot & \cdot & \cdot & \cdot & \cdot \\ \cdot & \cdot & \cdot & \cdot & \cdot & \cdot & \cdot \end{bmatrix}. \tag{4.35}$$

Theorem 4.1 Let $\{X_t\}$ be the ARMA(p,q) process defined by (4.28) and let $\{m(n),\ n = 1, 2, \cdots\}$ be a sequence of integers satisfying the conditions, $m < n$, $m \to \infty$ and $m = o(n^{1/3})$. Then in $\mathbb{R}^\infty$,

$$n^{1/2}(\hat{\phi}_{m1} + \pi_1, \cdots, \hat{\phi}_{mm} + \pi_m, 0, 0, \cdots) \Rightarrow N(\underset{\sim}{0}, \sigma^2\Gamma_\infty^{-1}),$$

where $N(\underset{\sim}{0}, \sigma^2\Gamma_\infty^{-1})$ denotes a Gaussian process with mean 0 and covariance function given by the entries of $\sigma^2\Gamma_\infty^{-1}$. In particular,

$$n^{1/2}(\hat{\phi}_{mj} + \pi_j) \Rightarrow N(0, \sum_{k=0}^{j-1} \pi_k^2),$$

where $\pi_0 = 1$. □

The proof of this result follows the ideas of Berk (1974) and can be found in Brockwell and Davis (1986).

Since the coefficients $\{\pi_j\}$ satisfy the relation

$$\pi_j = -\sum_{i=1}^{\min(j,q)} \theta_i \pi_{j-i} - \phi_j, \ j = 1, 2, \cdots, p+q, \tag{4.36}$$

where $\pi_0 = 1$ and $\phi_j = 0$ for $j > p$, one method for obtaining initial estimates of $\underset{\sim}{\phi} = (\phi_1, \cdots, \phi_p)'$ and $\underset{\sim}{\theta} = (\theta_1, \cdots, \theta_q)'$ is to choose m large and then solve (4.36) for $\underset{\sim}{\phi}$ and $\underset{\sim}{\theta}$ with $\underset{\sim}{\pi}_m$ replaced by

$-\hat{\underset{\sim}{\phi}}_m$. (A different approach to the estimation of $\underset{\sim}{\phi}$ and $\underset{\sim}{\theta}$ using the equations (4.36) is described by Fuller (1976), p.359.)

Notice that if $\{X_t\}$ is an AR(p) process then this procedure estimates $\underset{\sim}{\phi}$ by $(\hat{\phi}_{m1}, \cdots, \hat{\phi}_{mp})'$ which by Theorem 4.1 is asymptotically normal with covariance matrix,

$$\begin{bmatrix} 1 & 0 & \dots & 0 & 0 \\ \pi_1 & 1 & \dots & 0 & 0 \\ \vdots & \vdots & \ddots & \vdots & \vdots \\ \pi_{p-2} & \pi_{p-3} & \dots & 1 & 0 \\ \pi_{p-1} & \pi_{p-2} & \dots & \pi_1 & 1 \end{bmatrix} \begin{bmatrix} 1 & \pi_1 & \dots & \pi_{p-2} & \pi_{p-1} \\ 0 & 1 & \dots & \pi_{p-3} & \pi_{p-2} \\ \vdots & \vdots & \ddots & \vdots & \vdots \\ 0 & 0 & \dots & 1 & \pi_1 \\ 0 & 0 & \dots & 0 & 1 \end{bmatrix}. \tag{4.37}$$

On the other hand, the Yule-Walker estimate of $\underset{\sim}{\phi}$ is asymptotically normal with covariance matrix $\sigma^2\Gamma_p^{-1}$ which is not quite the same as the matrix in (4.37).

An alternative method for finding initial parameter estimates, particularly useful for moving average models, is based on estimating the coefficients $\underset{\sim}{\theta}_m = (\theta_{m1}, \cdots, \theta_{mm})'$ in the innovations algorithm. Let

$$\hat{\underset{\sim}{\theta}}_m = (\hat{\theta}_{m1}, \cdots, \hat{\theta}_{mm})' \tag{4.38}$$

denote the estimate of $\underset{\sim}{\theta}_m$ obtained from the recursion (2.4) with $\kappa(i,j) = \hat{\gamma}(i-j)$, $i,j = 1, \cdots, m+1$. As the next theorem shows, $\hat{\underset{\sim}{\theta}}_m$ is a consistent estimator of $\underset{\sim}{\psi}_m = (\psi_1, \psi_2, \cdots, \psi_m)'$.

Theorem 4.2 Let $\{X_t\}$ be the ARMA(p,q) process in (4.28) and let $\{m(n),\ n = 1, 2, \cdots\}$ be a sequence of integers such that $m < n$, $m \to \infty$ and $m = o(n^{1/3})$. Then in $\mathbb{R}^\infty$,

$$n^{1/2}(\hat{\theta}_{m1} - \psi_1, \hat{\theta}_{m2} - \psi_2, \cdots, \hat{\theta}_{mm} - \psi_m, 0, 0, \cdots) \Rightarrow N(0, T'T)$$

where $N(0, T'T)$ denotes a Gaussian process with mean zero and covariance function given by the entries in $T'T$. In particular,

$$n^{1/2}(\hat{\theta}_{mj} - \psi_j) \Rightarrow N(0, \sum_{k=0}^{j-1} \psi_k^2),$$

where $\psi_0 = 1$. □

The proof of this theorem is given in Brockwell and Davis (1986).

Initial estimates of $\underset{\sim}{\phi}$ and $\underset{\sim}{\theta}$ are now found by solving

$$\hat{\theta}_{mj} = \theta_j + \sum_{i=1}^{\min(j,p)} \phi_i \hat{\theta}_{m,j-i}, \qquad j = 1, 2, \cdots, p+q, \tag{4.39}$$

for $\underset{\sim}{\phi}$ and $\underset{\sim}{\theta}$ where $\hat{\theta}_{m,0} = 1$, $\theta_j = 0$ for $j > q$ and m satisfies the assumptions of Theorem 4.2. This procedure is parallel to the method based on (4.36) with the roles of $\underset{\sim}{\pi}_m$ and $\underset{\sim}{\psi}_m$, $\hat{\underset{\sim}{\phi}}_m$ and $\hat{\underset{\sim}{\theta}}_m$, and $\underset{\sim}{\phi}$ and $\underset{\sim}{\theta}$ interchanged. (A slightly different approach is to estimate $\underset{\sim}{\phi}$ from the lagged Yule-Walker equations and then solve the equations in (4.39), $j = 1, \cdots, q$, for $\underset{\sim}{\theta}$.)

Example 4.3 The MA(q) process For a pure moving average process, (4.39) always has a solution for $\underset{\sim}{\theta} = (\theta_1, \cdots, \theta_q)'$ given by $(\hat{\theta}_{m1}, \cdots, \hat{\theta}_{mq})'$. In this case, we have by Theorem 4.2 that

$n^{1/2}(\hat{\theta}_{m1}-\theta_1,\cdots,\hat{\theta}_{mq}-\theta_q)$ is asymptotically normal with zero mean and covariance matrix whose (i,j) element is equal to $\sum_{k=0}^{\min(i,j)-1}\theta_k\theta_{k+|i-j|}$, where $\theta_0=1$ and $\theta_j=0$ for $j>q$. There are several advantages of this initial parameter estimation procedure over the moment method of estimation as suggested by Wilson (1969). First, an estimate of $\underset{\sim}{\theta}$ using Wilson's method does not always exist. Second, if a moment estimate of $\underset{\sim}{\theta}$ does exist then the asymptotic variance of the estimator is difficult to compute for moving averages of order greater than one. Moreover the estimator we propose has substantially better asymptotic efficiency than the moment estimator. For example in the MA(1) case, the asymptotic efficiency of the moment estimator of θ_1 relative to the estimator $\hat{\theta}_{m1}$ is $e_1(\theta_1)=(1+\theta_1^2+4\theta_1^4+\theta_1^6+\theta_1^8)^{-1}(1-\theta_1^2)^2\leq 1$ (the moment estimate of θ_1 is $[1-(1-4\hat{\rho}^2(1))^{1/2}]/2\hat{\rho}(1)$ where $\hat{\rho}(1)=\hat{\gamma}(1)/\hat{\gamma}(0)$). In particular, $e_1(.25)=.82$, $e_1(.50)=.37$, $e_1(.75)=.06$. The efficiency of the estimator $\hat{\theta}_{m1}$ relative to the maximum likelihood estimator is $e_2(\theta_1)=(1-\theta_1^2)\leq 1$, with $e_2(.25)=.94$, $e_2(.50)=.75$, $e_2(.75)=.44$.

Example 4.4 The ARMA(p,p) process For an ARMA(p,p) process we have

$$\underset{\sim}{\psi}=M\underset{\sim}{\beta}$$

where $\underset{\sim}{\psi}=(\psi_1,\cdots,\psi_{2p})'$, $\underset{\sim}{\beta}=(\theta_1,\cdots,\theta_p,\phi_1,\cdots,\phi_p)'$ and

$$M=\begin{bmatrix}
1 & 0 & \cdot & \cdot & \cdot & 0 & 1 & 0 & 0 & \cdot & \cdot & 0\\
0 & 1 & & & & \cdot & \psi_1 & 1 & 0 & & & \cdot\\
\cdot & & \cdot & & & \cdot & \psi_2 & \psi_1 & 1 & & & \cdot\\
\cdot & & & \cdot & & \cdot & \cdot & \cdot & & & \cdot & \cdot\\
\cdot & & & & \cdot & 0 & \cdot & \cdot & & & & 0\\
0 & \cdot & \cdot & \cdot & 0 & 1 & \psi_{p-1} & \psi_{p-2} & & \cdot & \cdot & 1\\
0 & \cdot & \cdot & \cdot & \cdot & 0 & \psi_p & \psi_{p-1} & & \cdot & \cdot & \psi_1\\
\cdot & & & & & \cdot & \cdot & \cdot & & & & \cdot\\
\cdot & & & & & \cdot & \cdot & \cdot & & & & \cdot\\
\cdot & & & & & \cdot & \cdot & \cdot & & & & \cdot\\
0 & \cdot & \cdot & \cdot & \cdot & 0 & \psi_{2p-1} & \psi_{2p-2} & & \cdot & \cdot & \psi_p
\end{bmatrix}.$$

Solving (4.39) we obtain

$$\hat{\underset{\sim}{\psi}}=\hat{M}\hat{\underset{\sim}{\beta}}$$

where $\hat{\underset{\sim}{\psi}}=(\hat{\theta}_{m1},\cdots,\hat{\theta}_{m,2p})'$ and the elements of M are replaced by their estimated counterparts. Consequently,

$$\begin{aligned}
n^{1/2}(\hat{\underset{\sim}{\beta}}-\underset{\sim}{\beta}) &= n^{1/2}\hat{M}^{-1}(\hat{\underset{\sim}{\psi}}-\underset{\sim}{\psi}-(\hat{M}-M)\underset{\sim}{\beta})\\
&\sim n^{1/2}M^{-1}(\hat{\underset{\sim}{\psi}}-\underset{\sim}{\psi}-(\hat{M}-M)\beta),
\end{aligned}$$

and since

$$(\hat{\underset{\sim}{\psi}}-\underset{\sim}{\psi})-(\hat{M}-M)\underset{\sim}{\beta}=\begin{bmatrix}
1 & 0 & \cdot & \cdot & \cdot & \cdot & \cdot & \cdot & 0 & 0\\
-\phi_1 & 1 & 0 & \cdot & \cdot & \cdot & \cdot & \cdot & \cdot & 0\\
\cdot & & \cdot & & & & & & & \cdot\\
\cdot & & & \cdot & & & & & & \cdot\\
-\phi_p & & & & \cdot & & & & & \cdot\\
0 & & & & & \cdot & & & & \cdot\\
\cdot & & & & & & \cdot & & & \cdot\\
0 & \cdot & 0 & -\phi_p & \cdot & \cdot & -\phi_1 & & 1 & 0\\
0 & 0 & \cdot & 0 & -\phi_p & \cdot & \cdot & & -\phi_1 & 1
\end{bmatrix}(\hat{\underset{\sim}{\psi}}-\underset{\sim}{\psi})$$

$$=A(\hat{\underset{\sim}{\psi}}-\underset{\sim}{\psi}),$$

we have by Theorem 4.2

$$n^{1/2}(\hat{\beta} - \beta) \Rightarrow N(0, M^{-1}A(B'B)(M^{-1}A)')$$

where

$$B' = \begin{bmatrix} 1 & 0 & \cdots & 0 & 0 \\ \psi_1 & 1 & \cdots & 0 & 0 \\ \psi_2 & \psi_1 & \cdot & \cdot & \cdot \\ \cdot & \cdot & \cdot & \cdot & \cdot \\ \cdot & \cdot & & \cdot & \cdot \\ \cdot & \cdot & & 1 & 0 \\ \psi_{2p-1} & \psi_{2p-2} & & \psi_1 & 1 \end{bmatrix}.$$

A straightforward calculation gives,

$$C = AB' = \begin{bmatrix} 1 & 0 & \cdot & \cdot & & \cdot & \cdot & 0 & 0 \\ \theta_1 & 1 & 0 & \cdot & & \cdot & \cdot & \cdot & 0 \\ \cdot & & \cdot & & & & & & \cdot \\ \cdot & & & \cdot & & & & & \cdot \\ \theta_p & & & & \cdot & & & & \cdot \\ 0 & & & & & \cdot & & & \cdot \\ \cdot & & & & & & \cdot & & \cdot \\ 0 & \cdot & 0 & \theta_p & \cdot & \cdot & \theta_1 & 1 & 0 \\ 0 & 0 & \cdot & 0 & \theta_p & \cdot & \cdot & \theta_1 & 1 \end{bmatrix}$$

and hence

$$n^{1/2}(\hat{\beta} - \beta) \Rightarrow (0, M^{-1}C(M^{-1}C)').$$

The fact that M is partitioned into four $p \times p$ submatrices (one of which is the identity matrix and another the zero matrix) greatly facilitates the computation of M^{-1}.

In the special case $p = q = 1$, writing $\phi = \phi_1$ and $\theta = \theta_1$, we have $\psi_1 = \phi + \theta$,

$$M = \begin{bmatrix} 1 & 1 \\ 0 & \psi_1 \end{bmatrix}, \quad M^{-1} = \psi_1^{-1}\begin{bmatrix} \psi_1 & -1 \\ 0 & 1 \end{bmatrix}, \quad C = \begin{bmatrix} 1 & 0 \\ \theta & 1 \end{bmatrix},$$

and hence

$$(M^{-1}C)(M^{-1}C)' = (\theta + \phi)^{-2}\begin{bmatrix} \phi^2 + 1 & \phi\theta - 1 \\ \phi\theta - 1 & \theta^2 + 1 \end{bmatrix}. \tag{4.40}$$

Consequently $n^{1/2}(\hat{\theta} - \theta, \hat{\phi} - \phi)'$ is asymptotically normal with mean zero and covariance matrix (4.40). The asymptotic efficiencies of $\hat{\theta}$ and $\hat{\phi}$ relative to the corresponding maximum likelihood estimators are $(1 - \theta^2)(1 + \phi\theta)^2/(1 + \phi^2)$ and $(1 - \phi^2)(1 + \phi\theta)^2/(1 + \theta^2)$ respectively.

The moment estimators $\tilde{\theta}$ and $\tilde{\phi}$, obtain by equating the theoretical and sample autocorrelations $\rho(i)$ and $\hat{\rho}(i)$ at lags 1 and 2, are frequently used as preliminary parameter estimators. The estimator $\tilde{\phi}$ is equal to $\hat{\rho}(2)/\hat{\rho}(1)$ and the estimator $\tilde{\theta}$ (if it exists) is the solution with absolute value less than or equal to 1 of the quadratic equation, $(1 + \tilde{\theta}^2 + 2\tilde{\phi}\tilde{\theta})\hat{\rho}(1) = (1 + \tilde{\phi}\tilde{\theta})(\tilde{\phi} + \tilde{\theta})$.

Table 4 shows the asymptotic efficiencies of $\hat{\theta}$ and $\tilde{\theta}$ relative to the maximum likelihood estimator. Over most of the parameter space, $\hat{\theta}$ is vastly superior to $\tilde{\theta}$, although there is a small region in the vicinity of $\theta = -\phi$ where $\tilde{\theta}$ is marginally better.

Table 5 gives the asymptotic efficiencies of $\hat{\phi}$ and $\tilde{\phi}$ relative to the maximum likelihood estimator. Here the comparison is not so clear cut, since there is one region of the parameter space where $\hat{\phi}$ is distinctly superior, and another where $\tilde{\phi}$ is distinctly superior. In order to obtain reasonably efficient estimators of both θ and ϕ, these observations suggest that we should first compute $(\hat{\theta}, \hat{\phi})$. If $(\hat{\theta}, \hat{\phi})$ falls in the region of the parameter space where $\tilde{\phi}$ is superior, then use $\tilde{\phi}$ in place of $\hat{\phi}$. In this way we obtain an estimator of ϕ whose asymptotic efficiency is the maximum of the two efficiencies listed in Table 5.

Preliminary estimators which are easy to compute and which have reasonably high efficiency are extremely useful as initial values for the iterative non-linear optimization required to find maximum likelihood or least squares estimators. Good initial estimates can greatly reduce the computational time, especially when fitting high order models. A second important use of preliminary estimates is in model identification, as illustrated in our analysis of the airline data below (see (g)).

TABLE 4
Asymptotic efficiencies of $(\hat{\theta}, \tilde{\theta})$

$\phi\backslash\theta$	0	.2	.4	.6	.8	1.0
1.0	(.5,0)	(.691,0)	(.823,0)	(.819,0)	(.583,0)	(0,0)
.8	(.610,.017)	(.788,.007)	(.892,003)	(.855,.001)	(.590,.000)	(0,0)
.6	(.735,.271)	(.885,.125)	(.950,.048)	(.870,.014)	(.580,.002)	(0,0)
.4	(.862,.833)	(.965,.544)	(.974,.247)	(.848,.075)	(.541,.011)	(0,0)
.2	(.962,.993)	(.998,.851)	(.942,.514)	(.772,.187)	(.466,.030)	(0,0)
0	(1,1)	(.960,.919)	(.840,.666)	(.640,.301)	(.360,.053)	(0,0)
-.2	(.962,.993)	(.851,.915)	(.684,.693)	(.477,.366)	(.244,.075)	(0,0)
-.4	(.862,.833)	(.700,.855)	(.511,.627)	(.319,.328)	(.144,.071)	(0,0)
-.6	(.735,.271)	(.547,.449)	(.357,.428)	(.193,.212)	(.072,.039)	(0,0)
-.8	(.610,.017)	(.413,.037)	(.237,.071)	(.106,.071)	(.028,.015)	(0,0)
-1.0	(.5,0)	(.307,0)	(.151,0)	(.051,0)	(.007,0)	(0,0)

TABLE 5
Asymptotic efficiencies of $(\hat{\phi}, \tilde{\phi})$

$\phi\backslash\theta$	0	.2	.4	.6	.8	1.0
1.0	(0,.5)	(0,.632)	(0,.720)	(0,.771)	(0,.794)	(0,.800)
.8	(.360,.709)	(.466,.737)	(.541,.739)	(.580,.722)	(.590,.696)	(.583,.666)
.6	(.640,.885)	(.772,.821)	(.848,.735)	(.870,.654)	(.855,.585)	(.819,.527)
.4	(.840,.975)	(.942,.866)	(.974,.706)	(.950,.571)	(.892,.468)	(.823,.391)
.2	(.960,.998)	(.998,.873)	(.965,.658)	(.885,.480)	(.788,.354)	(.691,.268)
0	(1,1)	(.961,.861)	(.862,.600)	(.735,.389)	(.610,.252)	(.500,.167)
-.2	(.960,.998)	(.851,.851)	(.700,.546)	(.547,.307)	(.413,.167)	(.307,.091)
-.4	(.840,.975)	(.684,.851)	(.511,.511)	(.357,.239)	(.237,.101)	(.151,.041)
-.6	(.640,.885)	(.477,.821)	(.319,.507)	(.193,.193)	(.106,.056)	(.051,.013)
-.8	(.360,.709)	(.244,.645)	(.144,.486)	(.072,.184)	(.028,.028)	(.007,.002)
-1.0	(0,.5)	(0,.340)	(0,.188)	(0,.075)	(0,.016)	(0,0)

(e) Goodness of fit tests In the course of either the maximum likelihood or least squares parameter estimation described in (b) the standardized residuals $(X_j - \hat{X}_j)/v_{j-1}^{1/2}$, $j = 1, \cdots, n$, are computed and stored. If the fitted model is appropriate these residuals should be independently and identically distributed standard normal random variables, and the usual tests for goodness of fit can be made on this basis. In particular the graph of the residuals themselves should be examined for evidence of deviation from expected behaviour of an i.i.d. standard normal sequence. The sample autocorrelations $\hat{\rho}_r(h)$ of the residuals should be (for large sample size) approximately i.i.d. normal random variables with standard deviation $n^{-1/2}$, giving the usual check on the magnitudes of the autocorrelations. Strictly speaking the low-lag autocorrelations should be somewhat smaller (Box and Pierce (1970)) since we are examining *estimated* rather than true residuals. The portmanteau statistic of Box and Pierce (1970), $Q = n \sum_{h=1}^{m} \hat{\rho}_r(h)^2$, will be distributed approximately as $\chi^2(m - p - q)$ degrees of freedom, giving an overall test for the autocorrelations. Significant deviations of the autocorrelations $\hat{\rho}_r(h)$ from their expected behaviour assuming the validity of the model may suggest a more appropriate model.

(f) Spectral estimation The spectral density of the process $\{X_t\}$ defined by (4.1) is

$$f(\lambda) = |\sum_{j=0}^{\infty} \psi_j e^{-ij\lambda}|^2 \sigma^2/(2\pi), \qquad -\pi \leq \lambda \leq \pi,$$

where $\psi(z) = \theta(z)/\phi(z)$, $|z| \leq 1$. In view of Theorem 4.2, a natural estimator of $f(\lambda)$ is therefore

$$\hat{f}_n(\lambda) = |\sum_{j=0}^{m} \hat{\theta}_{mj} e^{-ij\lambda}|^2 \hat{v}_m/(2\pi), \tag{4.41}$$

where $\hat{\theta}_{m0} = 1$,

$$\hat{v}_m = \hat{\gamma}(0) - \sum_{j=1}^{m} \hat{\theta}_{mj}^2 \hat{v}_{m-j}$$

estimates the one-step prediction mean squared error,

$$v_m = E(X_{m+1} - \hat{X}_{m+1})^2 = \gamma(0) - \underset{\sim}{\phi}'_m \Gamma_m \underset{\sim}{\phi}_m,$$

and $\underset{\sim}{\phi}'_m = (\phi_{m1}, \cdots, \phi_{mm})$. It is easy to see from (2.4) that

$$\underset{\sim}{\phi}'_m \Gamma_m \underset{\sim}{\phi}_m = \sum_{j=1}^{m} \theta_{mj}^2 v_{m-j},$$

and that

$$\hat{v}_m = \hat{\gamma}(0) - \hat{\underset{\sim}{\phi}}'_m \hat{\Gamma}_m \hat{\underset{\sim}{\phi}}_m,$$

where $\hat{\underset{\sim}{\phi}}_m = \hat{\Gamma}_m^{-1} \hat{\underset{\sim}{\gamma}}_m$.

It was shown by Berk (1974) that if $\{m(n)\}$ is chosen in such a way that $m(n) = o(n^{1/3})$ and $n^{1/2} \sum_{j=m+1}^{\infty} |\psi_j| \to 0$, then

$$(n/m)^{1/2}(\hat{v}_m - \sigma^2) \xrightarrow{P} 0. \tag{4.42}$$

Moreover a straightforward calculation shows that as $n \to \infty$,

$$m^{-1} e^*(\lambda)(T'T)_m e(\mu) \to \begin{cases} 0 & \text{if } \mu \neq \lambda, \\ |\sum_{j=0}^{\infty} \psi_j e^{-ij\lambda}|^2 & \text{if } \mu = \lambda, \end{cases}$$

where $e(\mu) = (e^{i\mu}, e^{2i\mu}, \cdots, e^{mi\mu})$, $e^*(\lambda) = (e^{-i\lambda}, e^{-2i\lambda}, \cdots, e^{-mi\lambda})$ and $(T'T)_m$ is the upper left $m \times m$ truncation of the matrix $T'T$ defined by (4.33). If we set

$$\hat{c}_m(\lambda) = \sum_{j=0}^{m} \hat{\theta}_{mj} \cos j\lambda,$$

$$c_m(\lambda) = \sum_{j=0}^{m} \theta_{mj} \cos j\lambda,$$

$$\hat{s}_m(\lambda) = \sum_{j=0}^{m} \hat{\theta}_{mj} \sin j\lambda$$

and

$$s_m(\lambda) = \sum_{j=0}^{m} \theta_{mj} \sin j\lambda,$$

we therefore expect from Theorem 4.2 that for $\lambda \neq 0$ or $\pi/2$, $(n/m)^{1/2}(\hat{c}_m(\lambda) - c_m(\lambda))$ and $(n/m)^{1/2}(\hat{s}_m(\lambda) - s_m(\lambda))$ are asymptotically independent and normally distributed with mean zero and variance $|\sum_{j=0}^{\infty} \psi_j e^{-ij\lambda}|^2/2$. A Taylor expansion and (4.42) then give

$$(4.43)\quad (n/m)^{1/2}(\hat{f}_n(\lambda) - f(\lambda))$$

$$= (n/m)^{1/2}\sigma^2(2\pi)^{-1}[2\hat{c}_m(\lambda)(\hat{c}_m(\lambda) - c_m(\lambda)) + 2\hat{s}_m(\lambda)(\hat{s}_m(\lambda) - s_m(\lambda))]$$

$$+ (n/m)^{1/2}[\sigma^2(c_m^2(\lambda) + s_m^2(\lambda))/2\pi - f(\lambda)] + o_p(1).$$

The first term on the right hand side of (4.43) is asymptotically normal with mean zero and variance $2f^2(\lambda)$. The other terms vanish as $n \to \infty$ by the choice of m. Hence $(n/m)^{1/2}(\hat{f}_n(\lambda) - f(\lambda))$ is asymptotically normal with mean zero and variance $2f^2(\lambda)$.

Notice that the estimate given by (4.41) has the same asymptotic distribution as the spectral density estimate proposed by Berk (1974).

(g) Numerical examples

Example 4.5 The first example illustrates the estimation of the coefficients of an MA(2) process using the technique described in (d). One thousand realizations, each of length 200, of the MA(2) process,

$$(4.44)\qquad X_t = Z_t + 0.9Z_{t-1} - 0.6Z_{t-2}, \qquad Z_t \sim N(0,1),$$

were generated. The equivalent invertible representation of (4.44) is

$$(4.45)\qquad X_t = W_t + 0.2972W_{t-1} - 0.3313W_{t-2},$$

where $\{W_t\}$ is an independent zero-mean normal sequence with variance 1.8112.

For each realization, the estimators $\hat{\theta}_{m1}$ and $\hat{\theta}_{m2}$ of $\theta_1 = 0.2972$ and $\theta_2 = -0.3313$ were calculated from the sample autocovariances and the recursion relations (2.4). Several different values of m were used. The asymptotic distribution of $n^{1/2}(\hat{\theta}_{m1} - \theta_1, \hat{\theta}_{m2} - \theta_2)$, under the conditions of Theorem 4.2, is

$$N\left(\underset{\sim}{0}, \begin{bmatrix} 1 & \theta_1 \\ \theta_1 & 1+\theta_1^2 \end{bmatrix}\right) = N\left(\underset{\sim}{0}, \begin{bmatrix} 1.000 & .297 \\ .297 & 1.088 \end{bmatrix}\right). \tag{4.46}$$

In Table 6 we show the sample means and sample covariances of $(\hat{\theta}_{m1} - \theta_1, \hat{\theta}_{m2} - \theta_2)'$ based on 1000 pairs $(\hat{\theta}_{m1}, \hat{\theta}_{m2})$ each derived from a realization of length $n = 200$. We also show the corresponding moments based on the asymptotic distribution of Theorem 4.2.

The asymptotic distribution provides a very good approximation to the sample moments over the range of values of m considered. More generally we have found that for moderate sample sizes (100–500), the asymptotic distribution provides an excellent approximation for m as large as $n/4$. Table 6 also shows that the estimators $\hat{\theta}_{m1}$ and $\hat{\theta}_{m2}$ perform very well compared with the corresponding maximum likelihood estimators, for which the covariances (based on the asymptotic distribution) are $\sigma_{11} = .0045$, $\sigma_{22} = .0045$ and $\sigma_{12} = .0020$.

TABLE 6

The sample means $\hat{\mu}_i$ and covariances σ_{ij} of $(\hat{\theta}_{mi} - \theta_i)$, $i, j = 1, 2$ and the corresponding values from the asymptotic distribution of Theorem 4.2

m	5	10	20	40	Thm 4.2
$\hat{\mu}_1$	-.0337	-.0147	-.0145	-.0158	0
$\hat{\mu}_2$	.0139	.0005	-.0005	-.0012	0
$\hat{\sigma}_{11}$	.00500	.00498	.00529	.00545	.00500
$\hat{\sigma}_{22}$	.00523	.00571	.00599	.00627	.00544
$\hat{\sigma}_{12}$	.00156	.00149	.00157	.00167	.00149

Example 4.6 We conclude by applying the modelling, estimation and prediction developed in previous sections to the International Airline Passenger data $\{Y_t, t = 1, \cdots, 144\}$ of Box and Jenkins (1976), p.531. As in the analysis of Box and Jenkins, the data was first transformed by taking natural logarithms, $L_t = \ln Y_t$, then applying the operator $(1 - B)(1 - B^{12})$ to produce a new series, stationary in appearance, and with rapidly decaying sample autocorrelation function. If we write

$$X_t = (1 - B)(1 - B^{12})L_{t+13}, \qquad t = 1, \cdots, 131,$$

then the sample autocorrelation function of X_t suggests a moving average model with zero coefficients for lags greater than 23. (Box and Jenkins fitted a multiplicative moving average model of order 13.)

A graph which is far more informative for model identification than the sample autocorrelation function is that of $\hat{\theta}_{mj}$, $1 \le j \le 30$, which is shown in Figure 1 for $m = 30$ and $m = 50$. (It is a trivial matter to compute $\hat{\theta}_{mj}$ from the sample autocorrelation function and Proposition 2.1.) Since the asymptotic variance of $\hat{\theta}_{mj}$ is $\sigma_j^2(\theta_1, \cdots, \theta_{j-1}) = (1 + \theta_1^2 + \cdots + \theta_{j-1}^2)/n$ from Theorem 4.2, we have also plotted the bounds $\pm 1.96\hat{\sigma}_j$ where $\hat{\sigma}_j = \sigma_j(\hat{\theta}_{m1}, \cdots, \hat{\theta}_{m,j-1})$. A value of $\hat{\theta}_{mj}$ outside these bounds suggests that the corresponding coefficient θ_j is non-zero. The graphs of $\hat{\theta}_{mj}$

FIGURE 1

Graphs of $\theta_{30,j}$ and $\theta_{50,j}, j = 1, \cdots, 30$

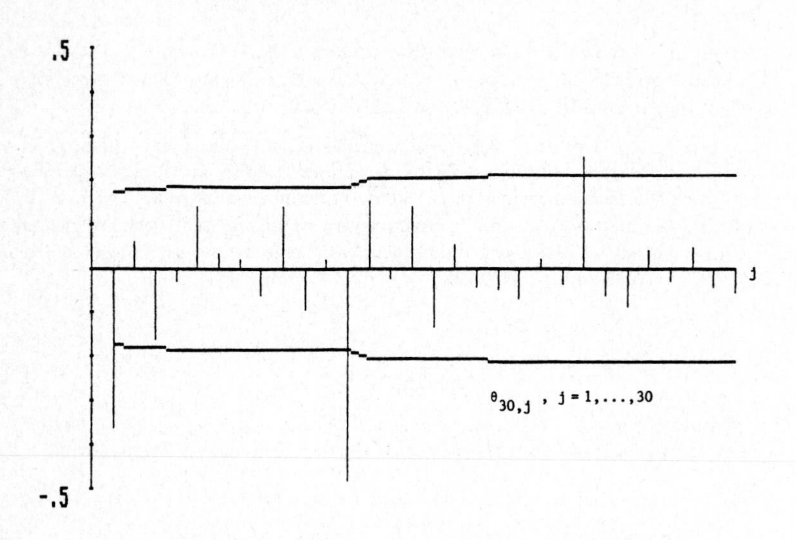

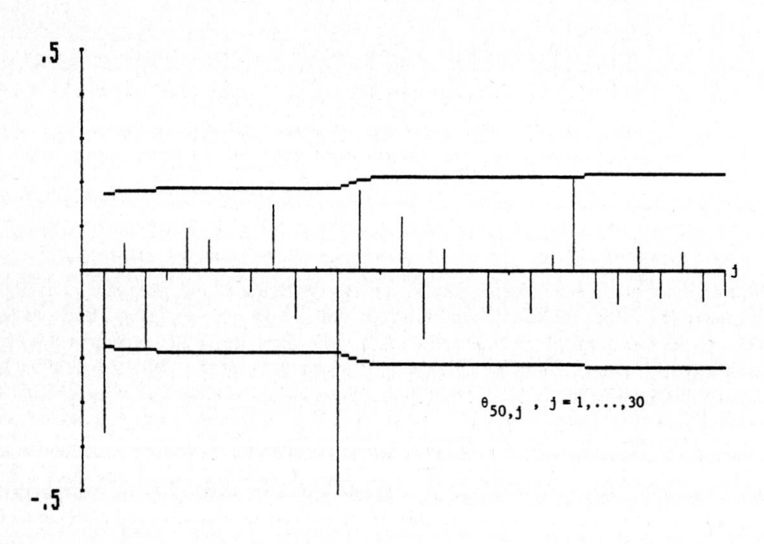

strongly suggest the model

$$X_t = Z_t + \theta_1 Z_{t-1} + \theta_3 Z_{t-3} + \theta_{12} Z_{t-12} + \theta_{23} Z_{t-23}, \tag{4.47}$$

where $\{Z_t\}$ is white noise. At the same time they provide us with the preliminary estimates $\hat{\theta}_j = \hat{\theta}_{30,j}$, $j = 1, 3, 12, 23$, where

$$\hat{\theta}_1 = -.357, \quad \hat{\theta}_3 = -.158, \quad \hat{\theta}_{12} = -.479 \quad \text{and} \quad \hat{\theta}_{23} = .254. \tag{4.48}$$

(There is very little difference between the values of $\hat{\theta}_{mj}$ for $30 \le m \le 50$.)

Using the maximum likelihood technique described in (b), estimates of the parameters $\theta_1, \theta_3, \theta_{12}$ and θ_{23} were then obtained, using as initial values in the optimization the preliminary estimates found in the preceding paragraph. The maximum likelihood model was found to be

$$X_t = Z_t - .372 Z_{t-1} - .214 Z_{t-3} - .537 Z_{t-12} + .232 Z_{t-23}, \tag{4.49}$$

where $\{Z_t\}$ is white noise with variance .00123. The Akaike information criterion for this model has the value, AIC $= -861.757$. The sample autocorrelation function of the standardized residuals $(X_t - \hat{X}_t)/v_t$, $, t = 1, \cdots, 131$, is indistinguishable from that of independent white noise on the basis of the goodness of fit tests described in (e).

The model for $\{X_t\}$ fitted by Box and Jenkins was

$$X_t = (1 - .396B)(1 - .614B^{12}) Z_t, \qquad \{Z_t\} \sim WN(0, .00134). \tag{4.50}$$

Although this model has two fewer parameters than (4.49), it gives a higher AIC value, viz. AIC $=$ -856.247. The sample autocorrelation function of the standardized residuals from the model (4.50) is compatible with that of white noise insofar as it passes the portmanteau test. There is however a rather large value, .219, at lag 23, which is well outside the .95 bounds, $\pm 1.96/\sqrt{131} = \pm .171$.

Maximum likelihood fitting was also carried out for the more general model,

$$X_t = Z_t + \theta_1 Z_{t-1} + \theta_3 Z_{t-3} + \theta_{12} Z_{t-12} + \theta_{13} Z_{t-13} + \theta_{23} Z_{t-23},$$

and for the models obtained by setting subsets of these parameters equal to zero. However the best such model on the basis of the AIC criterion was found to be (4.49), the one which was first suggested by our identification technique. Notice also that the preliminary estimated values (4.48) are quite close to the maximum likelihood estimates in (4.49).

The coefficients $\theta_{mj}, m \le 131$, and the residuals $X_t - \hat{X}_t$, $t \le 131$, from the model (4.49) are computed and stored in the course of the maximum likelihood estimation procedure. Since the coefficients θ_{mj} and one-step mean squared errors v_m are found to have attained (to three significant digits) their asymptotic values when $m = 131$, we can compute the predictors $P_{131} X_t$, $t \ge 132$ from the asymptotic expression,

$$P_{131} X_t = \sum_{j=t-131}^{23} \theta_j (X_{t-j} - \hat{X}_{t-j}), \tag{4.51}$$

with $\theta_1 = -.372$, $\theta_3 = -.214$, $\theta_{12} = -.537$, $\theta_{23} = .232$ and $\theta_j = 0$ for $j \ne 1, 2, 12$ or 23. The predicted values of $X_{132}, \cdots, X_{143}$ are shown in Table 7, along with the corresponding predicted values of $L_{145}, \cdots, L_{156}$. The mean squared errors of the predictors are accurately represented by their asymptotic forms, which are also shown in the table.

TABLE 7
Predictors and their mean squared errors, based on $L_1, \cdots, L_{144}$

t	Predictor of X_{t-13}	MSE	Predictor of L_t	MSE
145	.0108	.00123	6.108	.00123
146	.0060	.00139	6.050	.00172
147	.0461	.00139	6.165	.00220
148	-.0424	.00144	6.218	.00241
149	-.0156	.00144	6.226	.00262
150	.0209	.00144	6.373	.00283
151	.0042	.00144	6.528	.00304
152	.0103	.00144	6.512	.00325
153	-.0007	.00144	6.335	.00347
154	-.0073	.00144	6.230	.00368
155	.0106	.00144	6.074	.00389
156	-.0005	.00144	6.175	.00449

References

Ansley, Craig F. (1979). An algorithm for the exact likelihood of a mixed autoregressive-moving average process. *Biometrika 66*, 59-65.

Berk, K.N. (1974). Consistent autoregressive spectral estimates. *Ann. Statist. 2*, 489-501.

Billingsley, P. (1968). *Convergence of Probability Measures*. John Wiley & Sons Inc., New York.

Box, G.E.P. and Jenkins, G.M. (1976). *Time Series Analysis, Forecasting and Control*, 2nd edition. Holden Day, San Francisco.

Box, G.E.P. and Pierce, D.A. (1970). Distribution of residual autocorrelations in autoregressive-integrated moving average time series models. *J. Amer. Statist. Assoc. 65*, 1509-1526.

Brockwell, P.J. and Davis, R.A. (1986). Preliminary estimation for ARMA processes. Technical Report, Dept. of Statistics, Colorado State University.

Fuller, W.A. (1976). *Introduction to Statistical Time Series*. John Wiley & Sons Inc., New York.

Gardner, G., Harvey, A.C. and Phillips, G.D.A. (1980). An algorithm for exact maximum likelihood estimation of autoregressive-moving average models by means of Kalman filtering. *Appl. Statistics 29*, 311-322.

Jones, R.H. (1980). Maximum likelihood fitting of ARMA models to time series with missing observations. *Technometrics 22*, 389-395.

Kailath, T. (1970). The innovations approach to detection and estimation theory. *Proceedings IEEE 58*, 680-695.

Rissanen, J. and Barbosa, L. (1969). Properties of infinite covariance matrices and stability of optimum predictors. *Information Sci. 1*, 221-236.

Rissanen, J. (1973). A fast algorithm for optimum linear predictors. *IEEE Transactions on Automatic Control AC-18*, 555.

Schweppe, F.C. (1965). Evaluation of likelihood functions for Gaussian signals. *IEEE Transactions on Information Theory IT-11*, 61-70.

Wilson, G.T. (1969). Factorization of the generating function of a pure moving average process. *SIAM J. Num. Analysis 6*, 1-7.

Original paper received: 30.06.86
Final paper received: 27.05.87

Paper recommended by D.C. Boes

PROBABILITY AND STATISTICS
Essays in Honor of Franklin A. Graybill
J.N. Srivastava (Editor)

Karl Pearson at Gresham College

J. Leroy Folks
Department of Statistics
Oklahoma State University
Stillwater, Oklahoma
U. S. A.

Abstract

Karl Pearson's association with Gresham College is important to statisticians because it was there that he gave his first lectures on statistics and it was there that some of his statistical concepts were born.

AMS 01

KEY WORDS: Thomas Gresham, Royal Exchange, Royal Society, University of London, posterior odds, normal distribution, standard deviation.

1. INTRODUCTION

Karl Pearson was the nineteenth lecturer of geometry at Gresham College, an institution dating from the sixteenth century which had a considerable effect upon the early development of science (See Hartley and Hinshelwood (1960)). Because of the college's effect upon science, it affected statistics indirectly but it also is linked directly to statistics through Karl Pearson.

Pearson was a lecturer at Gresham College during the years of his transition from applied mathematics to statistics and some of his statistical reasoning was presented first at Gresham College. Also, Karl Pearson's lectures on statistics at Gresham College are important to anyone interested in the teaching of statistics.

2. THE ORIGINS OF GRESHAM COLLEGE.

Thomas Gresham (1519-1579) was apprenticed to his uncle, Sir John Gresham in the Mercer's Company in 1543 and was Master of the Company in 1569, 1573, and 1579. He served Henry VIII, Edward VI, and Elizabeth I and was knighted by her in 1559. Gresham offered to build the Royal Exchange if the city provided the site and the first stone was laid in 1566. After the original building was destroyed in the fire of 1666, the Royal Exchange was rebuilt on the same site and in Christopher Wren's plan for the new City of London, the Royal Exchange was to be the center of the city.

Gresham's will, dated July 5, 1575 called for the establishment of lectures in "divynitye, astronomy, musick, geometry, law, physicke, and rethoricke." The lectures were to be given free of charge to all who would attend. Sir Thomas Gresham died in 1579 and Lady

Gresham in 1596. Upon her death the City of London Corporation and the Mercer's Company obtained possession of the Royal Exchange and Gresham's home on Bishopgate Street. A joint committee was appointed by the Corporation and the Mercer's Company to manage the joint trusts of the will. Soon after the death of Lady Gresham, the seven lectureships were established and Gresham's home became known as Gresham College. The lecturers became known as professors and lived at Gresham College until 1768, when the site was taken for a new excise office. The lectures were given at the Royal Exchange until its destruction in 1838. In 1842 a new Gresham College building was erected at the intersection of Gresham and Basinghall Streets. The present Gresham College building, dating from 1913, houses the Business Research Centre of the City of London University.

The Royal Society evolved from discussions among the Gresham professors and was chartered at Gresham College by Charles II in 1662. The College escaped the great fire of 1666 and the Royal Society remained there until 1710.

To be appointed a Gresham lecturer was one of the highest honors in the land and the list of early lecturers includes the names of many notables. Karl Pearson (1891) himself wrote a paper recognizing the greatness of the early Gresham professors. Henry Briggs, the first geometry lecturer was involved in the study of logarithms as early as 1615 and constructed the first table with base 10. Lawrence Rooke, geometry lecturer from 1657 through 1662 was instrumental in obtaining the Royal Society charter. Isaac Barrow, geometry lecturer from 1662 through 1664 was considered inferior only to Newton in geometry and was elected a fellow of the Royal Society in 1663. Perhaps the most famous was Robert Hooke, geometry lecturer from 1665 through 1703, who invented the watch, a marine barometer, and several new lamps. He is known for his research on elasticity and strength of materials. Partly because scholars were scholars in many fields, professors were not always chosen for the field in which they are most famous. For example, Christopher Wren, architect, was Professor of Astronomy from 1567 through 1570; William Petty, physician and anatomist, was appointed Professor of Music in 1650; and in 1657 Jonathan Goddard, physician, was appointed Professor Rhetoric. Like many wills, Gresham's will contained some awkward stipulations. It prohibited marriage of lecturers. This restriction was removed by an act of Parliament in 1767.

By the late nineteenth century, Gresham College continued but was no longer the center of scientific activity. Gresham's trust still provided one hundred pounds a year for each of the seven professors. This was not a trivial sum in 1890, amounting to about $500 in U. S. currency. For this stipend they were expected to deliver four lectures per term in each of three terms.

In 1884 Karl Pearson was appointed Professor of Applied Mathematics and Mechanics at University College, London. When B. M. Cowie, Gresham Professor Geometry since 1854, resigned in 1890, Karl Pearson applied for the Gresham College position.

3. KARL PEARSON'S ASSOCIATION WITH GRESHAM COLLEGE

Although Gresham College had lost much of its former greatness, it is evident that a Gresham professorship was still to be prized. Pearson (1891) indicated his own high esteem for the title "Professor of Geometry at Gresham College." His application included a letter of application, a list of publications, and testimonials (letters of recommendation). The Grand Gresham Committee considered applications from twenty applicants and reduced the list to six, including Pearson. The six were asked to give probationary lectures and Karl Pearson presented his thirty minute probationary lecture at Gresham College late on a Friday afternoon on December 12, 1890. This lecture entitled "The Applications of Geometry to Practical Life" was later reprinted in Nature. Person developed the theme that the Egyptians, unlike the Greeks, were interested in geometry not for its own sake but for its use in commerce and industry, in measuring land, and in planning buildings and pyramids. In his lecture he says that for a long time to come the College should be more concerned with spreading and utilizing existing knowledge than with pure research. As an indication of how he expected to develop the lectures in geometry, he says in another place that the graphical representation of statistics is extremely important to problems of insurance and commerce and that conclusions may be reached more readily from the geometrical than from the numerical representation of statistics. In view of Pearson's comments it should not have been a surprise that his third term of four lectures was on the Geometry of Statistics.

At a Grand Gresham Committee meeting on December 15, 1890, the list of applicants was narrowed by a show of hands to Karl Pearson and W. H. Wagstaff. Subsequently Karl Pearson was elected. He held this post concurrently with his chair at University College.

In the 1890's there were several attempts to establish a new University of London. A draft charter referred to the university as Albert University and later as Gresham University. Still later there was a proposal for a Victoria University. Thomas Gresham's trust and lectureships would have been incorporated into the new university. Professors at University College, including Pearson and Weldon, signed a statement opposing the proposed Gresham University charter. On February 4, 1892, Pearson wrote to the Gresham committee offering to resign his professorship if the Gresham trust were reconstituted to allow formation of Gresham University. Later in 1892 it became clear that the Gresham Committee could not alter Gresham's will to contribute to a new teaching university and that the lectures would have to continue as popular lectures.

In the Gresham Committee minutes, there are more references to Karl Pearson during his tenure than to any other lecturer. On January 20, 1891, Karl Pearson's eclection was reported. On November 13, 1891, he was reelected. On May 4, 1892, his letter offering to resign was read. On July 22, 1892, a letter from Pearson was read expressing the sort of disappointment that other teachers have felt when their efforts seem not to be appreciated. Having spent seven pounds over his allowance of ten pounds, he wrote that he would not try to illustrate the next course of lectures. Apparently he was disappointed by the attendance and felt that the committee might think the money poorly spent. Some of his modern counterparts may have similar results to report for their attempts at educational innovation. The committee voted to reimburse Pearson. On December 6, 1892, Pearson was reelected Lecturer on Geometry.

In February, 1893, Pearson wrote to the Committee to inform them that he would not be able to give the lectures in April because of ill health and to tender the resignation of his lectureship. Pearson was asked to appoint a deputy lecturer and to reconsider his resignation. Pearson's April lectures were given by John Venn, W. F. R. Weldon, W. A. Whitworth, and Robert S. Ball.

On June 9, 1893, a letter from Pearson was read to the Committee requesting permission to change lectures from October to November. The request was granted. On June 26, 1893, a letter from Pearson was read asking for an allowance to illustrate lectures. He was given an allowance of six pounds, seventeen shillings, and eight pence for the preceding year and an allowance not to exceed ten pounds for the ensuing year. On November 22, 1893, Pearson was appointed for the following year. On March 8, 1894, a letter was read from Pearson resigning the Gresham lectureship because of health reasons. The resignation was accepted and Pearson's association with Gresham College came to an end after two lectures in May of 1894.

4. THE GRESHAM COLLEGE LECTURES

These lectures, given over three and one-half years, predate the beginning of statistics courses and the biometry laboratory at University College. Pearson was thinking hard about making existing ideas of probability useful. Syllabi of the lectures were published by E. S. Pearson (1938) and much information (including some complete lectures) is available in the University College Library. In today's terminology, Pearson's lectures would fall under five headings: philosophy of science, statistical graphics, theory of probability, probability distributions, and frequency distributions.

It is clear that Pearson cared about being an effective teacher and we can still learn from him. He prepared his lectures and syllabi carefully and announced the topics in advance of his next set of lectures. He used lime light or lantern illustrations. This is noteworthy when we realize that only in very recent years have some statisticians ventured beyond blackboard and chalk at annual meetings. He involved the students in generating data: guessing the height of people, ages of people, etc., and all sorts of real data in addition to the usual cards and marbles data. He urged students to give or send their questions to him so that he could answer their questions at the next course of lectures.

The students responded. They came from all over London: Wimbledon, Harrow, Peckham, Woolwich, etc. as well as from outside London. Despite Pearson's evident disappointment which he expressed in his letter to the Gresham Committee, the students reacted positively with enthusiasm to his efforts and letters poured in with questions and comments about the lectures.

Finally, there are three points in the lectures which should be noted. First, it is quite clear that Pearson asked his audience to vote for the number of black and white balls given some results of drawing. Thus he was asking them to formulate their posterior odds ratio. Second, the lecture on November 21, 1893 includes the expression normal curve. While this surely is not the first use of the word normal in this sense, it may be the first use in English. Third, the lecture on January 31, 1893 includes the expression standard deviation. This surely is one of the first uses of this expression.

Acknowledgment

I am indebted to Ms. Imray (retired) and Mrs. Sutton, archivists at Mercer's Hall for their very kind assistance.

References

Adamson, Ian (1975). Foundation and Early History of Gresham College, London, 1596 - 1704, Ph.D. Dissertation, St. John's College, Cambridge.

Burgon, J. W. (1839). The Life and Times of Sir Thomas Gresham, London. Referenced in Pearson's papers.

Featherstone, Ernest (1952). Sir Thomas Gresham and His Trusts, Blades, East, and Blades, Limited : London.

Hartley, Harold, and Cyril Hinshelwood (1960). Gresham College and the Royal Society, Headley Brothers, Ltd. : London.

Pearson, E. S. (1938). Karl Pearson, An Appreciation of Some Aspects of His Life and Work, Cambridge University Press: Cambridge, England.

Pearson, E. S. (1966). The Neyman-Pearson Story: 1926 - 34, in Research Papers in Statistics, Festchrift for J. Neyman, edited by F. N. David, John Wiley : New York.

Pearson, Karl (1891). "The Application of Geometry to Practical Life," Pearson's probationary Gresham College lecture, Nature, 43, 273 - 276.

Ward, John (1740). The Lives of Professors of Gresham College. London. Referenced in Pearson's papers.

Winckworth, Peter (1966). A History of the Gresham Lecturers, printed by Vail & Co., Ltd. for the City University : London.

Original paper received: 29.10.85
Final paper received: 17.09.87

Paper recommended by P.W. Mielke

PROBABILITY AND STATISTICS
Essays in Honor of Franklin A. Graybill
J.N. Srivastava (Editor)

ON AN INTERMITTENT STOCHASTIC PROCESS

J. Gani
Statistics Program
University of California
Santa Barbara, CA 93106
U.S.A.

P. Todorovic
Department of Statistics
University of Kentucky
Lexington, KY 40506
U.S.A.

We consider the random walk along a straight line of a particle whose velocity is a non-negative intermittent stochastic process $\xi(t)$ taking positive values on random time intervals. Under certain simple assumptions, we evaluate the Laplace-Stieltjes transform of the bivariate marginal distribution functions (d.f.'s) of $\xi(t)$. We then investigate the asymptotic behaviour under translation of the time parameter of this transform; the marginal distributions under translation converge pointwise to the marginal d.f.'s of a strictly stationary stochastic process. Finally, we show that the intermittent process is ergodic, and sketch some applications.

Key words: random walk, renewal processes, asymptotic behaviour, ergodicity.

AMS classification: 60 G10, 60K05.

1. INTRODUCTION

In various scientific investigations, one must often consider stochastic models involving functionals of an intermittent stochastic process which vanishes on random time intervals. For instance, the rainfall intensity at a given point is such a process. In this paper we study a process of this kind; to motivate our work we shall begin by giving an example.

Consider a particle moving along a straight line as follows: At time $t = 0$ the particle is at point $x = 0$, where it remains for a random sojourn time. At the end of this time, it begins to move with a random uniform velocity V_1 to a new position on the line. The duration of the motion is a random variable. The particle remains in this new position for a random time, and then begins to move again with a random uniform velocity V_2 until it reaches a new position; the motion of the particle along the line continues indefinitely in this manner. If we take $\xi(t)$ to be the velocity of the particle at a time $t \geq 0$, then we shall be concerned in this paper with the process $\{\xi(t);\ t \geq 0\}$. This provides a more physically realistic model of various phenomena (see e.g., [1] and [6]) than an ordinary random walk, where the particle is assumed to move from one point in space to another instantly.

Let Z_t, $t \geq 0$ be a stochastic process with the state space $\{0,1\}$ such that $Z_0 \equiv 0$ and defined for $t > 0$ as follows:

$$Z_t = \begin{cases} 1 & \text{if the particle moves at time } t, \\ 0 & \text{if the particle is at rest at time } t. \end{cases}$$

Let $N_j(t)$ be a counting random function on $R_+ = [0,\infty)$ defined as:

$$N_j(t) = \#\{0 < s \leq t;\ Z_{s-} \neq Z_s = 1 - j\}$$

where $j = 0,1$. If we write

$$N(t) = N_0(t) + N_1(t) \tag{1.1}$$

then we can easily see that

$$Z_t = \tfrac{1}{2}[1-(-1)^{N(t)}]. \tag{1.2}$$

In what follows, particular points of the process $N_j(t)$ will be denoted by $\{\tau_{j,n}\}_1^\infty$ where $\tau_{j,0} \equiv 0$ for $j = 0,1$ and

$$\tau_{j,n} = \inf\{t; N_j(t) = n\} \qquad n = 1,2,\ldots \tag{1.3}$$

The points of $N(t)$ are $\{\tau_n\}_1^\infty$. It is clear that

$$\tau_{j,n} = \tau_{2n+j-1} \qquad j = 0,1; \quad n \geq 1. \tag{1.4}$$

Figure 1 illustrates a possible realization of Z_t.

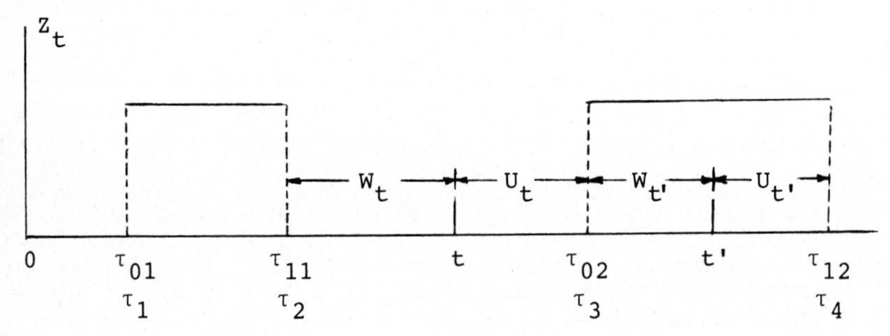

Figure 1. A possible realization of Z_t

According to (1.2), Z_t is right continuous; we also assume that it has left-hand limits.

In the following, two other processes are also required:

$$U_t = \tau_{N(t)+1} - t \text{ and } W_t = t - \tau_{N(t)}. \tag{1.5}$$

They are the lengths of the time intervals from any point $t \geq 0$ to the next and last jumps of Z_t, respectively.

2. NOTATION AND PRELIMINARIES

Let $\xi(t)$ be the velocity of the particle at time $t > 0$. It is clear that

$$\xi(t) = Z_t V_{N_1(t)+1} \cdot \tag{2.1}$$

We proceed to investigate various properties of $\xi(t)$ under the following assumptions:

(i) $\{V_n\}_1^\infty$ is an i.i.d. sequence of random variables independent of $\{N_0(t),\ N_1(t)\}$.

(ii) $\{(Z_t, U_t);\ t \geq 0$ is a homogeneous Markov process.

Pyke and Schauffele [3] have shown that the second condition implies that $\{Z_t;\ t \geq 0\}$ is a semi-Markov process. It follows that $\{N_0(t),\ N_1(t)\}$ is a Markov renewal process. In other words, following Pyke [2],

$$\{(\tau_{0,n+1} - \tau_{0,n})\}_0^\infty$$

is a delayed renewal process while

$$\{(\tau_{1,n+1} - \tau_{1,n})\}_0^\infty$$

is an ordinary renewal process.

Some additional notations are now required: let us write

$$\Lambda_j(t) = E\{N_j(t)\} \quad j = 0,1, \quad \Lambda(t) = \Lambda_0(t) + \Lambda_1(t) \tag{2.2}$$

$$B(s) = P\{\tau_{01} \leq s\}, \quad \beta_1 = \int_0^\infty s dB(s) \tag{2.3}$$

$$D(s) = P\{\tau_{11} - \tau_{01} \leq s\}, \quad \beta_2 = \int_0^\infty s dD(s) \tag{2.4}$$

$$G(s) = P\{\tau_{11} \leq s\}, \quad \int_0^\infty s dG(s) = \beta_1 + \beta_2 \tag{2.5}$$

$$\phi(\theta) = E\{e^{-\theta V_1}\} \quad \theta \geq 0, \quad P_{0j}(t) = P\{Z_t = j | Z_0 = 0\} \quad j=0,1. \tag{2.6}$$

The next two results of Pyke [2] are well known

$$\begin{aligned} P_{00}(t) &= [1 - B(t)] * [\Lambda_1(t) + 1] \\ \lim_{t\to\infty} P_{00}(t) &= \frac{\beta_1}{\beta_1 + \beta_2} . \end{aligned} \tag{2.7}$$

The first equation follows easily from (1.2) and (1.4). For

$$\begin{aligned} P_{00}(t) = P\{Z_t = 0\} &= \sum_{n=0}^{\infty} P\{N(t) = 2n\} \\ &= \sum_{k=0}^{\infty} P\{\tau_{1,k} \leq t\} - \sum_{k=0}^{\infty} P\{\tau_{1,k} + (\tau_{0,k+1} - \tau_{1,k}) \leq t\} \\ &= 1 - B(t) + \{1 - B(t)\} * \Lambda_1(t) . \end{aligned} \tag{2.8}$$

The limiting result follows from the key renewal theorem of Smith [4].

Remark 2.2

It is easy to verify that the following two relations hold:

$$\Lambda_1(t) = \Lambda_0(t) * D(t) \quad (2.9)$$

and

$$\Lambda_1(t) * B(t) = \Lambda_0(t) - B(t). \quad (2.10)$$

From (2.8) and (2.10) we have

$$P_{00}(t) = 1 + \Lambda_1(t) - \Lambda_0(t), \quad P_{01}(t) = \Lambda_0(t) - \Lambda_1(t). \quad (2.11)$$

Let $L(\theta;t)$ be the Laplace-Stieltjes (L-S) transform of $\xi(t)$, then

$$L(\theta;t) = \phi(\theta) + [1 - \phi(\theta)]P_{00}(t). \quad (2.12)$$

For, since

$$L(\theta;t) = E\{e^{-\theta\xi(t)}\} = P\{Z_t = 0\} + P\{Z_t = 1\}\phi(\theta),$$

the assertion follows directly. In addition, from (2.7) we have that

$$\lim_{t\to\infty} L(\theta;t) = \phi(\theta) + [1 - \phi(\theta)]\frac{\beta_1}{\beta_1 + \beta_2}. \quad (2.13)$$

3. SECOND ORDER CHARACTERISTICS OF $\xi(t)$

In this section we investigate some second order characteristics of the process $\xi(t)$. We begin by evaluating its bivariate L-S transform

$$L(\theta_1,\theta_2;t_1,t_2) = E(\exp\{-\sum_{j=1}^{2} \theta_j\xi(t_j)\}). \quad (3.1)$$

Proposition 3.1

For any $0 \le t_1 \le t_2 < \infty$

$$L(\theta_1,\theta_2;\, t_1,t_2) = P\{Z_{t_1} = Z_{t_2} = 0\} + P\{Z_{t_1} = 1,\, Z_{t_2} = 0\}\phi(\theta_1)$$

$$+ \phi(\theta_2)P\{Z_{t_1} = 0,\, Z_{t_2} = 1\} + P\{Z_{t_1} = 0,\, N(t_1) = N(t_2)\}\phi(\theta_1 + \theta_2)$$

$$+ \phi(\theta_1)\phi(\theta_2)P\{Z_{t_1} = Z_{t_2} = 1,\, N(t_1) < N(t_2)\}. \quad (3.2)$$

Proof:

To prove the proposition, let us write

$$L(\theta_1,\theta_2;t_1,t_2) = E\{e^{-\sum_{r=1}^{2}\theta_r\xi(t_r)} \sum_{i=0}^{1}\sum_{j=0}^{1} I\{Z_{t_1} = i,\, Z_{t_2} = j\}\}.$$

From this equation (2.1), assumption (i) and the fact that

$$P\{Z_{t_1} = Z_{t_2} = 1,\ N_1(t_1) = N_1(t_2)\} = P\{Z_{t_1} = 1,\ N_1(t_1) = N_1(t_2)\}, \tag{3.3}$$

the assertion follows.

The next proposition gives the joint distribution of Z_t and U_t.

<u>Proposition 3.2</u>

For any $t > 0$ and $y > 0$

$$P\{Z_t = 1,\ U_t \le y\} = \int_0^t [D(t + y - s) - D(t - s)]\, d\Lambda_0(s). \tag{3.4}$$

Proof:

From (1.2), (1.4) and (1.5) we obtain

$$\begin{aligned} P\{Z_t = 1,\ U_t \le y\} &= \sum_{n=1}^{\infty} P\{N(t) = 2n - 1,\ \tau_{2n} \le t + y\} \\ &= \sum_{n=1}^{\infty} P\{\tau_{2n-1} \le t \le \tau_{2n},\ \tau_{2n} \le t + y\} \\ &= \sum_{n=1}^{\infty} P\{\tau_{2n-1} \le t,\ \tau_{2n} \le t + y\} - \sum_{n=1}^{\infty} P\{\tau_{2n} \le t\} \\ &= \sum_{n=1}^{\infty} P\{\tau_{0,n} \le t,\ \tau_{1,n} \le t + y\} - \Lambda_1(t) \\ &= \sum_{n=1}^{\infty} P\{\tau_{0,n} \le t,\ \tau_{0,n} + (\tau_{1,n} - \tau_{0,n}) \le t\} - \Lambda_1(t) \\ &= \sum_{n=1}^{\infty} \int_0^t P\{\tau_{1,n} - \tau_{0,n} \le t - s\} dP\{\tau_{0,n} \le s\} - \Lambda_1(t). \end{aligned}$$

This, together with (2.9), proves the proposition.

<u>Remark 3.1</u>

In a similar fashion, one can also show that

$$P\{Z_t = 0,\ U_t \le y\} = \int_0^t B(t + y - s)\, d\Lambda_1(s) + B(t + y) - \Lambda_0(t). \tag{3.5}$$

Let us write $H(y;t) = P\{U_t \le y\}$; combining (3.4) and (3.5) we obtain:

$$H(y;t) = \int_0^t [B(t+y-s)\, d\Lambda_1(s) + D(t+y-s)\, d\Lambda_0(s)] + B(t+y) - \Lambda_0(t) - \Lambda_1(t). \tag{3.6}$$

<u>Proposition 3.3</u>

For all $0 \le t_1 \le t_2 < \infty$

$$P\{Z_{t_1}=0,\ N_1(t_1)=N_1(t_2)\}=1-B(t_2)+\int_0^{t_1}[1-B(t_2-s)d\Lambda_1(s)]. \tag{3.7}$$

Proof:

Write

$$P\{Z_{t_1}=0,\ N_1(t_1)=N_1(t_2)\}=P\{Z_{t_1}=0\}-P\{Z_{t_1}=0,\ N_1(t_1)<N_1(t_2)\}$$

$$=P\{Z_{t_1}=0\}-P\{Z_{t_1}=0,\ U_{t_1}\le t_2-t_1\}.$$

From this, (2.11) and (3.5) the assertion follows.

Proposition 3.4

For every $0 \le t_1 \le t_2 < \infty$

$$P\{Z_{t_1} = 1,\ N_1(t_1) < N_1(t_2)\} = \int_0^{t_1}[D(t_2 - s) - D(t_1 - s)]d\Lambda_0(s). \tag{3.8}$$

Proof:

Bearing in mind that

$$P\{Z_{t_1} = 1, N_1(t_1) < N_1(t_2)\} = P\{Z_{t_1} = 1,\ U_{t_1} \le t_2 - t_1\}$$

$$= P\{U_{t_1} \le t_2 - t_1\} - P\{Z_{t_1} = 0,\ U_{t_1} \le t_2 - t_1\},$$

and recalling (3.5) and (3.6), the assertion is readily proved.

Proposition 3.5

For any $0 \le t_1 \le t_2 < \infty$

$$P\{Z_{t_1} = Z_{t_2} = 1,\ N_1(t_1) < N_1(t_2)\} = P\{Z_{t_1} = Z_{t_2} = 0, N_1(t_1)<N_1(t_2)\}. \tag{3.9}$$

The proof is simple and is left to the reader.

Proposition 3.6

For any $0 \le t_1 \le t_2 < \infty$

$$P\{Z_{t_1}=Z_{t_2}=1,\ N_1(t_1)<N_1(t_2)\} = \int_0^{t_1}\Lambda_1(t_2-s)d\Lambda_0(s) + \int_0^{t_1}\Lambda_1(t_2-s)d\Lambda_1(s)$$
$$-\int_0^{t_1} \Lambda_1(t_2-s) * D(t_2-s)d\Lambda_0(s). \tag{3.10}$$

Proof:

$$P\{Z_{t_1}=Z_{t_2}=1, N_1(t_1)<N_1(t_2)\} = \sum_{n=1}^{\infty}\sum_{j=n+1}^{\infty} P\{\tau_{0n}\le t_1<\tau_{1n}, \tau_{0j}\le t_2<\tau_{1j}\}$$

$$= \sum_{n=1}^{\infty}\sum_{j=n+1}^{\infty} (P\{\tau_{0n}\le t_1, \tau_{0j}\le t_2\} - P\{\tau_{1n}\le t_1, \tau_{0j}\le t_2\}$$

$$-P\{\tau_{0n}\le t_1, \tau_{1j}\le t_2\} + P\{\tau_{1n}\le t_1, \tau_{1j}\le t_2\}$$

a) Consider the first elements of the sum

$$P\{\tau_{0n}\le t_1, \tau_{0j}\le t_2\} = P\{\tau_{0n}\le t_1, \tau_{0n}+(\tau_{0j}-\tau_{0n})\le t_2\}$$

$$= \int_0^{t_1} P\{\tau_{0j}-\tau_{0n}\le t_2 - s\}dP\{\tau_{0n}\le s\}$$

$$= \int_0^{t} P\{\tau_{1,j-n}\le t_2 - s\}dP\{\tau_{0n}\le s\}.$$

Hence

$$\sum_{n=1}^{\infty}\sum_{j=n+1}^{\infty} P\{\tau_{0n}\le t_1, \tau_{0j}\le t_2\} = \int_0^{t_1}\Lambda_1(t_2-s)d\Lambda_0(s).$$

b) Proceeding to the next elements

$$P\{\tau_{1n}\le t_1, \tau_{0j}\le t_2\} = P\{\tau_{1n}\le t_1, \tau_{1n}+(\tau_{1,j-1}-\tau_{1n})+(\tau_{0j}-\tau_{1,j-1})\le t_2\}$$

$$= \int_0^{t_1} P\{(\tau_{1,j-1}-\tau_{1n})+(\tau_{0j}-\tau_{1,j-1})\le t_2-s\}dP\{\tau_{1n}\le s\}$$

$$= \int_0^{t_1}\int_0^{t_2-s} P\{\tau_{1,j-n-1}\le t_2-s-u\}dB(u)dP\{\tau_{1n}\le s\}.$$

From this we have:

$$\sum_{n=1}^{\infty}\sum_{j=n+1}^{\infty} P\{\tau_{1n}\le t_1, \tau_{0j}\le t_2\} = \int_0^{t_1}\int_0^{t_2-s} dB(u)d\Lambda_1(s) + \int_0^{t_1}\int_0^{t_2-s}\Lambda_1(t_2-s-u)dB(u)d\Lambda_1(s)$$

$$= \int_0^{t_1} B(t_2-s)d\Lambda_1(s) + \int_0^{t_1}\Lambda_1(t_2-s)*B(t_2-s)d\Lambda_1(s).$$

c) The third elements are

$$P\{\tau_{0n}\le t_1, \tau_{1j}\le t_2\} = P\{\tau_{0n}\le t_1, \tau_{0n}+(\tau_{0j}-\tau_{0n})+(\tau_{1j}-\tau_{0j})\le t_2\}$$

$$= \int_0^{t_1}\int_0^{t_2-s} P\{\tau_{1,j-n}\le t_2-s-u\}dD(u)dP\{\tau_{0n}\le s\}.$$

Therefore

$$\sum_{n=1}^{\infty}\sum_{j=n+1}^{\infty} P\{\tau_{0n}\le t_1, \tau_{1j}\le t_2\} = \int_0^{t_1}\Lambda_1(t_2-s)*D(t_2-s)d\Lambda_0(s).$$

d) Finally

$$P\{\tau_{1n}\le t_1, \tau_{1j}\le t_2\} = P\{\tau_{1n}\le t_1, \tau_{1n}+(\tau_{1j}-\tau_{1n})\le t_2\}$$

$$= \int_0^{t_1} P\{\tau_{1,j-n}\le t_2-s\}dP\{\tau_{1n}\le s\}.$$

Hence,

$$\sum_{n=1}^{\infty}\sum_{j=n+1}^{\infty} P\{\tau_{1n} \le t_1,\ \tau_{1j} \le t_2\} = \int_0^{t_1} \Lambda_1(t_2-s)\,d\Lambda_1(s).$$

From this, the proof follows.

4. EVALUATION OF $E\{\xi(t_2)|\xi(t_1)\}$

In this section we are concerned with the conditional expectation $E\{\xi(t_2)|\xi(t_1)\}$. To determine this function suppose that $E\{V_1\}=v$ exists. We can now prove the following result:

Proposition 4.1

For any $0 < t_1 \le t_2 < \infty$ we have (a.e.) that

$$\begin{aligned} E\{\xi(t_2)|\xi(t_1)\} = {} & P\{N_1(t_1) = N_1(t_2)|Z_{t_1} = 1\}\xi(t_1) \\ & +[P\{Z_{t_2}=1,\ N_1(t_1)<N_1(t_2)|Z_{t_1}=1\}Z_{t_1} \\ & +P\{Z_{t_2}=0|Z_{t_1}=0\}(1-Z_{t_1})]v. \qquad (4.1) \end{aligned}$$

Proof:

To show that this assertion holds, notice first of all that $\sigma\{\xi(t)\}$ is an atomic σ-algebra whose only atom is the set $\{Z_t=0\}$. Thus, one can write:

$$E\{\xi(t_2)|\xi(t_1)\}=E\{\xi(t_2)|Z_{t_1}=0\}I_{\{Z_{t_1}=0\}}+E\{\xi(t_2)|\xi(t_1)\}I_{\{Z_{t_1}=1\}}. \qquad (4.2)$$

From assumption (i),

$$E\{\xi(t_2)|Z_{t_1}=0\} = vP\{Z_{t_2}=1|Z_{t_1}=0\}. \qquad (4.3)$$

On the other hand we have the following relation:

$$E\{\xi(t_2)|\xi(t_1)\} = E\{E\{\xi(t_2)|Z_{t_1},\ V_{N_1(t_1)+1}\}|\xi(t_1)\}.$$

Let us evaluate

$$\begin{aligned} E\{\xi(t_2)|Z_{t_1},\ V_{N_1(t_1)+1}\} = {} & E\{\xi(t_2)(I_{\{N_1(t_1)=N_1(t_2)\}} \\ & +I_{\{N_1(t_1)<N_1(t_2)\}})|Z_{t_1},\ V_{N_1(t_1)+1}\} \\ = {} & E\{V_{N_1(t_1)+1}Z_{t_2}I_{\{N_1(t_1)=N_1(t_2)\}}|Z_{t_1},\ V_{N_1(t_1)+1}\} \\ & +E\{V_{N_1(t_2)+1}Z_{t_2}I_{\{N_1(t_1)<N(t_2)\}}|Z_{t_1},V_{N_1(t_1)+1}\} \end{aligned}$$

$$= V_{N_1(t_1)+1} P\{Z_{t_2} = 1, N_1(t_1) = N_1(t_2) | Z_{t_1}\}$$

$$+ vP\{Z_{t_2} = 1, N_1(t_1) < N_1(t_2) | Z_{t_1}\}.$$

From this, (4.2), (4.3) and the fact that $\{Z_t = 1\} \in \sigma\{\xi(t)\}$ it follows that

$$E\{\xi(t_2) | \xi(t_1)\} = vP\{Z_{t_2} = 1 | Z_{t_1} = 0\} I_{\{Z_{t_1} = 0\}}$$

$$+ \xi(t_1) P\{Z_{t_2} = 1, N_1(t_1) = N_1(t_2) | Z_{t_1} = 1\}$$

$$+ vP\{Z_{t_2} = 1, N_1(t_1) < N_1(t_2) | Z_{t_1} = 1\} Z_{t_1}.$$

Finally, taking into account the result (3.3), the proposition follows.

Corollary 4.1

If $E\{V_1\} = 0$ we have from (2.11) and the proposition 3.4 that (a.e.)

$$E\{\xi(t_2) | \xi(t_1)\} = \frac{\int_0^{t_1} \{1 - D(t_2 - s)\} d\Lambda_0(s)}{P_{01}(t_1)} \cdot \xi(t_1).$$

5. ASYMPTOTIC BEHAVIOUR

With the knowledge of the explicit form of the bivariate L-S transform $L(\theta_1, \theta_2; t_1, t_2)$ it is possible to make a statement about its approach to equilibrium under translation of the parameter t_1. To be more specific, we examine the asymptotic behaviour of $L(\theta_1, \theta_2; t_1, t_2)$ when $d = t_2 - t_1$ is fixed, and $t_1 \to \infty$. To do so we shall need the following lemma.

Lemma 5.1

For any $y \geq 0$

$$\lim_{t \to \infty} P\{Z_t = 1, U_t \leq y\} = \frac{1}{\beta_1 + \beta_2} \int_0^y [1 - D(u)] du. \quad (5.1)$$

Proof:

A proof of this can be found in Pyke [2]. Another can be easily obtained from (3.4). Writing the right-hand side of this relation as

$$\Lambda_0(t) - \Lambda_1(t) - [1 - D(t + y)] * \Lambda_0(t + y) + \int_t^{t+y} [1 - D(t + y - s)] d\Lambda_0(s)$$

we find from this and the key renewal theorem (see also (2.8) and (2.11)) that (5.1) follows.

Remark 5.1

In the same fashion one can show

$$\lim_{t\to\infty} P\{Z_t = 0,\ U_t \le y\} = \frac{1}{\beta_1+\beta_2} \int_0^y [1 - B(u)]du \tag{5.2}$$

and that

$$\lim_{t_1\to\infty} P\{Z_{t_1} = 0,\ N_1(t_1) = N_1(t_2)\} = \frac{1}{\beta_1+\beta_2} \{\beta_1 - \int_0^d [1-B(u)]du\}. \tag{5.3}$$

On the other hand, taking into account the limiting result (2.7) and (5.3) we obtain immediately that

$$\lim_{t_1\to\infty} P\{Z_{t_1} = 1,\ N_1(t_1) = N_1(t_2)\} = \frac{1}{\beta_1+\beta_2} \{\beta_2 - \int_0^d [1 - D(u)]du\}. \tag{5.4}$$

<u>Proposition 5.1</u>

For $d = t_2 - t_1$ fixed

$$\lim_{t_1\to\infty} P\{Z_{t_1} = Z_{t_2} = 1,\ N_1(t_1) < N_1(t_2)\} = \frac{1}{\beta_1+\beta_2} \int_0^d [1-D(u)] * [P_{01}(u)]du. \tag{5.5}$$

Proof:

It is easy to verify that

$$\int_0^{t_1} \Lambda_1(t_2-s)d\Lambda_0(s) = \Lambda_1(t_2) * \Lambda_0(t_2) - \int_{t_1}^{t_2} \Lambda_1(t_2-s)d\Lambda_0(s),$$

and

$$\int_0^{t_1} B(t_2-s)d\Lambda_1(s) + \int_0^{t_1} \Lambda_1(t_2-s) * B(t_2-s)d\Lambda_1(s)$$

$$= [1+\Lambda_1(t_2)] * \Lambda_1(t_2) * B(t_2) - \int_{t_1}^{t_2} [1+\Lambda_1(t_2-s)] * B(t_2-s)d\Lambda_1(s),$$

while

$$\int_0^{t_1} \Lambda_1(t_2-s) * D(t_2-s)d\Lambda_0(s) = \Lambda_1(t_2) * D(t_2) * \Lambda_0(t_2) - \int_{t_1}^{t_2} \Lambda_1(t_2-s) * D(t_2-s)d\Lambda_0(s),$$

and

$$\int_0^{t_1} \Lambda_1(t_2-s)d\Lambda_1(s) = \Lambda_1(t_2) * \Lambda_1(t_2) - \int_{t_1}^{t_2} \Lambda_1(t_2-s)d\Lambda_1(s).$$

From this, (2.9) and (2.10), we obtain

$$P\{Z_{t_1} = Z_{t_2} = 1,\ N_1(t_1) < N_1(t_2)\} = \int_{t_1}^{t_2} [\Lambda_0(t_2-s)d\Lambda_1(s) - \Lambda_1(t_2-s)d\Lambda_0(s)$$

$$- \Lambda_1(t_2-s)d\Lambda_1(s) + \Lambda_1(t_2-s) * D(t_2-s)d\Lambda_0(s)].$$

Thus, for large t_1

$$P\{Z_{t_1} = Z_{t_2} = 1,\ N_1(t_1) < N_1(t_2)\} \approx \frac{1}{\beta_1+\beta_2} \int_{t_1}^{t_2} [\Lambda_0(t_2-s) - \Lambda_1(t_2-s)$$

$$- \Lambda_1(t_2-s) * \{1-D(t_2-s)\}]ds.$$

From this, after a simple transformation we have for $t_1 \to \infty$

$$P\{Z_{t_1} = Z_{t_2} = 1,\ N_1(t_1) < N_1(t_2)\} = \frac{1}{\beta_1+\beta_2} \int_0^d [P_{01}(u) - \Lambda_1(u) * \{1-D(u)\}]du,$$

which completes the proof. Note from (2.11) that $\Lambda_0(u) - \Lambda_1(u) = P_{01}(u)$.

Corollary 5.1

From proposition 3.5 it follows likewise that

$$\lim_{t_1\to\infty} P\{Z_{t_1} = Z_{t_2} = 0,\ N_1(t_1) < N_1(t_2)\} = \frac{1}{\beta_1+\beta_2} \int_0^d [1-D(u)] * [P_{01}(u)]du. \tag{5.6}$$

Corollary 5.2

From (5.4) and (5.5) we obtain, taking into account relations (3.3) and (2.11) that

$$\lim_{t_1\to\infty} P\{Z_{t_1} = Z_{t_2} = 1\} = \frac{1}{\beta_1+\beta_2} \{\beta_2 - \int_0^d P_{00}(u) * [1-D(u)]du\}. \tag{5.7}$$

On the other hand, (5.3) and (5.6) give

$$\lim_{t_1\to\infty} P\{Z_{t_1} = Z_{t_2} = 0\} = \frac{1}{\beta_1+\beta_2} \{\beta_1 - \int_0^d \{[1-B(u)][1-D(u)] * P_{01}(u)\}du\}. \tag{5.8}$$

The results of this section can be extended to show that for all $n \geq 1$

$$L(\theta_1,\dots,\theta_n;t_1,\dots,t_n) \to L^*(\theta_1,\dots,\theta_n;d_1,\dots,d_{n-1}) \tag{5.9}$$

pointwise as $t_1 \to \infty$, where $d_i = t_{i+1} - t_i$, $i = 1,\dots,(n-1)$. The function on the left-hand side of (5.8) is the L-S transform of $(\xi(t_1),\dots,\xi(t_n))$; on the other hand, L^* is a marginal n-dimensional L-S transform of a strictly stationary stochastic process. Thus, roughly speaking, for large values of t the process $\xi(t)$ becomes strictly stationary.

6. ERGODICITY OF $\xi(t)$

In this section we shall show that the process $\xi(t)$ is also ergodic.

Proposition 6.1

$$\lim_{t\to\infty} \frac{1}{t} \int_0^t \xi(s)ds = \lim_{t\to\infty} \frac{1}{t} \int_0^t E\{\xi(s)\}ds = \frac{\beta_2 v}{\beta_1+\beta_2}$$

where $|v| = |E\{V_1\}| < \infty$.

Proof:

Writing

$$M(t) = V_{N_1(t)+1}[t - \tau_{0,N_1(t)+1}]\xi(t),$$

we have

$$\int_0^t \xi(s)ds = \begin{cases} 0 & \text{if} \quad \tau_{01} > t, \\ b_0(t) & \text{if} \quad \tau_{01} < t,\ N_1(t) = 0, \\ X_{N_1(t)} + M(t) & \text{if} \quad N_1(t) \geq 1, \end{cases}$$

where $M(t) = b_k(t) = [t - \tau_{0,k+1}]V_{k+1}$ and $X_{N_1(t)} = X_k$ on the set $\{N_1(t) = k\}$; here

$$X_k = \sum_{i=1}^{k} (\tau_{1i} = \tau_{0i})V_k \qquad X_0 \equiv 0.$$

Hence,

$$\frac{1}{t}\int_0^t \xi(s)ds = \{\frac{1}{N_1(t)} X_{N_1(t)}\} \frac{N_1(t)}{t} + \frac{M(t)}{t}.$$

From the renewal theorem, as $t \to \infty$

$$\frac{N_1(t)}{t} \to \frac{1}{\beta_1+\beta_2} \text{ (a.e.)}, \quad \frac{1}{N_1(t)} X_{N_1(t)} \to \beta_2 v \text{ (a.e.)}.$$

Finally, as $t \to \infty$,

$$\frac{|M(t)|}{t} \leq \frac{N_1(t)}{t} \sum_{n=1}^{\infty} \frac{1}{n}|V_{n+1}|(\tau_{1,n+1} - \tau_{0,n+1}) I_{\{N_1(t)=n\}} \to 0 \text{ (a.e.)}.$$

This proves the first part of the proposition. On the other hand, from (3.6),

$$E\{\xi(t)\} = v[\Lambda_0(t) - \Lambda_1(t)] = v[1 - D(t)] * \Lambda_0(t).$$

It is simple to verify that

$$\frac{1}{t}\int_0^t [1 - D(u)] * \Lambda_0(u)du \to \frac{\beta_2}{\beta_1+\beta_2}.$$

From this, the assertion follows.

7. SOME APPLICATIONS

Various functionals of $\xi(t)$ are of interest in practical applications. Of particular importance is the stochastic process

$$X(t) = \int_0^t \xi(s)ds, \tag{7.1}$$

where $X(t)$ represents the position of the particle at time t. In problems such as longitudinal dispersion in porous media or sediment transport, the velocities $\{V_i\}_1^\infty$ form an i.i.d. sequence of positive r.v.'s, so that $X(t)$ is a nondecreasing random process. In such a case it is more convenient to deal with its first arrival time T_x

into the set $[x,\infty)$ than with $X(t)$ itself. From the relation

$$\{X(t) \leq x\} = \{T_x > t\} \tag{7.2}$$

one can then easily obtain the d.f. of $X(t)$. For simplicity, we shall use the following notations: $R_1 = \tau_{0,1}$ and $R_n = \tau_{0,n} - \tau_{1,n-1}$, $U_n = \tau_{1,n} - \tau_{0,n}$, $n = 1,2,\ldots$. If we also write

$$X_n = \sum_{i=1}^{n} U_i V_i \text{ and } \eta(x) = \sup\{n;\ X_n \leq x\},$$

then we have:

<u>Proposition 7.1</u>.

For every $x > 0$

$$T_x = \sum_{i=1}^{1+\eta(x)} R_i + \sum_{i=0}^{\eta(x)} U_i + \frac{x - \sum_{i=1}^{\eta(x)} U_i V_i}{V_{\eta(x)+1}}$$

and $T_0 \equiv 0$.

Proof:

It is clear that

$$x = \sum_{i=0}^{N(T_x)} (R_i + U_i)V_i + [T_x - \tau_{N(T_x)}]V_{\eta(T_x)+1}$$

$$- \sum_{i=1}^{N(T_x)+1} R_i V_i = \sum_{i=0}^{N(T_x)} U_i V_i - R_{N(T_x)+1} V_{N(T_x)+1}$$

$$+ [T_x - \tau_{N(T_x)}]V_{N(T_x)+1} .$$

Taking into account that $N(T_x) = \eta(x)$, we see that

$$\frac{x - \sum_{i=0}^{\eta(x)} U_i V_i}{V_{\eta(x)+1}} + R_{\eta(x)+1} + \tau_{\eta(x)} \equiv T_x .$$

This proves the assertion.

<u>Corollary 7.1</u>.

In the special case where one may assume that V_i is a constant V for all $i = 1,2,\ldots$ it follows from (7.3) that

$$T_x = \sum_{i=1}^{\eta(x)+1} R_i + \frac{x}{V} . \tag{7.4}$$

Assuming that $\{R_i\}_1^\infty$ and $\{U_i\}_1^\infty$ are independent r.v.'s one may easily determine the d.f. of T_x, and from (7.2) the d.f. of X(t) follows immediately. This d.f. is of interest in a particular application, which is outlined below.

Consider the steady flow of a fluid (say fresh water) through a column packed with some porous material and assume that a set of tagged particles is introduced in the flow. The transport of these particles in the direction of the (mean) flow is called "longitudinal dispersion". A tagged particle in the flow undergoes a particular kind of random walk. It progresses in a series of steps of random length, with a rest period of random duration between two consecutive steps. The velocity of the particle in the longitudinal direction can be modeled by an intermittent stochastic process. In the case of a statistically homogeneous and isotropic porous medium one can use the process $\{\xi(t); t \geq 0\}$.

The distance along the longitudinal direction traveled by the particle during [0,t] is given by (7.1), starting at time $t = 0$ with the particle at the cross-section $x = 0$ of the column. Therefore, if at time $t = 0$ the tagged particles are released in the cross-section $x = 0$ of the column and no absorption by the medium takes place, the d.f. of (7.1) represents the longitudinal concentration function of these particles in the flow at any time $t > 0$ after their release. For more information on this and related problems see Todorovic [5], Gani and Todorovic [1], Todorovic and Gani [6].

ACKNOWLEDGMENT

We should like to thank Professor R. Pyke for his valuable comments on an earlier version of this paper. Both authors are grateful for the support of the Office of Naval Research through Contract No. N00014-84-K-0568 during the writing of this paper.

REFERENCES

[1] Gani, J. and Todorovic, P., A model for the transport of solid particles in a fluid flow. Stoch. Proc. Applns 14 (1983) 1-17.

[2] Pyke, R., Markov renewal processes; definitions and preliminary properties. Ann. Math. Statist. 32 (1961) 1231-1242.

[3] Pyke, R. and Schauffele, R. A., Limit theorems for Markov renewal processes. Ann. Math. Statist. 35 (1964) 1746-1764.

[4] Smith, W. L., Asymptotic renewal theorems. Proc. Roy. Soc. Edinburgh A64 (1954) 9-48.

[5] Todorovic, P., Stochastic modeling of longitudinal dispersion in a porous medium. Math. Scientist 5 (1980) 45-54.

[6] Todorovic, P. and Gani, J., A model of longitudinal dispersion in non-homogeneous porous media. Stoch. Anal. Applns 1 (1983) 327-339.

Original paper received: 11.02.85
Final paper received: 25.02.86

Paper recommended by M.M. Siddiqui

PROBABILITY AND STATISTICS
Essays in Honor of Franklin A. Graybill
J.N. Srivastava (Editor)

ON A UNIFICATION OF BIAS REDUCTION AND NUMERICAL APPROXIMATION

H. L. Gray

Department of Statistical Science
Southern Methodist University
Dallas, Texas
U.S.A.

In this paper a slight, but important, extension of the generalized jackknife is given. By showing that the error in a numerical approximation is simply the degenerate case of bias in an estimator, it is demonstrated that a large body of theory of numerical analysis can be encompassed by the jackknife theory. Numerous examples are given which include parameter estimation, spectral approximation, spectral estimation and the approximation of tail probabilities.

INTRODUCTION

In [2] the relationship between the so called e_n-transformation or ε-algorithm and the generalized jackknife statistic was discussed and many new results concerning the latter were obtained. In this paper we expand that discussion somewhat and establish a much more general relationship between the jackknife and general numerical methods. That is, by considering numerical approximations as degenerate estimators it is pointed out that the error in a numerical approximation is a special case of the concept of bias in an estimator. In this way it is seen that a large body of the theory of numerical analysis can be considered as just a special case of the generalized jackknife, i.e. the degenerate case. From this point of view, it follows that the generalized jackknife can be considered as the natural extension of one of the more fundamental ideas in the theory of numerical approximation.

Although for the most part this paper is expository and the tools employed are not new, they are utilized in such a way as to suggest a more general applicability of the bias reduction technique employed in the generalized jackknife. In order to demonstrate the validity of this last remark, we first define the generalized jackknife, and then a simple example is given which exemplifies it as a bias reduction method. Following this example, it is shown that such well known results as Simpson's rule, Romberg quadrature, Weddle's rule, Newton's rule, Newton-Cotes, Lagrange interpolation, and the e_n-transform are all simple applications of the generalized jackknife and are in fact even more simple applications of the generalized jackknife than the first, admittedly trivial, example. Finally two somewhat more complicated examples are given, approximating and estimating the spectral density and approximating tail probabilities.

The following definition is a rather simple, but significant, extension of the one given in [6] for the generalized jackknife.

Definition 1.
Let $\hat{\theta}_1(n), \hat{\theta}_2(n), \ldots, \hat{\theta}_{k+1}(n)$ be k+1 estimators (possibly degenerate) each of which depend on n and let a_{ij} and C_j, $j=1,2,\ldots,k+1$, be real or complex numbers. Then the generalized jackknife $G(\cdot;a_{ij},C_j)$ is defined by

$$G(\hat{\theta}_1,\hat{\theta}_2,\ldots,\hat{\theta}_{k+1};a_{ij},C_j) = \frac{H_{k+1}(\hat{\theta}_j;a_{ij})}{H_{k+1}(C_j;a_{ij})} \quad , \tag{1}$$

where $H_{k+1}(Z_j;a_{ij}) = \begin{vmatrix} Z_1 & Z_2 & \cdots & Z_{k+1} \\ a_{11} & a_{12} & \cdots & a_{1,k+1} \\ \vdots & & & \\ a_{k,1} & & \cdots & a_{k,k+1} \end{vmatrix}$

and $H_{k+1}(C_j;a_{ij}) \neq 0$.

Now let us define $B_j(n,\theta)$ by

$$E[\hat{\theta}_j(n)] = C_j\theta + B_j(n,\theta) \ . \tag{2}$$

We will only have need for two sets of C_j's in (1), namely $\{C_j \equiv 1\}$ and the set $\{C_1 = 1,\ C_j \equiv 0,\ j \geq 2\}$. Hence we will restrict ourselves hereafter to those sets of C_j. In either event, $B_1(n,\theta)$ is the bias in the estimator $\theta_1(n)$. One should note that if θ_1 is degenerate that

$$E[\hat{\theta}_1(n)] = \hat{\theta}_1(n) = C_1\theta + B_1(n,\theta) = \theta + B_1(n,\theta) \tag{3}$$

and hence

$$\hat{\theta}_1 - \theta = B_1(n,\theta) \ . \tag{4}$$

In this case (the degenerate one) $B_1(n,\theta)$ is usually referred to as the error in $\hat{\theta}_1$. Thus the problem of reducing the error in an approximation can be looked at as a special case of the problem of reducing the bias in an estimator. We will therefore generally refer to the quantity in (4) as bias but it should be understood that when the θ_i are degenerate the word "error" is more conventional.

Although the introduction of C_j in Definition 1 adds some utility, it does not effect the bias reduction property of the generalized jackknife. That is, the following theorem still obtains.

Theorem 1. Let $\hat{\theta}_1, \hat{\theta}_2, \ldots, \hat{\theta}_{k+1}$ be k+1 estimators, such that

$$E[\hat{\theta}_j - C_j\theta] = \sum_{i=1}^{\infty} a_{ij}(n)b_i(\theta) \ . \tag{5}$$

Then for every set of C_j and $a_{ij} = a_{ij}(n)$ such that (1) is defined,

we have

$$E[G(\hat{\theta}_1,\hat{\theta}_2,\ldots,\hat{\theta}_{k+1};a_{ij},C_j)] = \theta + \frac{H_{k+1}(\varepsilon_j;a_{ij}(n))}{H_{k+1}(C_j;a_{ij}(n))} \ , \tag{6}$$

where

$$\varepsilon_j = \sum_{i=k+1}^{\infty} a_{ij}(n)b_i(\theta) \ .$$

The proof of the above theorem is the same as its counterpart in (4). It holds for all C_j but, as stated previously, we will restrict ourselves to the sets of C_j already mentioned.

Corollary. If $a_{ij}(n) \equiv 0$ when $i = k+1, \ldots$, then $G(\hat\theta_1, \hat\theta_2, \ldots, \hat\theta_{k+1}; a_{ij}, C_j)$ is an unbiased estimator for θ, i.e.,

$$E\left[G(\hat\theta_1, \hat\theta_2, \ldots, \hat\theta_{k+1}; a_{ij}, C_j) \right] = \theta \ . \tag{7}$$

In order to actually make use of Theorem 1 it is necessary to obtain k+1 estimators and their bias expansions. There is fortunately a standard way to obtain these estimators in the usual statistical setting. In the approximation area, their method of obtainment is on the surface more varied, but at the proper level of understanding, no different. In the next section, we review the standard approach for selecting the $\hat\theta_i$ when they are not degenerate. We then demonstrate how this bias reduction technique extends to the degenerate case.

ANALYSIS

In general the problem of selecting k+1 estimators in (1) is more a problem of selecting one estimator, and "perturbing" it properly to obtain the other k estimators, than selecting k+1 distinct estimators. In the case of the jackknife, for example, one usually selects the $\hat\theta_j$ as follows. Let $\hat\theta$ be a given estimator such that (5) holds when $j = 1$ and let $C_j \equiv 1$. Then define

$$\hat\theta_1 = \hat\theta(X_1, X_2, \ldots, X_n)$$

$$\hat\theta_2 = \overline{\hat\theta^{i}} = \overline{\hat\theta}(X_1, X_2, \ldots, X_{i-1}, X_{i+1}, \ldots, X_n) \tag{8}$$

$$\hat\theta_3 = \overline{\hat\theta^{\,i_1, i_2}} = \overline{\hat\theta}\,(X_1, \ldots, X_{i_1-1}, X_{i_1+1}, \ldots, X_{i_2-1}, X_{i_2+1}, \ldots, X_n),$$

etc., where the bar denotes the average over the indicated possible subsamples. Clearly if (5) holds for $j = 1$, then it holds for $j = 2, 3, \ldots, k+1$ and because of (8), in this case we can shorten the notation to

$$G^{(k)}(\hat\theta; a_i(n-j+1)) = \frac{H_{k+1}(\hat\theta; a_i(n-j+1))}{H_{k+1}(1; a_i(n-j+1))} \tag{9}$$

$i = 1, 2, \ldots, k$, $j = 1, 2, \ldots, k+1$ or simply $G^{(k)}(\hat\theta)$, which is the common notation for the generalized jackknife.

The procedure is the same in the degenerate case. That is, select an approximation $\hat\theta(h)$ (here we have denoted n by h since in the degenerate case the quantity perturbed is not the sample size) and obtain the other k approximations by varying h. The following simple examples should clarify the preceeding notions. One should keep in mind that in every example given, the "estimator" obtained is a form of the generalized jackknife.

Example 1.

Let $\bar X^3$ be an estimator for μ^3 based on the random sample $X_1, X_2, \ldots, X_n$ from a distribution with mean μ and variance σ^2. Then

$$E[\bar X^3] = \mu^3 + \frac{3\mu\sigma^2}{n} + \frac{1}{n^2} E[(X-\mu)^3] \ .$$

Therefore from the corollary we use k = 2 and find

$$G^{(2)}(\bar{X}^3) = \frac{H_3(\bar{X}^3;a_{ij})}{H_3(1;a_{ij})} = \frac{\begin{vmatrix} \bar{X}^3 & \overline{(\bar{X}^i)^3} & \overline{(\bar{X}^{i,j})^3} \\ \frac{1}{n} & \frac{1}{n-1} & \frac{1}{n-2} \\ \frac{1}{n^2} & \frac{1}{(n-1)^2} & \frac{1}{(n-2)^2} \end{vmatrix}}{\begin{vmatrix} 1 & 1 & 1 \\ \frac{1}{n} & \frac{1}{n-1} & \frac{1}{n-2} \\ \frac{1}{n^2} & \frac{1}{(n-1)^2} & \frac{1}{(n-2)^2} \end{vmatrix}}$$

$$= \bar{X}^3 - \frac{3\bar{X}}{n(n-1)} \sum_{i=1}^{n} (X_i-\bar{X})^2 + \frac{1}{n(n-1)(n-2)} \sum_{i=1}^{n} (X_i-\bar{X})^3 ,$$

and

$$E[G^{(2)}(\bar{X}^3) = \mu^3.$$

Example 2. (Simpson's Rule)

As possibly a more fundamental example we consider the problem of approximating

$$\theta = \int_a^b f(x)dx \tag{10}$$

by the trapezoidal rule, T(h), i.e.,

$$\hat{\theta}(h) = T(h) = \frac{h}{2}\{f(a)+2f(a+h) + \ldots + 2f(a+(m-1)h) + f(a+mh)\}, \tag{11}$$

where $a + mh = b$. Now if f is analytic over [a,b] the bias in T(h) can be shown to be given by

$$T(h) - \theta = b_1h^2 + b_2h^4 + \ldots + b_mh^{2m} + R_m , \tag{12}$$

where the b_i do not depend on h and $R_m = O(h^{2m+2})$. Equation (12) is referred to as the Euler-Maclaurin summation formula. In light of Equation (12) and Theorem 1 there are a number of ways we could select a second approximation to reduce the bias in T(h). One of the more natural choices is T(h/2). From (12)

$$T(\tfrac{h}{2}) = \theta + \frac{b_1h^2}{4} + \frac{b_2h^4}{16} + \ldots + \frac{b_mh^{2m}}{4^m} + R_m , \tag{13}$$

and the first order generalized jackknife is then given by

$$G(T(h),T(\tfrac{h}{2});2^{-2i(j-1)},1) = \frac{\begin{vmatrix} T(h) & T(\frac{h}{2}) \\ 1 & \frac{1}{4} \end{vmatrix}}{\begin{vmatrix} 1 & 1 \\ 1 & \frac{1}{4} \end{vmatrix}}$$

$$= \frac{4}{3}[T(\tfrac{h}{2}) - \tfrac{1}{4}T(h)]$$

$$= \frac{h}{6}[f(a) + 4f(a+\tfrac{h}{2}) + 2f(a+h) + 4f(a + \tfrac{3h}{2}) + \ldots + f(a+mh)]. \quad (14)$$

Equation 14 is of course better known as Simpson's rule, a result familiar to all students of elementary calculus. Numerous additional quadrature formulas could be obtained in precisely the same way. For example Weddle's rule, Newton-Cotes method, Newton's rule, etc. could all be obtained by simply jackknifing T(h) using different partitions of the interval to produce the required additional approximations. The next example demonstrates this notion more fully.

Example 3 (Romberg Quadrature)

Consider further the problem of reducing the bias in the trapezoidal approximation of the integral in (10). Example 2 established that Simpson's rule, as well as several common quadrature methods, are in fact first order jackknives of the trapezoid rule. The observation can be extended to the higher order jackknife. The k+1 approximations required could be obtained by again using (12) and noting that

$$T(\tfrac{h}{2^j}) = \theta + \frac{b_1 h^2}{2^{2j}} + \frac{b_2 h^4}{2^{4j}} + \ldots + \frac{b_m h^{2m}}{2^{2mj}} + R_m, \quad (15)$$

$$j = 0,1,2,\ldots,k.$$

Then the k-th order jackknife is given by

$$G(\hat{\theta}_1,\ldots,\hat{\theta}_{k+1};a_{ij},1) = \frac{H_{k+1}(T(\frac{h}{2^{j-1}});\ 2^{-2i(j-1)})}{H_{k+1}(1;2^{-2i(j-1)})}$$

$$= \frac{\begin{vmatrix} T(h) & T(\frac{h}{2}) & \ldots & T(\frac{h}{2^k}) \\ 1 & 2^{-2} & \ldots & 2^{-2k} \\ \vdots & & & \\ 1 & 2^{-2k} & \ldots & 2^{-2k(k+1)} \end{vmatrix}}{\begin{vmatrix} 1 & 1 & \ldots & 1 \\ 1 & 2^{-2} & \ldots & 2^{-2k} \\ \vdots & & & \\ & -2k & & -2k(k+1) \end{vmatrix}}. \quad (16)$$

The approximation in (16) can be written in a recursive form and evaluated very efficiently. Thus the step size, h, can be progressively reduced until $G(\hat{\theta}_1,\ldots,\hat{\theta}_{k+1};a_{ij},1)$ gives the desired accuracy for θ. For these reasons it is a very popular method of numerical quadrature and is more commonly referred to as Romberg quadrature.

The first few examples in numerical analysis we have given are numerical approximation of integrals. This was because of the Euler-Maclaurin formula and it was not meant to suggest that the application of the jackknife principle is limited to that arena. The key of course is the bias expansion. The remaining examples are all somewhat distinct.

Example 4 (Lagrange Interpolation)

Consider the problem of approximating f(x) given several values of f in the neighborhood of x, i.e., $f(x_1),f(x_2),\ldots,f(x_{k+1})$. Now suppose f has a Taylor expansion about x, valid in an interval containing the x_i. Then

$$f(x_1) = f(x) + (x_1-x)f'(x) + \ldots + \frac{(x_1-x)^k}{k!} f^{(k)}(x) + R_k(x,x_1)$$

$$f(x_2) = f(x) + (x_2-x)f'(x) + \ldots + \frac{(x_2-x)^k}{k!} f^{(k)}(x) + R_k(x,x_2)$$

$$\vdots$$

$$f(x_{k+1}) = f(x) + (x_{k+1}-x)f'(x) + \ldots + \frac{(x_{k+1}-x)^k}{k!} f^{(k)}(x) + R(x,x_{k+1}).$$

Thus we have for k-th order jackknife,

$$G[f(x_1),\ldots,f(x_{k+1});(x_j-x),1] = \frac{\begin{vmatrix} f(x_1) & f(x_2) & \ldots & f(x_{k+1}) \\ (x_1-x) & (x_2-x) & \ldots & (x_{k+1}-x) \\ (x_1-x)^k & (x_2-x)^k & \ldots & (x_{k+1}-x)^k \end{vmatrix}}{\begin{vmatrix} 1 & 1 & \ldots & 1 \\ (x_1-x) & (x_2-x) & \ldots & (x_{k+1}-x) \\ \vdots & & & \\ (x_1-x)^k & (x_2-x)^k & \ldots & (x_{k+1}-x)^k \end{vmatrix}}$$

which, when expanded, is the well known Lagrange formula for interpolation.

Example 5 (e_p-Transformation or ε-algorithm)

This example was discussed in some detail in [2] regarding the relationship between the jackknife and the e_p-transform. We repeat the example here with some modification since the picture is now more clear. In the following we see that the relationship between the e_p-transform and the jackknife is a fundamental one, but no more so than the relation between jackknifing and Lagrange interpolation or Romberg integration or even Simpson's rule.

Consider the problem of approximating

$$\theta = \sum_{k=C}^{\infty} a_k$$

by the natural approximation

$$\theta(m) = \sum_{k=C}^{m} a_k \ . \tag{17}$$

Now suppose that the error in (17) eventually satisfies a p-th order homogeneous linear difference equation with constant, but unknown, coefficients, i.e., for m>M

$$\Delta^p(\theta(m)-\theta) + \alpha_1\Delta^{p-1}(\theta(m)-\theta) + \ldots + \alpha_p(\theta(m)-\theta) = 0 \ . \tag{18}$$

Rewriting (18) we have (using a backward difference),

$$\theta(m) = \theta + b_1 a_m + b_2 a_{m-1} + \ldots + b_p a_{m-p+1} \ . \tag{19}$$

Of course, since the b_i are unknown, θ cannot be calculated directly from (19).

However, the generalized jackknife of $\theta(m)$ can be computed by noting that

$$\theta(m-j) = \theta + b_1 a_{m-j} + b_2 a_{m-j-1} + \ldots + b_p a_{m-j-p+1} \ . \tag{20}$$

$$j = 0,1,\ldots,p \ .$$

Then from (1) and (20), adopting the notation of (9), we have

$$G^{(p)}(\theta(m);a_{m-i-j+2},1) = \frac{\begin{vmatrix} \theta(m) & \theta(m-1) & \ldots & \theta(m-p) \\ a_m & a_{m-1} & \ldots & a_{m-p} \\ \vdots & & & \\ a_{m-p+1} & & \ldots & a_{m-2p+1} \end{vmatrix}}{\begin{vmatrix} 1 & 1 & \ldots & 1 \\ a_m & a_{m-1} & \ldots & a_{m-p} \\ \vdots & & & \\ a_{m-p+1} & a_{m-p} & \ldots & a_{m-2p+1} \end{vmatrix}} , \tag{21}$$

and for m-p+1 > M it follows from (7) and Theorem 1 that

$$G^{(p)}(\theta(m);a_{m-i-j+2},1) \equiv \theta = \sum_{k=C}^{\infty} a_k \ . \tag{22}$$

In the form (21), the generalized jackknife is known as the e_p-transformation. The transformation has been used extensively in numerical analysis for increasing the rate of convergence of a sequence. It can be evaluated very efficiently by the so called ε-algorithm. In the numerical example which follows we make use of the assumption of Equation (19). However, it should be stressed that the major application of the e_p-transformation is to sequences for which (19) is only "approximately" true, see [5].

Let

$$\theta(m) = \sum_{k=1}^{m} \frac{k \cos\pi(k-1)}{2^{k-1}}$$

If one notes that for all m, $\frac{m \cos\pi(m-1)}{2^{m-1}}$ is a solution of a second order difference equation with constant coefficients then it follows that $\theta(m)-\theta$ satisfies (19) for $p = 2$ and $m > 3$. Thus we obtain

$$\sum_{k=1}^{\infty} \frac{k\cos\pi(k-1)}{2^{k-1}} = G^{(2)}(\theta(4);a_{kj},1) = \frac{\begin{vmatrix} \frac{1}{4} & \frac{3}{4} & 0 \\ -\frac{1}{2} & \frac{3}{4} & -1 \\ \frac{3}{4} & -1 & 1 \end{vmatrix}}{\begin{vmatrix} 1 & 1 & 1 \\ -\frac{1}{2} & \frac{3}{4} & -1 \\ \frac{3}{4} & -1 & 1 \end{vmatrix}} = \frac{4}{9} .$$

The sum of the infinite series has therefore been obtained by jackknifing $\theta(4)$, where

$$\theta(4) = \sum_{k=1}^{4} \frac{k\cos\pi(k-1)}{2^{k-1}} .$$

In this example the a_m were real. Although not widely known, if the a_m are complex, the e_p-transform, as defined by (19), is still valid and enjoys many of the same properties, see [4]. This observation will be used in the next example.

Example 6 (Spectral Density of an ARMA process)

Let X_t be ARMA(p,q) with autocorrelation ρ, i.e.,

$$X_t - \phi_1 X_{t-1} - \dots - \phi_p X_{t-p} = a_t - \phi_1 a_{t-1} - \dots - \phi_q a_{t-q} ,$$

where a_t is zero mean white noise. Then for $m > q$

$$\rho(m) - \phi_1\rho(m-1) - \dots - \phi_p\rho(m-p) = 0 . \tag{23}$$

Moreover for $|f| \le .5$ the spectral density of X_t, $S(f)$, is by definition

$$S(f) = \sum_{-\infty}^{\infty} e^{-2\pi ifk}\rho(k)$$

$$= 1 + 2 \operatorname{Re} \sum_{k=1}^{\infty} e^{-2\pi i f k} \rho(k) , \tag{24}$$

since $\rho(k) = \rho(-k)$. Now let

$$S_m(f;j) = \sum_{k=-j}^{m} e^{-2\pi i f k} \rho(k) \tag{25}$$

and

$$S(f;j) = \sum_{k=-j}^{\infty} e^{-2\pi i f k} \rho(k) . \tag{26}$$

From (24) for $j > 0$

$$S(f) = 1 + 2 \operatorname{Re}(S(f;j) - S_0(f;j)) . \tag{27}$$

A natural approximation for $S(f)$ is then given by the partial sum

$$S_m(f) = 1 + 2 \operatorname{Re}(S_m(f;j) - S_0(f;j)) . \tag{28}$$

But from (23) it is easy to show that for $m > q$

$$S_m(f;j) = S(f;j) + b_1\alpha_m(f) + b_2\alpha_{m-1}(f) + \dots + b_p\alpha_{m-p+1}(f), \tag{29}$$

where

$$\alpha_m(f) = e^{-2\pi i f m} \rho(m) .$$

But (29) is exactly the same form as (19), and as in (21), for $m > q+p-1$, we obtain

$$G^{(p)}(\theta_m; \alpha_{m-i-j+1}, 1) = S(f;j) , \tag{30}$$

where

$$\theta_m = S_m(f;j)$$

and j is chosen positive and sufficiently large that $G^{(p)}$ is defined, i.e., for $m = q + p$, $j = p - q$. Then

$$S(f) = 1 + 2 \operatorname{Re}(G^{(p)}(\theta_m; \alpha_{m-i-j+2}, 1) - S_0(f;j)) . \tag{31}$$

Thus the spectral density of an ARMA(p,q) process has been obtained by jackknifing the finite sum in (28). It should be noted that we could have taken $j=1$ in (27) and avoided what might seem as needless confusion introduced by the notation of (27). The astute reader may have noticed however that introducing j in (27) and using the fact that $\rho(m) = \rho(-m)$ allows us to compute the jackknife from fewer values of $\rho(m)$. This is important when $\rho(m)$ must be estimated. In that event, it can be shown that if ρ is estimated by the sample autocorrelation, then (31) yields the method of moments ARMA spectral estimator. On the other hand if $q = 0$ and the estimates of ρ that arise from (23) using the Burg (Marple) estimates for the ϕ_i are used, then the resulting spectral estimator is the Burg (Marple) spectral estimator. This shows that ARMA spectral estimation method is a bias reduction method and in that sense suggests that it is more closely related to tapering than windowing.

Example 7 (Tail Probabilities)

In this example we will show how the jackknife can be used to obtain an approximation to a tail probability.

Let

$$F(\infty;t) = \int_t^\infty f(u)du \quad . \tag{32}$$

Then a natural approximation to $F(\infty;t)$ is

$$F(x;t) = \int_t^x f(u)du \tag{33}$$

and the error, $E(x)$, is given by

$$E(x) = F(x;t) - F(\infty;t) = -\int_x^\infty f(u)du \ . \tag{34}$$

Suppose that for some n and some set of constants, $\{a_i\}$, that

$$E^{(n)}(x) + a_n E^{(n-1)}(x) + \ldots + a_1 E(x) = 0 \ , \tag{35}$$

i.e. that the error satisfies a linear differential equation with constant coefficients of order n. Then, substituting $F(x;t) - F(\infty;t)$ in (35) and rearranging we have (analogous to (20))

$$F(x;t) = F(\infty;t) + b_1 f(x) + b_2 f^{(1)}(x) + \ldots + b_n f^{(n-1)}(x) \quad , \tag{36}$$

where

$$b_m = \frac{a_{m+1}}{a_1} \ , \quad m = 1,2,\ldots,n-1, \quad b_n = -\frac{1}{a_1} \quad .$$

But now we have a bias expansion for our approximation and it can be used in a variety of ways to determine a generalized jackknife. For example, from (36), for $m = 1,2,\ldots$, we can write

$$f^{(m-1)}(x) = b_1 f^{(m)}(x) + \ldots + b_n f^{(m+n-1)}(x) \ . \tag{37}$$

From (36) and (37) the n-th order jackknife is then (shortening the notation)

$$G^{(n)}[F(x;t)] = \frac{\begin{vmatrix} F(x;t) & f(x) & \ldots & f^{(n-1)}(x) \\ f(x) & f^{(1)}(x) & \ldots & f^{(n)}(x) \\ \vdots & & & \\ f^{(n-1)}(x) & f^{(n)}(x) & \ldots & f^{(2n-1)}(x) \end{vmatrix}}{\begin{vmatrix} 1 & 0 & \ldots & 0 \\ f(x) & f^{(1)}(x) & \ldots & f^{(n)}(x) \\ \vdots & & & \\ f^{(n-1)}(x) & f^{(n)}(x) & \ldots & f^{(2n-1)}(x) \end{vmatrix}} \ . \tag{38}$$

Note that this is the first example of the jackknife which makes use of our extended definition, i.e. in this case $C_1 = 1$ and $C_j = 0$, $j \geq 2$.

From Theorem 1 it then follows that

$$G^{(n)}[F(x;t)] \equiv \int_t^\infty f(u)du$$

for every t. In fact we can take x = t to obtain

$$\int_t^\infty f(u)du = G^{(n)}[F(x;x)] = \frac{\begin{vmatrix} 0 & f(x) & \cdots & f^{(n-1)}(x) \\ f(x) & f^{(1)}(x) & \cdots & f^{(n)}(x) \\ \vdots & & & \vdots \\ f^{(n-1)}(x) & \cdots & & f^{(2n-1)}(x) \end{vmatrix}}{\begin{vmatrix} f^{(1)}(x) & \cdots & f^{(n)}(x) \\ \vdots & & \\ f^{(n)}(x) & \cdots & f^{(2n-1)}(x) \end{vmatrix}} . \quad (39)$$

The result of (39) can be shown to be of value for a much larger class of functions than those satisfying (35) since the jackknife need not eliminate all of the bias to be of value. It can be shown under rather general conditions that the generalized jackknife of (39) converges to the tail probability with n, i.e.,

$$\lim_{n\to\infty} G^{(n)}[F(x,x)] = \int_x^\infty f(u)du, \quad (40)$$

see (3). Thus the generalized jackknife of the rather natural approximation of the tail probability produces an approximating function to that probability that converges to the tail probability as the order of the jackknife increases.

In order to exemplify (39) and (40), let

$$f(x;\alpha,\beta) = \begin{cases} \dfrac{1}{\Gamma(\alpha+1)\beta^{\alpha+1}} x^\alpha e^{-x/\beta} & , \quad x \ge 0 \\ 0 & , \quad x < 0 \end{cases}$$

and consider

$$F(\infty;t) = \int_t^\infty f(x;\alpha,\beta)dx .$$

Table 1 tabulates $G^{(n)}[F(x;x)]$ as a function of n when $\beta = 2$ and $\alpha = (m/2) - 1$, m an integer, i.e. in the chi-square case. However the table is representative of the behavior of the jackknife approximation for a variety of distributions. The convergence with n can clearly be seen in the table. If one notes that for m even, f satisfies a homogeneous linear differential equation with constant coefficients, then that behavior is also clear in the table. For example $G^{(n)}(F(x;x))$ is exact for m = 4 and $n \ge 2$. If m = 6, the approximation is no longer exact at n = 2, but is exact for $n \ge 3$. In any event the approximation improves as n increases and as t increases. The latter behavior is due to the fact that we took x = t. That is, taking x = t will limit the usefulness of the approximation to the tails of the distribution.

CONCLUDING REMARKS

We have demonstrated that if we consider the error in a numerical approximation as a degenerate random variable then numerical error is simply a special case of statistical bias. With this connection made, general methods for bias reduction should translate into general methods for error reduction and vice-versa. We have shown this to be the case and demonstrated the value of the observation in both arenas. Although the extension of the generalized jackknife given here is a simple one, it is an important one, as the final example shows.

REFERENCES

1. Gray, H. L., Atchison, T. A. and McWilliams, G. V. (1971), "Higher Order G-Transformations,", SIAM J. Numer. Anal., Vol. 8, No. 2, 365-81.

2. Gray, H. L., Watkins, T. A. and Adams, J. E. (1972), "On the jackknife statistic, its extensions and its relation to e_n-transformation," Annals of Math. Stat. Vol. 43, No. 1, 1-30.

3. Gray, H. L., Lewis, T. O. (1971), "Approximation of Tail Probabilities by Means of the B_n-Transformation," JASA Vol. 66, No. 336, pp. 897-899.

4. Morton, M. J. and Gray, H. L. (1984), "The G-Spectral Estimator," JASA, Vol. 79, No. 387, pp. 692-701.

5. Shanks, D. (1955), "Non-linear transformation of divergent and slowly convergent sequences", J. Math. Phys.. V. 34, pp. 1-42, MR 28 #1736.

6. Schucany, W. R., Gray, H. L. and Owen, D. B. (1971), "On Bias Reduction in Estimation," J. Amer. Stat. Assoc., Vol. 66, No. 335, 524-33.

Original paper received: 07.03.85
Final paper received: 01.09.86

Paper recommended by D.C. Boes

PROBABILITY AND STATISTICS
Essays in Honor of Franklin A. Graybill
J.N. Srivastava (Editor)

INVARIANT INFERENCE FOR VARIANCE COMPONENTS

David A. Harville

Department of Statistics
Iowa State University
Ames, Iowa 50011
U.S.A.

Consider a general mixed linear model in which an observable random column vector $\underset{\sim}{y}$ follows a multivariate normal distribution with mean vector $\underset{\sim}{X}_0\underset{\sim}{\beta}_0$ and variance-covariance matrix $\theta_{k+1}\underset{\sim}{I} + \sum_{i=1}^{k} \theta_i\underset{\sim}{X}_i\underset{\sim}{X}_i'$. Let $\underset{\sim}{X}_i^* = (\underset{\sim}{X}_0, \ldots, \underset{\sim}{X}_i)$. A function $g(\underset{\sim}{y})$ of $\underset{\sim}{y}$ is defined to be $\underset{\sim}{X}_i^*$-invariant if $g(\underset{\sim}{y} + \underset{\sim}{X}_i^*\underset{\sim}{u}) \equiv g(\underset{\sim}{y})$ for every $\underset{\sim}{u}$. Point estimators of θ_{i+1}, ..., θ_{k+1} that are $\underset{\sim}{X}_i^*$-invariant can be obtained via the method of fitting constants or via modified versions of the maximum likelihood, minimum variance quadratic unbiased, and Bayesian approaches. A general bootstrap-like method can be used to obtain $\underset{\sim}{X}_i^*$-invariant approximate confidence intervals for $\theta_{i+1}, \ldots, \theta_{k+1}$.

1. INTRODUCTION

In the customary mixed linear models, the expected value $\underset{\sim}{\mu}$ of an $n \times 1$ observable random vector $\underset{\sim}{y}$ is assumed to be of the general form $\underset{\sim}{\mu} = \underset{\sim}{X}\underset{\sim}{\beta}$, where X is a given $n \times m$ matrix and $\underset{\sim}{\beta}$ is an $m \times 1$ vector of unknown, unconstrained parameters. Further, the elements of the variance-covariance matrix of $\underset{\sim}{y}$ are linear combinations of unknown parameters called variance components.

In devising procedures for inferences about the variance components, consideration is customarily restricted to translation invariant procedures, that is, to procedures that depend on $\underset{\sim}{y}$ solely through the values of one or more translation invariant statistics. [A (possibly vector-valued) function $g(\underset{\sim}{y})$ of $\underset{\sim}{y}$ is said to be translation invariant if $g(\underset{\sim}{y} + \underset{\sim}{X}\underset{\sim}{u}) = g(\underset{\sim}{y})$ for every $m \times 1$ vector $\underset{\sim}{u}$ (and for all values of $\underset{\sim}{y}$).] In particular, most point estimators of the variance components are translation invariant, including the method-of-fitting-constants estimators, the maximum likelihood (ML) and restricted maximum likelihood (REML) estimators, and the most common versions of the minimum variance and minimum norm quadratic unbiased estimators. One virtue of translation-invariant procedures is that their statistical properties do not depend on $\underset{\sim}{\beta}$.

Consider the special case of an unbalanced (normal-theory) one-way random-effects model

$$y_{ij} = \mu + a_i + e_{ij} \qquad (i = 1, \ldots, k;\ j = 1, \ldots, n_i) \tag{1.1}$$

where μ is an unknown, unconstrained parameter and $a_1, \ldots, a_k$ and $e_{11}, e_{12}, \ldots, e_{kn_k}$ are independently distributed normal random variables with variances $\sigma_a^2 \geq 0$ and $\sigma_e^2 > 0$, respectively. Under model (1.1), a function $g(y_{11}, y_{12}, \ldots, y_{kn_k})$

of $y_{11}, y_{12}, \ldots, y_{kn_k}$ is translation invariant if $g(y_{11} + u, y_{12} + u, \ldots, y_{kn_k} + u) = g(y_{11}, y_{12}, \ldots, y_{kn_k})$ for every scalar u (and for all values of $y_{11}, y_{12}, \ldots, y_{kn_k}$).

Harville (1969) considered, under model (1.1), a more restrictive type of invariance, which he called α-invariance to distinguish it from translation invariance (called μ-invariance by Harville). A function $g(y_{11}, y_{12}, \ldots, y_{kn_k})$ of $y_{11}, y_{12}, \ldots, y_{kn_k}$ is said to be α-invariant if $g(y_{11} + u_1, y_{12} + u_1, \ldots, y_{kn_k} + u_k) = g(y_{11}, y_{12}, \ldots, y_{kn_k})$ for all scalars $u_1, \ldots, u_k$ (and for all values of $y_{11}, y_{12}, \ldots, y_{kn_k}$). Note that $g(y_{11}, y_{12}, \ldots, y_{kn_k})$ is α-invariant under the random-effects model (1.1) if and only if it is translation invariant under the corresponding fixed-effects model. For purposes of making inferences about the residual variance σ_e^2, we may wish to consider only α-invariant procedures. The appeal of such procedures is that their statistical properties do not depend on σ_a^2 (or, since α-invariance implies μ-invariance, on μ).

In the present paper, we extend the concept of α-invariance to a general mixed linear model by introducing a hierarchy of levels of invariance and then consider how the various levels may be achieved in making inferences about a subset of the variance components.

After introducing (in Section 2) some alternative versions of the general mixed linear model, we formally define (in Section 3) the various levels of invariance and then characterize (in Sections 3 and 4) these levels both in general terms and as applied to inference procedures that depend on the data only through the values of one or more quadratic forms. In Sections 5, 6, and 7, we investigate the extent to which three techniques for estimating variance components are invariant or can be modified so as to incorporate various levels of invariance. These techniques are the method of fitting constants, REML, and minimum variance quadratic unbiased estimation. In Section 8, we consider how the Bayesian approach to inference is affected by the imposition of invariance requirements of various levels of severity. Finally, in Section 9, we describe a general bootstrap-like procedure for constructing approximate confidence intervals, list various ways in which this procedure can be used to obtain confidence intervals for variance components, and consider the extent to which these intervals are invariant.

2. GENERAL MIXED LINEAR MODEL

One model for the n × 1 observable random vector $\underset{\sim}{y}$ is the "general mixed linear model"

$$\underset{\sim}{y} = \underset{\sim}{X}_0\underset{\sim}{\beta}_0 + \underset{\sim}{X}_1\underset{\sim}{b}_1 + \cdots + \underset{\sim}{X}_k\underset{\sim}{b}_k + \underset{\sim}{b}_{k+1} ,$$

where $\underset{\sim}{X}_i$ is an $n \times m_i$ known matrix ($i = 0, \ldots, k$), $\underset{\sim}{\beta}_0$ is an $m_0 \times 1$ vector of unknown, unconstrained parameters, and $\underset{\sim}{b}_i$ is an $m_i \times 1$ unobservable vector of "random effects" whose distribution is $N(\underset{\sim}{0}, \theta_i\underset{\sim}{I})$, that is, multivariate normal with mean vector $\underset{\sim}{0}$ and variance-covariance matrix $\theta_i\underset{\sim}{I}$ ($i = 1, \ldots, k+1$). It is assumed that $\underset{\sim}{b}_1, \ldots, \underset{\sim}{b}_{k+1}$ are distributed independently and that $\theta_1, \ldots, \theta_{k+1}$ are unknown parameters that satisfy $\theta_i \geq 0$ ($i = 1, \ldots, k$) and $\theta_{k+1} > 0$. For convenience, we refer to this model as Model Ia. There is an extensive literature on inference about the "variance components" $\theta_1, \ldots, \theta_{k+1}$ and about linear functions of the elements of the vector β_0 of "fixed effects".

For i = 0, ..., k, define $m_i^* = m_0 + \ldots + m_i$ and $X_i^* = (X_0, \ldots, X_i)$, and let $r_i^* = \text{rank}(X_i^*)$. Note that $X_i^* = (X_{i-1}^*, X_i)$ (i = 1, ..., k). Let $r_0 = r_0^*$, $r_i = r_i^* - r_{i-1}^*$ (i = 1, ..., k), and $r_{k+1} = n - r_k^*$, and define X_{k+1} to be the n × n identity matrix. Subsequently, we assume that

$$r_1 > 0, \ldots, r_{k+1} > 0 . \tag{2.1}$$

In what follows, we consider inferences about the last k-i+1 of the variance components, that is, about $\theta_{i+1}, \ldots, \theta_{k+1}$. Note that the ordering of the variance components is somewhat, though — because of assumption (2.1) — not completely, arbitrary. Thus, in the case of the customary additive, random-effects model for a two-way crossed classification, we can take the elements of b_1 to be the effects of the first factor, or, alternatively, of the second factor, whereas, in the case of the customary random-effects model for a two-way nested classification, assumption (2.1) is satisfied only if we take the elements of b_1 to be the elements of the first factor.

Besides Model Ia, we shall have occasion to refer to various other models, some of which we now introduce. We define Model Ib(i) to be a model for y that is identical to Model Ia, except that the first i vectors $b_1, \ldots, b_i$ of random effects are replaced by unconstrained vectors $\beta_1, \ldots, \beta_i$ of fixed effects (that is, unknown parameters). Thus, in Model Ib(i),

$$y = X_i^*\beta_i^* + X_{i+1}b_{i+1} + \ldots + X_k b_k + b_{k+1} ,$$

where $\beta_i^{*\prime} = (\beta_0', \ldots, \beta_i')$. Note that, if in Model Ia we condition on $b_1, \ldots, b_i$, we obtain (as the values of $b_1, \ldots, b_i$ range over all possible vectors) a family of distributions for y identical to that encompassed by Model Ib(i).

For i = 0, ..., k, let $\theta_i^* = (\theta_{i+1}, \ldots, \theta_{k+1})'$. Define $\Sigma_i = \theta_{k+1}I + \sum_{s=i+1}^{k} \theta_s X_s X_s'$ (i = 0, ..., k-1) and $\Sigma_k = \theta_{k+1}I$. Take Ω_i to be a set of (k-i+1)-dimensional vectors consisting of all values of θ_i^* for which the k-i+1 matrices $\Sigma_i, \ldots, \Sigma_k$ are all positive definite (i = 0, ..., k). Let $\theta = \theta_0^* = (\theta_1, \ldots, \theta_{k+1})'$ and $\Sigma = \Sigma_0 = \theta_{k+1}I + \sum_{s=1}^{k} \theta_s X_s X_s'$. Also, let $\Omega = \Omega_0$, so that Ω consists of all values of θ for which the matrices Σ and $\Sigma_1, \ldots, \Sigma_k$ are all positive definite.

Under Model Ia, $y \sim N(X_0\beta_0, \Sigma)$, and the parameter space for θ is

$$\Omega^+ = \{\theta : \theta_{k+1} > 0, \theta_j \geq 0 \ (j = 1, \ldots, k)\}.$$

Note, however, that the matrix Σ is positive definite for some values of the vector θ that include one or more negative elements.

We define Model IIa to be a model for y in which it is assumed only that $y \sim N(X_0\beta_0, \Sigma)$ and in which the parameter space for θ is taken to be the set Ω. Note that the parameter space Ω for θ in Model IIa includes the parameter space Ω^+ for θ in Model Ia as a subset. In general, Ω contains values of θ for which one or more of the first k elements are negative. Thus, in Model IIa, the parameters $\theta_1, \ldots, \theta_k$ are not necessarily interpretable as variances.

Similarly, we define Model IIb(i) to be a model for y that is the same as Model Ib(i), except that the parameter space for θ_i^* is enlarged from

$$\Omega_i^+ = \{\theta_i^*: \theta_{k+1} > 0, \theta_j \geq 0 \ (j = i+1, \ldots, k)\}$$

to the set Ω_i. That is, in Model IIb(i), we assume that $y \sim N(X_i^* \beta_i^*, \Sigma_i)$ with $\theta_i^* \in \Omega_i$.

3. INVARIANCE

We shall say that a (possibly vector-valued) function $g(y)$ of y is _X_i^*-invariant_ if $g(y + X_i^* u) = g(y)$ for every $m_i^* \times 1$ vector u (and for every value of y). Clearly, an X_i^*-invariant function is, for $j = 0, \ldots, i-1$, also X_j^*-invariant.

Take F_i to be any matrix of constants such that $F_i' X_i^* = 0$ and $\text{rank}(F_i) = n - r_i^*$. Note that one such matrix is the matrix $I - X_i^*(X_i^{*\prime} X_i^*)^- X_i^{*\prime}$. (We write A^- for an arbitrary generalized inverse of a matrix A, that is, A^- is any matrix satisfying $AA^-A = A$.)

Let $z_i = F_i' y$. The vector z_i satisfies the definition (e.g., Ferguson 1967, p. 243) of a maximal invariant (with respect to transformations of y of the general form $T(y) = y + X_i^* u$). It follows (e.g., Ferguson 1967, pp. 244-245) that a function $g(y)$ of y is X_i^*-invariant if and only if it is expressible as a function of z_i, that is, if and only if $g(y) = h(F_i' y)$ for some function $h(\cdot)$. Observing that $z_i \sim N(0, F_i' \Sigma_i F_i)$ under each of Models Ia, IIa, Ib(i), and IIb(i), we conclude that, under Model Ia or IIa, the distribution of any X_i^*-invariant statistic $g(y)$ does not vary with β_0 or with $\theta_1, \ldots, \theta_i$ and that, under Model Ib(i) or IIb(i), the distribution of $g(y)$ does not vary with $\beta_0, \beta_1, \ldots, \beta_i$.

In our discussion of inference about the (k-i+1)-dimensional subset $\theta_{i+1}, \ldots, \theta_{k+1}$ of the parameters of Model Ia or IIa, we emphasize procedures that use only X_i^*-invariant statistics. The appeal of such procedures is that their statistical properties do not depend on β_0 or on $\theta_1, \ldots, \theta_i$. Moreoever, even if the joint distribution of $b_1, \ldots, b_i$ deviated from the assumed form, it would still be the case that $z_i \sim N(0, F_i' \Sigma_i F_i)$. Consequently, such a deviation would not affect the statistical properties of an X_i^*-invariant procedure. Further, the conditional (given $b_1, \ldots, b_i$) properties of an X_i^*-invariant procedure do not depend on the values of $b_1, \ldots, b_i$ (or on β_0). (Note, however, that, if the assumption that $b_1, \ldots, b_i$ are distributed independently of $b_{i+1}, \ldots, b_{k+1}$ were dropped, then in general the conditional properties of such a procedure would depend on the values of $b_1, \ldots, b_i$.)

4. QUADRATIC FORMS

In devising procedures for inferences about variance components, consideration is sometimes restricted to procedures that depend on y only through the values of one or more quadratic forms. We now discuss X_i^*-invariance as applied to quadratic forms.

Let A represent an arbitrary $n \times n$ symmetric matrix of constants. The quadratic form $y' A y$ is X_i^*-invariant if and only if

$$\mathbf{A}\mathbf{X}_i^* = \mathbf{0}$$

(e.g., LaMotte 1973, p. 313).

If the quadratic form $\mathbf{y}'\mathbf{A}\mathbf{y}$ in $\mathbf{y}$ is $\mathbf{X}_i^*$-invariant, then it can be re-expressed as a quadratic form in $\mathbf{z}_i$, as is known and as we now demonstrate. (One almost immediate consequence is that the class of all quadratic forms in $\mathbf{z}_i$ is the same as the class of all $\mathbf{X}_i^*$-invariant quadratic forms in $\mathbf{y}$.)

Suppose that $\mathbf{y}'\mathbf{A}\mathbf{y}$ is $\mathbf{X}_i^*$-invariant. Then, $\mathbf{A}\mathbf{X}_i^* = \mathbf{0}$, implying that $\mathbf{A} = \mathbf{R}_i\mathbf{F}_i'$ for some matrix $\mathbf{R}_i$ and hence that

$$\mathbf{A} = \mathbf{R}_i\mathbf{F}_i'\mathbf{F}_i(\mathbf{F}_i'\mathbf{F}_i)^-\mathbf{F}_i' = \mathbf{A}\mathbf{F}_i(\mathbf{F}_i'\mathbf{F}_i)^-\mathbf{F}_i' = \mathbf{A}'\mathbf{F}_i(\mathbf{F}_i'\mathbf{F}_i)^-\mathbf{F}_i' = \mathbf{F}_i(\mathbf{F}_i'\mathbf{F}_i)^-\mathbf{F}_i'\mathbf{A}\mathbf{F}_i(\mathbf{F}_i'\mathbf{F}_i)^-\mathbf{F}_i'.$$

We conclude that

$$\mathbf{y}'\mathbf{A}\mathbf{y} = \mathbf{z}'(\mathbf{F}_i'\mathbf{F}_i)^-\mathbf{F}_i'\mathbf{A}\mathbf{F}_i(\mathbf{F}_i'\mathbf{F}_i)^-\mathbf{z} .$$

We now give expressions for the mean and variance of the arbitrary quadratic form $\mathbf{y}'\mathbf{A}\mathbf{y}$ and determine the circumstances under which these quantities do not vary with $\boldsymbol{\beta}_0$ or with $\theta_1, \ldots, \theta_i$ (in the case of Model Ia or IIa) or $\boldsymbol{\beta}_1, \ldots, \boldsymbol{\beta}_i$ [in the case of Model Ib(i) or IIb(i)].

Applying standard results on the mean and variance of a quadratic form (e.g., Searle 1971, pp. 55 and 57), we find that, under Model Ia or IIa,

$$E(\mathbf{y}'\mathbf{A}\mathbf{y}) = \boldsymbol{\beta}_0'\mathbf{X}_0'\mathbf{A}\mathbf{X}_0\boldsymbol{\beta}_0 + \textstyle\sum_{j=1}^{k+1} \theta_j \mathrm{tr}(\mathbf{X}_j'\mathbf{A}\mathbf{X}_j) , \qquad (4.1)$$

$$\mathrm{var}(\mathbf{y}'\mathbf{A}\mathbf{y}) = 4\textstyle\sum_{j=1}^{k} \theta_j\boldsymbol{\beta}_0'(\mathbf{X}_j'\mathbf{A}\mathbf{X}_0)'\mathbf{X}_j'\mathbf{A}\mathbf{X}_0\boldsymbol{\beta}_0 + 2\sum_{s=1}^{k+1}\sum_{j=1}^{k+1} \theta_j\theta_s \mathrm{tr}[(\mathbf{X}_s'\mathbf{A}\mathbf{X}_j)'\mathbf{X}_s'\mathbf{A}\mathbf{X}_j] \qquad (4.2)$$

and that, under Model Ib(i) or IIb(i),

$$E(\mathbf{y}'\mathbf{A}\mathbf{y}) = \boldsymbol{\beta}_i^{*\prime}\mathbf{X}_i^{*\prime}\mathbf{A}\mathbf{X}_i^*\boldsymbol{\beta}_i^* + \textstyle\sum_{j=i+i}^{k+1} \theta_j \mathrm{tr}(\mathbf{X}_j'\mathbf{A}\mathbf{X}_j) , \qquad (4.3)$$

$$\mathrm{var}(\mathbf{y}'\mathbf{A}\mathbf{y}) = 4\textstyle\sum_{j=i+1}^{k+1} \theta_j\boldsymbol{\beta}_i^{*\prime}(\mathbf{X}_j'\mathbf{A}\mathbf{X}_i^*)'\mathbf{X}_j'\mathbf{A}\mathbf{X}_i^*\boldsymbol{\beta}_i^*$$
$$+ 2 \textstyle\sum_{s=i+1}^{k+1}\sum_{j=i+1}^{k+1} \theta_j\theta_s \mathrm{tr}[\mathbf{X}_s'\mathbf{A}\mathbf{X}_j)'\mathbf{X}_s'\mathbf{A}\mathbf{X}_j] . \qquad (4.4)$$

If $\mathbf{y}'\mathbf{A}\mathbf{y}$ is $\mathbf{X}_i^*$-invariant, then expressions (4.1) and (4.3) for $E(\mathbf{y}'\mathbf{A}\mathbf{y})$, as determined under Model Ia (or IIa) and Model Ib(i) [or IIb(i)], respectively, both simplify to

$$E(\mathbf{y}'\mathbf{A}\mathbf{y}) = \textstyle\sum_{j=i+1}^{k+1} \theta_j \mathrm{tr}(\mathbf{X}_j'\mathbf{A}\mathbf{X}_j) . \qquad (4.5)$$

Expression (4.5) does not, of course, involve $\boldsymbol{\beta}_0$ and $\theta_1, \ldots, \theta_i$ or $\boldsymbol{\beta}_1, \ldots, \boldsymbol{\beta}_i$. Note that expression (4.3) also simplifies to expression (4.5) under the weaker condition $\mathbf{X}_i^{*\prime}\mathbf{A}\mathbf{X}_i^* = \mathbf{0}$ and that expression (4.1) simplifies to expression (4.5) under the even less restrictive conditions

$$\mathbf{X}_0'\mathbf{A}\mathbf{X}_0 = \mathbf{0}, \quad \mathrm{tr}(\mathbf{X}_j'\mathbf{A}\mathbf{X}_j) = 0 \quad (j = 1, \ldots, i) .$$

It is clear from expressions (4.2) and (4.4) that, for $\mathrm{var}(\mathbf{y}'\mathbf{A}\mathbf{y})$ not to depend on $\boldsymbol{\beta}_0$ or on $\theta_1, \ldots, \theta_i$ (in the case of Model Ia or IIa) or $\boldsymbol{\beta}_1, \ldots, \boldsymbol{\beta}_i$ [in the case of Model Ib(i) or IIb(i)], it is necessary and sufficient that $\mathbf{A}\mathbf{X}_i^* = \mathbf{0}$, that is, that $\mathbf{y}'\mathbf{A}\mathbf{y}$ be $\mathbf{X}_i^*$-invariant. For an $\mathbf{X}_i^*$-invariant quadratic form $\mathbf{y}'\mathbf{A}\mathbf{y}$, exppressions (4.2) and (4.4) both simplify to

$$\mathrm{var}(\underset{\sim}{y}'\underset{\sim}{A}\underset{\sim}{y}) = 2\sum_{s=i+1}^{k+1}\sum_{j=i+1}^{k+1} \theta_j\theta_s \mathrm{tr}[(\underset{\sim}{X}_s'\underset{\sim}{A}\underset{\sim}{X}_j)'\underset{\sim}{X}_s'\underset{\sim}{A}\underset{\sim}{X}_j] .$$

5. METHOD OF FITTING CONSTANTS

In conjunction with Model Ia, a long-standing method for obtaining (point) estimates of the k+1 variance components $\theta_1, \ldots, \theta_{k+1}$ is the method of fitting constants, also known as Henderson's (1953) Method 3. The method-of-fitting-constants estimator of θ_{i+1} is $\underset{\sim}{X}_i^*$-invariant (i = 0, ..., k), as we now show.

Let $\underset{\sim}{P}_i = \underset{\sim}{X}_i^*(\underset{\sim}{X}_i^{*\prime}\underset{\sim}{X}_i^*)^-\underset{\sim}{X}_i^{*\prime}$ (i = 0, ..., k). Define

$$Q_i = \underset{\sim}{y}'(\underset{\sim}{P}_i - \underset{\sim}{P}_{i-1})\underset{\sim}{y} \quad (i = 1, \ldots, k), \quad Q_{k+1} = \underset{\sim}{y}'(\underset{\sim}{I} - \underset{\sim}{P}_k)\underset{\sim}{y}.$$

Note that the quadratic form Q_i is $\underset{\sim}{X}_{i-1}^*$-invariant (i = 1, ..., k+1) and that $E(Q_{k+1}) = r_{k+1}\theta_{k+1}$ and, for i = 1, ..., k,

$$E(Q_i) = \sum_{j=i}^{k} \lambda_{ji}\theta_j + r_i\theta_{k+1}$$

where $\lambda_{ji} = \mathrm{tr}[\underset{\sim}{X}_j'(\underset{\sim}{P}_i - \underset{\sim}{P}_{i-1})\underset{\sim}{X}_j]$ (j = i, ..., k) .

In the method of fitting constants, the estimates of the k+1 variance components are obtained by equating the observed values of the k+1 quadratic forms to their expected values and by solving these equations for $\theta_1, \ldots, \theta_{k+1}$. Starting with $\hat{\theta}_{k+1}$, the estimates $\hat{\theta}_{k+1}, \hat{\theta}_k, \ldots, \hat{\theta}_1$ of $\theta_{k+1}, \theta_k, \ldots, \theta_1$, respectively, can be computed successively from the formulas

$$\hat{\theta}_{k+1} = Q_{k+1}/r_{k+1}, \quad \hat{\theta}_i = (Q_i - \sum_{j=i+1}^{k} \lambda_{ji}\hat{\theta}_j - r_i\hat{\theta}_{k+1})/\lambda_{ii} \quad (i = 1, \ldots, k) .$$

It is clear that the method-of-fitting-constants estimator of θ_{i+1} is unbiased under Model Ia [and also under Models IIa, Ib(i), and IIb(i)] and that it is $\underset{\sim}{X}_i^*$-invariant (i = 0, ..., k). (In general, the unbiasedness of the method-of-fitting-constants estimator is achieved at a price — except in special cases, these estimators can, with positive probability, produce negative estimates.)

6. ML AND REML

Suppose that $\underset{\sim}{y}$ follows Model Ia or IIa. The ML technique can be used to obtain estimates of $\theta_1, \ldots, \theta_{k+1}$ and of estimable linear functions of the elements of $\underset{\sim}{\beta}_0$. The ML estimators of $\theta_1, \ldots, \theta_{k+1}$ are $\underset{\sim}{X}_0$-invariant (e.g., Harville 1977, p. 325). However, for i = 1, ..., k, the ML estimators of $\theta_{i+1}, \ldots, \theta_{k+1}$ are not, in general, $\underset{\sim}{X}_i^*$-invariant.

Patterson and Thompson (1971), following Thompson (1962), proposed a technique for estimating $\theta_1, \ldots, \theta_{k+1}$ that is sometimes called REML. We now describe their technique and introduce some variations.

The matrix $\underset{\sim}{F}_i$, introduced in Section 3, can be chosen to have exactly $n-r_i^*$ columns. For any such choice, the matrix $\underset{\sim}{F}_i'\underset{\sim}{\Sigma}\underset{\sim}{F}_i = \underset{\sim}{F}_i'\underset{\sim}{\Sigma}_i\underset{\sim}{F}_i$, which is the variance-covariance matrix of $\underset{\sim}{z}_i$, is nonsingular, and the distribution of $\underset{\sim}{z}_i$ has a probability density function which is given by

$$f_i(\underset{\sim}{z}_i;\ \underset{\sim}{\theta}_i^*) = (2\pi)^{-(n-r_i^*)/2} |\underset{\sim}{F}_i'\underset{\sim}{\Sigma}_i\underset{\sim}{F}_i|^{-1/2}\exp[-(1/2)\underset{\sim}{z}_i'(\underset{\sim}{F}_i'\underset{\sim}{\Sigma}_i\underset{\sim}{F}_i)^{-1}\underset{\sim}{z}_i] .$$

One choice for $\underset{\sim}{F}_i$ that has $n-r_i^*$ columns is $\underset{\sim}{F}_i = \underset{\sim}{A}_i$, where $\underset{\sim}{A}_i$ is any $n \times (n-r_i^*)$ matrix such that $A_i A_i' = \underset{\sim}{I} - \underset{\sim}{P}_i$ and $\underset{\sim}{A}_i'\underset{\sim}{A}_i = \underset{\sim}{I}$. For $\underset{\sim}{F}_i = \underset{\sim}{A}_i$, the probability density function of $\underset{\sim}{z}_i$ can be re-expressed as

$$f_i(\underset{\sim}{F}_i'\underset{\sim}{y};\ \underset{\sim}{\theta}_i^*) = (2\pi)^{-(n-r_i^*)/2} |\underset{\sim}{X}_i^{+\prime}\underset{\sim}{X}_i^+|^{1/2} |\underset{\sim}{\Sigma}_i|^{-1/2} |\underset{\sim}{X}_i^{+\prime}\underset{\sim}{\Sigma}_i^{-1}\underset{\sim}{X}_i^+|^{-1/2}$$
$$\cdot \exp[-(1/2)(\underset{\sim}{y} - \underset{\sim}{X}_i^*\hat{\underset{\sim}{\beta}}_i^*)'\underset{\sim}{\Sigma}_i^{-1}(\underset{\sim}{y} - \underset{\sim}{X}_i^*\hat{\underset{\sim}{\beta}}_i^*)] , \tag{6.1}$$

where $\underset{\sim}{X}_i^+$ is an $n \times r_i^*$ matrix whose columns are any r_i^* linearly independent columns of $\underset{\sim}{X}_i^*$ and where $\hat{\underset{\sim}{\beta}}_i^*$ is any solution to $\underset{\sim}{X}_i^{*\prime}\underset{\sim}{\Sigma}_i^{-1}\underset{\sim}{X}_i^*\hat{\underset{\sim}{\beta}}_i^* = \underset{\sim}{X}_i^{*\prime}\underset{\sim}{\Sigma}_i^{-1}\underset{\sim}{y}$. Upon observing that the distribution of $\underset{\sim}{z}_i$ is the same under Model Ib(i) or IIb(i) as under Model Ia or IIa, result (6.1) follows immediately from result (3) of Harville (1974).

The function $\ell_i(\cdot;\ \underset{\sim}{z}_i)$ defined by $\ell_i(\underset{\sim}{\theta}_i^*;\ \underset{\sim}{z}_i) = f_i(\underset{\sim}{z}_i;\ \underset{\sim}{\theta}_i^*)$ is the likelihhood function when the $n-r_i^*$ elements of $\underset{\sim}{z}_i$ are regarded as the totality of the data. Note that, aside from a multiplicative constant, the probability density function $f_i(\cdot;\ \underset{\sim}{\theta}_i^*)$ and hence the likelihood function $\ell_i(\cdot;\ \underset{\sim}{z}_i)$ is the same for any two choices of the matrix $\underset{\sim}{F}_i$ having $n-r_i^*$ columns.

The REML estimator $\hat{\underset{\sim}{\theta}} = (\hat{\theta}_1, \ldots, \hat{\theta}_{k+1})'$ is, by definition, a value of $\underset{\sim}{\theta}$ that maximizes the likelihood function $\ell_0(\underset{\sim}{\theta};\ \underset{\sim}{z}_0)$ associated with the vector $\underset{\sim}{z}_0$. (If a maximum does not exist, the estimator could be set equal to a value of $\underset{\sim}{\theta}$ for which $\ell_0(\underset{\sim}{\theta}_0;\ \underset{\sim}{z}_0)$ is sufficiently close to its supremum.) As in the case of the ML estimators, the REML estimators $\hat{\theta}_1, \ldots, \hat{\theta}_{k+1}$ are X_0-invariant, but, for i = 1, ..., k, $\hat{\theta}_{i+1}, \ldots, \hat{\theta}_{k+1}$ are not, in general, X_i^*-invariant. There are reasons — discussed, for example, by Harville (1977, sec. 4.3) — for preferring REML to ML as a general technique for estimating $\theta_1, \ldots, \theta_{k+1}$.

There is a variation on the REML technique that could be used to obtain X_i^*-invariant estimators of $\theta_{i+1}, \ldots, \theta_{k+1}$. Instead of taking the estimators of $\theta_{i+1}, \ldots, \theta_{k+1}$ to be the last k-i+1 elements of the value of $\underset{\sim}{\theta}$ that maximizes the likelihood function $\ell_0(\underset{\sim}{\theta};\ \underset{\sim}{z}_0)$ associated with the vector $\underset{\sim}{z}_0$, as in REML, we could take the estimators to be the elements of the value of $\underset{\sim}{\theta}_i^*$ that maximizes the likelihood function $\ell_i(\underset{\sim}{\theta}_i^*;\ \underset{\sim}{z}_i)$ associated with the vector $\underset{\sim}{z}_i$. For convenience, we refer to the latter estimators of $\theta_{i+1}, \ldots, \theta_{k+1}$ as the REML-i estimators. Clearly, the REML-i estimators are $\underset{\sim}{X}_i^*$-invariant. Note that the REML-k estimator of θ_{k+1} is identical to the method-of-fitting-constants estimator $\underset{\sim}{y}'(\underset{\sim}{I}-\underset{\sim}{P}_k)\underset{\sim}{y}/r_{k+1}$.

As an alternative to the REML-i estimator of $\underset{\sim}{\theta}_i^*$, we could form a composite estimator $\tilde{\underset{\sim}{\theta}}_i^* = (\tilde{\theta}_{i+1}, \ldots, \tilde{\theta}_{k+1})'$, where $\tilde{\theta}_{j+1}$ is the REML-j estimator of θ_{j+1} (j = i, ..., k). Note that (unlike the REML-i estimator of θ_{j+1}) $\tilde{\theta}_{j+1}$ is $\underset{\sim}{X}_j^*$-invariant as well as X_i^*-invariant (j = i+1, ..., k). In the case of Model IIa, one apparent drawback of $\tilde{\underset{\sim}{\theta}}_i^*$ as an estimator of θ_i^* is that it is not necessarily confined to the space Ω_i.

We now introduce another estimator of $\underset{\sim}{\theta}_i^*$ — one which, like the composite estimator, gives an $\underset{\sim}{X}_j^*$-invariant estimator of θ_{j+1} (j = i, ..., k). This estimator seems to have more intuitive appeal than the composite estimator, has less extensive computational requirements, and is constrained, by definition, to the space Ω_i^+ or Ω_i. The estimator is $\bar{\underset{\sim}{\theta}}_i^* = (\bar{\theta}_{i+1}, \ldots, \bar{\theta}_{k+1})'$ where $\bar{\theta}_{k+1} = \underset{\sim}{y}'(\underset{\sim}{I} - \underset{\sim}{P}_k)\underset{\sim}{y}/r_{k+1}$ is the REML-k estimator of θ_{k+1} and where, for j = k-1, ..., i, $\bar{\theta}_{j+1}$ is a value at which the function $\ell_j[(\theta_{j+1}, \bar{\theta}_{j+2}, \ldots, \bar{\theta}_{k+1})'; \underset{\sim}{z}_j]$ attains its maximum for θ_{j+1} such that $(\theta_{j+1}, \bar{\theta}_{j+2}, \ldots, \bar{\theta}_{k+1})'$ is contained in Ω_j^+ or Ω_j. For convenience, we refer to the estimators $\bar{\theta}_{i+1}, \ldots, \bar{\theta}_{k+1}$ as the <u>sequential REML estimators</u>.

We now outline a sequential procedure for computing $\bar{\theta}_{i+1}, \ldots, \bar{\theta}_{k+1}$. After k-j+1 steps of the sequential procedure, $\bar{\theta}_{j+1}, \ldots, \bar{\theta}_{k+1}$ will have been computed, and the next [(k-j+2)-nd] step is to compute $\bar{\theta}_j$.

Consider the model

$$\underset{\sim}{y} = \underset{\sim}{X}_{j-1}^* \underset{\sim}{\beta}_{j-1}^* + \underset{\sim}{X}_j \underset{\sim}{b}_j + \underset{\sim}{d}_j , \qquad (6.2)$$

where $\underset{\sim}{b}_j$ $N(\underset{\sim}{0}, \theta_j \underset{\sim}{I})$ (as before), where $\underset{\sim}{d}_j$ is an n × 1 unobservable random vector that is distributed as $N(\underset{\sim}{0}, \underset{\sim}{\Sigma}_j)$ independently of $\underset{\sim}{b}_j$, and where (unlike before) the values of $\theta_{j+1}, \ldots, \theta_{k+1}$ are regarded as known. Take the parameter space for θ_j in this model to consist either of those values for which $\theta_j \geq 0$ or those values for which Σ_{j-1} is positive definite (depending upon whether $\bar{\theta}_j$ is being computed for Model Ia or Model IIa).

The distribution of $\underset{\sim}{z}_{j-1}$ under model (6.2) is $N(\underset{\sim}{0}, \underset{\sim}{F}_{j-1}' \underset{\sim}{\Sigma}_{j-1} \underset{\sim}{F}_{j-1})$, the same as under Model Ia or IIa. Consequently, a REML estimator of the parameter θ_j in model (6.2) is (according to the definition of REML) a value of θ_j that maximizes $\ell_{j-1}[(\theta_j, \theta_{j+1}, \ldots, \theta_{k+1})'; \underset{\sim}{z}_{j-1}]$. We conclude that the problem of computing $\bar{\theta}_j$ is equivalent to the problem of computing a REML estimate of θ_j for model (6.2) (when, in the latter problem, we set $\theta_{j+1} = \bar{\theta}_{j+1}, \ldots, \theta_{k+1} = \bar{\theta}_{k+1}$). The problem of computing a REML estimate of θ_j for model (6.2) is very similar to the computational problem considered by Dempster et al. (1984) and by Callanan (1985), and the procedures described by them can be readily adapted for use in computing $\bar{\theta}_j$.

7. MINIMUM VARIANCE QUADRATIC UNBIASED ESTIMATION

We now consider an approach to the estimation of the parameters $\theta_1, \ldots, \theta_{k+1}$ (in Model Ia or IIa) known as MIVQUE (Rao, 1971; LaMotte, 1973). The acronym MIVQUE is used both for minimum variance quadratic unbiased estimation and for minimum variance quadratic unbiased estimator. We restrict attention to the most common version of MIVQUE, in which the estimator of θ_i is taken to be that quadratic, unbiased, $\underset{\sim}{X}_0^*$-invariant estimator whose variance is smaller than that of any other such estimator at some specified point in the parameter space Ω^+ or Ω. Our discussion of MIVQUE parallels our discussion (in Section 6) of REML.

We use the symbol $\underset{\sim}{\omega} = (\omega_1, \ldots, \omega_{k+1})'$ to represent any specified point in Ω^+ or

Ω, allowing us to distinguish such a point from the "true" parameter vector $\underset{\sim}{\theta}$. Let $\underset{\sim}{\omega}_i^* = (\omega_{i+1}, \ldots, \omega_{k+1})'$, so that $\underset{\sim}{\omega}_i^*$ represents a specified point in Ω_i^+ or Ω_i. Define $\underset{\sim}{V}_i = \underset{\sim}{V}(\underset{\sim}{\omega}_i^*) = \omega_{k+1}\underset{\sim}{I} + \sum_{s=i+1}^{k} \omega_s \underset{\sim}{X}_s \underset{\sim}{X}_s'$ $(i = 0, \ldots, k-1)$ and $\underset{\sim}{V}_k = \underset{\sim}{V}_k(\omega_k^*) = \omega_{k+1}\underset{\sim}{I}$. Note that $\underset{\sim}{\omega} = \underset{\sim}{\omega}_0^*$.

Let $\underset{\sim}{H}_i = \underset{\sim}{H}_i(\underset{\sim}{\omega}_i^*) = \underset{\sim}{V}_i^{-1} - \underset{\sim}{V}_i^{-1}\underset{\sim}{X}_i^*(\underset{\sim}{X}_i^{*\prime}\underset{\sim}{V}_i^{-1}\underset{\sim}{X}_i^*)^{-}\underset{\sim}{X}_i^{*\prime}\underset{\sim}{V}_i^{-1}$. Define $\underset{\sim}{G}_i = \underset{\sim}{G}_i(\underset{\sim}{\omega}_i^*)$ to be a $(k-i-1) \times (k-i+1)$ symmetric matrix whose jsth element is $\mathrm{tr}(\underset{\sim}{X}_{i+j}'\underset{\sim}{H}_i\underset{\sim}{X}_{i+s}\underset{\sim}{X}_{i+s}'\underset{\sim}{H}_i\underset{\sim}{X}_{i+j})$. Similarly, define $q_i = q_i(\underset{\sim}{\omega}_i^*)$ to be a $(k-i+1) \times 1$ vector whose jth element is $\underset{\sim}{y}'\underset{\sim}{H}_i\underset{\sim}{X}_{i+j}\underset{\sim}{X}_{i+j}'\underset{\sim}{H}_i\underset{\sim}{y}$.

Assuming that there exist quadratic, unbiased, X_0^*-invariant estimators of $\theta_1, \ldots, \theta_{k+1}$, there exists a unique solution $\hat{\underset{\sim}{\theta}} = \hat{\underset{\sim}{\theta}}(\underset{\sim}{\omega})$ of the linear system

$$\underset{\sim}{G}_0\hat{\underset{\sim}{\theta}} = \underset{\sim}{q}_0$$

(e.g., LaMotte, 1973). Denote the elements of $\hat{\underset{\sim}{\theta}}$ by $\hat{\theta}_1 = \hat{\theta}_1(\underset{\sim}{\omega}), \ldots, \hat{\theta}_{k+1} = \hat{\theta}_{k+1}(\underset{\sim}{\omega})$, respectively. The estimator $\hat{\theta}_i$ is a MIVQUE of θ_i; it is a quadratic, unbiased, X_0^*-invariant estimator, and its variance is smaller at $\underset{\sim}{\theta} = \underset{\sim}{\omega}$ than that of any other such estimator (LaMotte, 1973).

The MIVQUEs of the k-i+1 parameters $\theta_{i+1}, \ldots, \theta_{k+1}$ are not, in general, X_i^*-invariant. Quadratic, $\underset{\sim}{X}_i^*$-invariant estimators of $\theta_{i+1}, \ldots, \theta_{k+1}$ can be obtained by applying MIVQUE to Model Ib(i) or IIb(i) instead of applying it directly to Model Ia or IIa. For convenience, we refer to these estimators as MIVQU-i estimators. Results discussed in Section 4 imply that [under Model Ia or IIa as well as under Model Ib(i) or IIb(i)] the MIVQU-i estimators are unbiased and their variances are smaller at $\underset{\sim}{\theta}_i^* = \underset{\sim}{\omega}_i^*$ than those of any other quadratic, unbiased, $\underset{\sim}{X}_i^*$-invariant estimators.

The MIVQU-i estimators of $\theta_{i+1}, \ldots, \theta_{k+1}$ are the first, ..., (k-i+1)st components, respectively, of the solution to the linear system $\underset{\sim}{G}_i\hat{\underset{\sim}{\theta}}_i = \underset{\sim}{q}_i$. (Assuming that there exist quadratic, unbiased, $\underset{\sim}{X}_i^*$-invariant estimators of $\theta_{i+1}, \ldots, \theta_{k+1}$, this linear system has a unique solution.)

For $j = i+1, \ldots, k$, the MIVQU-i estimators of $\theta_{j+1}, \ldots, \theta_{k+1}$, like the MIVQUEs, are not, in general, $\underset{\sim}{X}_j^*$-invariant. Consider, however, the composite estimators of $\theta_{i+1}, \ldots, \theta_{k+1}$ obtained by taking the estimator $\tilde{\theta}_{j+1}$ of θ_{j+1} to be the MIVQU-j estimator $(j = i, \ldots, k)$. The estimators $\tilde{\theta}_{i+1}, \ldots, \tilde{\theta}_{k+1}$ are quadratic and unbiased, and, for $j = i, \ldots, k$, $\tilde{\theta}_{j+1}$ is $\underset{\sim}{X}_j^*$-invariant.

We now describe another MIVQUE-related procedure for estimating $\theta_{i+1}, \ldots, \theta_{k+1}$. The estimator of θ_{j+1} produced by this procedure — like the composite MIVQUE estimator $\tilde{\theta}_{j+1}$ — is $\underset{\sim}{X}_j^*$-invariant $(j = 1, \ldots, k)$. The estimator of θ_{k+1} is taken to be $\bar{\theta}_{k+1} = \underset{\sim}{y}'(\underset{\sim}{I}-\underset{\sim}{P}_k)\underset{\sim}{y}/r_{k+1}$. For $j = k, \ldots, i+1$, the estimator $\bar{\theta}_j$ of θ_j is taken to be that $\underset{\sim}{X}_{j-1}^*$-invariant estimator which, if the model were model (6.2) and if $\bar{\theta}_{j+1}, \ldots, \bar{\theta}_{k+1}$ were regarded as known values of $\theta_{j+1}, \ldots, \theta_{k+1}$, respectively, would be an unbiased estimator of general form $c + \underset{\sim}{y}'\underset{\sim}{A}\underset{\sim}{y}$ and would have a smaller variance at $\theta_j = \omega_j$ than any other X^*_{j-1}-invariant estimator of general form $c + y'Ay$.

By, for example, using the results of LaMotte (1973), it can be shown that

$$\bar{\theta}_j = [y'H^*_{j-1}X_jX_j'H^*_{j-1}y - \sum_{s=j+1}^{k} \bar{\theta}_s \mathrm{tr}(X_j'H^*_{j-1}X_sX_s'H^*_{j-1}X_j) - \bar{\theta}_{k=1}\mathrm{tr}(X_j'H^{*2}_{j-1}X_j)]/\mathrm{tr}[(X_j'H^*_{j-1}X_j)^2] , \tag{7.1}$$

where H^*_{j-1} is the matrix obtained from $H_{j-1} = H_{j-1}(\omega^*_{j-1})$ by replacing $\omega_{j+1}, \ldots, \omega_{k+1}$ with $\bar{\theta}_{j+1}, \ldots, \bar{\theta}_{k+1}$, respectively ($j = k, \ldots, i+1$). (To insure that the sequence $\bar{\theta}_k, \ldots, \bar{\theta}_{i+1}$ is well-defined, it may be necessary to modify formula (7.1) so as to eliminate values of $\bar{\theta}_j$ for which the vector $(\bar{\theta}_j, \ldots, \bar{\theta}_{k+1})'$ lies outside the space Ω^+_{j-1} or Ω_{j-1}.) For convenience, we refer to the estimators $\bar{\theta}_{i+1}, \ldots, \bar{\theta}_{k+1}$ as sequential MIVQUEs.

An iterative version of MIVQUE has sometimes been proposed. Iterative MIVQUE is known (in cases where the REML estimate of θ lies in the interior of the parameter space Ω^+ or Ω) to be essentially equivalent to REML. Iterative versions of the MIVQU-i and sequential MIVQUE procedures can be obtained by analogy and can be shown to be essentially equivalent (in much the same sense that iterative MIVQUE is essentially equivalent to REML) to the REML-i and sequential REML procedures.

8. BAYESIAN INFERENCE

Take τ_0 to be the $r_0 \times 1$ vector of linear parametric functions defined by $X_0^+\tau_0 = X_0\beta_0$. Then, parameterizing in terms of τ_0 rather than β_0, the likelihood function associated with y is

$$\ell(\theta, \tau_0; y) = (2\pi)^{-n/2}|\Sigma|^{-1/2}\exp[-(1/2)(y - X_0^+\tau_0)'\Sigma^{-1}(y - X_0^+\tau_0)] .$$

In the Bayesian approach to inference about $\theta_{i+1}, \ldots, \theta_{k+1}$, a prior distribution is assigned to θ and τ_0. Together, the prior distribution and the likelihood function $\ell(\theta, \tau_0; y)$ determine the posterior distribution of θ and τ_0. In general, Bayesian inferences about $\theta_{i+1}, \ldots, \theta_{k+1}$, which are based on their (marginal) posterior distribution, are not X_i^*-invariant or even X_0^*-invariant.

Inferences about $\theta_{i+1}, \ldots, \theta_{k+1}$ that are X_i^*-invariant could be obtained via a modified Bayesian approach, to be called the Bayesian-i approach, in which the likelihood function is taken to be $\ell_i(\theta_i^*; z_i)$ rather than $\ell(\theta, \tau_0; y)$. We now give some results that are helpful in comparing the Bayesian-0 approach (to inferences about θ) with the standard Bayesian approach [that in which the likelihood function is taken to be $\ell(\theta, \tau_0; y)$] and in comparing the Bayesian-i approach (to inferences about $\theta_{i+1}, \ldots, \theta_{k+1}$) with the Bayesian-(i-1) approach ($i = 1, \ldots, k$).

Taking $F_0 = A_0$, we have that

$$\ell_0(\theta; z_0) = |X_0^{+\prime}X_0^+|^{1/2}\int_{-\infty}^{\infty} \cdots \int_{-\infty}^{\infty} \ell(\theta, \tau_0; y)d\tau_0$$

(Harville, 1974). Thus, if a priori τ_0 were distributed diffusely and independently of θ, the Bayesian-0 approach to inference about θ would be essentially equivalent to the standard Bayesian approach.

Basically the same conclusion can be reached in a different way. If a priori (in the standard Bayesian approach) $\underset{\sim}{\tau}_0$ were distributed as $N(\underset{\sim}{0}, \theta_0 \underset{\sim}{I})$ independently of $\underset{\sim}{\theta}$, then the joint posterior probability density function of $\underset{\sim}{\tau}_0$ and $\underset{\sim}{\theta}$ would be proportional to

$$g(\underset{\sim}{\theta}, \underset{\sim}{\tau}_0; \underset{\sim}{y}) = h(\underset{\sim}{\theta})(2\pi\theta_0)^{-r_0/2}\exp[-\underset{\sim}{\tau}_0'\underset{\sim}{\tau}_0/(2\theta_0)]\ell(\underset{\sim}{\theta}, \underset{\sim}{\tau}_0; \underset{\sim}{y})$$

where $h(\cdot)$ is the prior probability density function of $\underset{\sim}{\theta}$. Further, the marginal posterior probability density function of θ would be proportional to

$$g^*(\underset{\sim}{\theta}; \underset{\sim}{y}) = \int_{-\infty}^{\infty} \cdots \int_{-\infty}^{\infty} g(\underset{\sim}{\theta}, \underset{\sim}{\tau}_0; \underset{\sim}{y})d\underset{\sim}{\tau}_0 = h(\underset{\sim}{\theta})q(\underset{\sim}{\theta}; \underset{\sim}{y})$$

where $q(\underset{\sim}{\theta}; \underset{\sim}{y}) = (2\pi)^{-n/2}|\underset{\sim}{\Sigma} + \theta_0\underset{\sim}{X}_0^+\underset{\sim}{X}_0^{+\prime}|^{-1/2}\exp[-(1/2)\underset{\sim}{y}'(\underset{\sim}{\Sigma} + \theta_0\underset{\sim}{X}_0^+\underset{\sim}{X}_0^{+\prime})^{-1}\underset{\sim}{y}]$.

Observe that $g^*(\underset{\sim}{\theta}; \underset{\sim}{y})$ can be rewritten as

$$g^*(\underset{\sim}{\theta}; \underset{\sim}{y}) = h(\underset{\sim}{\theta})(2\pi\theta_0)^{-r_0/2}|\underset{\sim}{X}_0^{+\prime}\underset{\sim}{X}_0^+|^{-1/2}[(2\pi\theta_0)^{r_0/2}|\underset{\sim}{X}_0^{+\prime}\underset{\sim}{X}_0^+|^{1/2}q(\underset{\sim}{\theta}; \underset{\sim}{y})].$$

Moreover,

$$\lim_{\theta_0\to\infty} [(2\pi\theta_0)^{r_0/2}|\underset{\sim}{X}_0^{+\prime}\underset{\sim}{X}_0^+|^{1/2}q(\underset{\sim}{\theta}, \underset{\sim}{y})] = \ell_0(\underset{\sim}{\theta}; \underset{\sim}{z}_0) \tag{8.1}$$

(e.g., Sallas 1979, pp. 93-94; Dempster, Rubin, and Tsutakawa 1981, sec. 5.4). This result suggests that, if θ_0 is large, then

$$g^*(\underset{\sim}{\theta}; \underset{\sim}{y}) \doteq h(\underset{\sim}{\theta})(2\pi\theta_0)^{-r_0/2}|\underset{\sim}{X}_0^{+\prime}\underset{\sim}{X}_0^+|^{-1/2}\ell_0(\underset{\sim}{\theta}; \underset{\sim}{z}_0) . \tag{8.2}$$

As before, we conclude that, if there were little prior information about $\underset{\sim}{\tau}_0$, then the Bayesian-0 approach to inference about $\underset{\sim}{\theta}$ would be nearly equivalent to the standard Bayesian approach.

We now derive an approximate relationship, somewhat comparable to (8.2), between the posterior distribution of $\theta_{i+1}, \ldots, \theta_{k+1}$ in the Bayesian-(i-1) approach and that in the Bayesian-i approach.

For convenience, take $\underset{\sim}{F}_{i-1} = \underset{\sim}{A}_{i-1}$. Let

$$\underset{\sim}{\Lambda}_i = \theta_{k+1}\underset{\sim}{I} + \sum_{s=i+1}^{k} \theta_s\underset{\sim}{A}_{i-1}'\underset{\sim}{X}_s\underset{\sim}{X}_s'\underset{\sim}{A}_{i-1} ,$$

and define $\underset{\sim}{N}_i$ to be an $(n-r_{i-1}^*) \times r_i$ matrix such that $\underset{\sim}{N}_i\underset{\sim}{N}_i' = \underset{\sim}{A}_{i-1}'\underset{\sim}{X}_i\underset{\sim}{X}_i'\underset{\sim}{A}_{i-1}$. Then $\underset{\sim}{z}_{i-1} \sim N(\underset{\sim}{0}, \underset{\sim}{\Lambda}_i + \theta_i\underset{\sim}{N}_i\underset{\sim}{N}_i')$.

Suppose that a priori θ_i and $\underset{\sim}{\theta}_i^*$ are distributed independently with probability density functions $d_i(\cdot)$ and $h_i(\cdot)$, respectively. Then, the joint posterior probability density function of θ_i and $\underset{\sim}{\theta}_i^*$ in the Bayesian-(i-1) approach is proportional to

$$g_{i-1}(\underset{\sim}{\theta}_i^*, \theta_i; \underset{\sim}{z}_{i-1}) = h_i(\underset{\sim}{\theta}_i^*)d_i(\theta_i)\ell_{i-1}(\underset{\sim}{\theta}_{i-1}^*; \underset{\sim}{z}_{i-1}) .$$

Note that

$$\ell_{i-1}(\underset{\sim}{\theta}_{i-1}^*; \underset{\sim}{z}_{i-1}) = (2\pi)^{-(n-r_{i-1}^*)/2}|\underset{\sim}{\Lambda}_i + \theta_i\underset{\sim}{N}_i\underset{\sim}{N}_i'|^{-1/2}$$
$$\cdot\exp[-(1/2)\underset{\sim}{z}_{i-1}'(\underset{\sim}{\Lambda}_i + \theta_i\underset{\sim}{N}_i\underset{\sim}{N}_i')^{-1}\underset{\sim}{z}_{i-1}]$$

and that $g_{i-1}(\underset{\sim}{\theta}_i^*, \theta_i; \underset{\sim}{z}_{i-1})$ can be rewritten as

$$g_{i-1}(\underset{\sim}{\theta}_i^*, \theta_i; \underset{\sim}{z}_{i-1}) = h_i(\underset{\sim}{\theta}_i^*)d_i(\theta_i)(2\pi\theta_i)^{-r_i/2}|\underset{\sim}{N}_i'\underset{\sim}{N}_i|^{-1/2}$$

$$\cdot[(2\pi\theta_i)^{r_i/2}|\underset{\sim}{N}_i'\underset{\sim}{N}_i|^{1/2}\ell_{i-1}(\underset{\sim}{\theta}^*_{i-1};\ \underset{\sim}{z}_{i-1})].$$

Let $\underset{\sim}{B}_i$ represent any $(n-r^*_{i-1}) \times (n-r^*_i)$ matrix such that $\underset{\sim}{B}_i\underset{\sim}{B}_i' = \underset{\sim}{I} - \underset{\sim}{N}_i(\underset{\sim}{N}_i'\underset{\sim}{N}_i)^{-1}\underset{\sim}{N}_i'$ and $\underset{\sim}{B}_i'\underset{\sim}{B}_i = \underset{\sim}{I}$. Then,

$$\begin{aligned}\underset{\sim}{A}_{i-1}\underset{\sim}{B}_i(\underset{\sim}{A}_{i-1}\underset{\sim}{B}_i)' &= \underset{\sim}{A}_{i-1}[\underset{\sim}{I} - \underset{\sim}{N}_i(\underset{\sim}{N}_i'\underset{\sim}{N}_i)^{-1}\underset{\sim}{N}_i']\underset{\sim}{A}_{i-1}' \\ &= \underset{\sim}{A}_{i-1}\{\underset{\sim}{I} - \underset{\sim}{A}_{i-1}'\underset{\sim}{X}_i[\underset{\sim}{X}_i'(\underset{\sim}{I}-\underset{\sim}{P}_{i-1})\underset{\sim}{X}_i]^-\underset{\sim}{X}_i'\underset{\sim}{A}_{i-1}\}\underset{\sim}{A}_{i-1}' \\ &= \underset{\sim}{I} - \underset{\sim}{P}_{i-1} - (\underset{\sim}{I}-\underset{\sim}{P}_{i-1})\underset{\sim}{X}_i[\underset{\sim}{X}_i'(\underset{\sim}{I}-\underset{\sim}{P}_{i-1})\underset{\sim}{X}_i]^-\underset{\sim}{X}_i'(\underset{\sim}{I}-\underset{\sim}{P}_{i-1}) \\ &= \underset{\sim}{I} - \underset{\sim}{P}_i ,\end{aligned}$$

and $(\underset{\sim}{A}_{i-1}\underset{\sim}{B}_i)'\underset{\sim}{A}_{i-1}\underset{\sim}{B}_i = \underset{\sim}{I}$. Consequently, we can choose

$$\underset{\sim}{F}_i = \underset{\sim}{A}_i = \underset{\sim}{A}_{i-1}\underset{\sim}{B}_i ,$$

in which case $\underset{\sim}{z}_i = \underset{\sim}{F}_i'\underset{\sim}{y} = \underset{\sim}{B}_i'\underset{\sim}{z}_{i-1}$.

Choosing $\underset{\sim}{F}_i = \underset{\sim}{A}_i = \underset{\sim}{A}_{i-1}\underset{\sim}{B}_i$ and essentially repeating the step that led to result (8.1) (with $\underset{\sim}{z}_{i-1}$, $\underset{\sim}{B}_i$, $\underset{\sim}{z}_i$, $\underset{\sim}{\Lambda}_i$, and $\underset{\sim}{N}_i$ in place of $\underset{\sim}{y}$, $\underset{\sim}{A}_i$, $\underset{\sim}{z}_0$, $\underset{\sim}{\Sigma}$, and $\underset{\sim}{X}_0^+$, respectively), we find that

$$\lim_{\theta_i \to \infty}[(2\pi\theta_i)^{r_i/2}|\underset{\sim}{N}_i'\underset{\sim}{N}_i|^{1/2}\ell_{i-1}(\underset{\sim}{\theta}^*_{i-1};\ \underset{\sim}{z}_{i-1})] = \ell_i(\underset{\sim}{\theta}^*_i;\ \underset{\sim}{z}_i) . \qquad (8.3)$$

The (marginal) posterior probability density function of $\underset{\sim}{\theta}^*_i$ in the Bayesian-(i-1) approach is proportional to

$$\begin{aligned}g^*_{i-1}(\underset{\sim}{\theta}^*_i;\ \underset{\sim}{z}_{i-1}) &= \int_0^\infty g_{i-1}(\underset{\sim}{\theta}^*_i,\ \theta_i;\ \underset{\sim}{z}_{i-1})d\theta_i \\ &= h_i(\underset{\sim}{\theta}^*_i)|\underset{\sim}{N}_i'\underset{\sim}{N}_i|^{-1/2}\int_0^\infty d_i(\theta_i)(2\pi\theta_i)^{-r_i/2}[(2\pi\theta_i)^{r_i/2}|\underset{\sim}{N}_i'\underset{\sim}{N}_i|^{1/2} \\ &\qquad \cdot\ell_{i-1}(\underset{\sim}{\theta}^*_{i-1};\ \underset{\sim}{z}_{i-1})]d\theta_i .\end{aligned}$$

If the prior probability density function $d_i(\theta_i)$ were such that the factor $d_i(\theta_i)(2\pi\theta_i)^{-r_i/2}$ were relatively small for small values of θ_i, then result (8.3) would suggest that

$$\begin{aligned}g^*_{i-1}(\underset{\sim}{\theta}^*_i;\ \underset{\sim}{z}_{i-1}) &\doteq h_i(\underset{\sim}{\theta}^*_i)\ell_i(\underset{\sim}{\theta}^*_i;\ \underset{\sim}{z}_i)|\underset{\sim}{N}_i'\underset{\sim}{N}_i|^{-1/2}\int_0^\infty d_i(\theta_i)(2\pi\theta_i)^{-r_i/2}\,d\theta_i \qquad (8.4)\\ &\propto h_i(\underset{\sim}{\theta}^*_i)\ell_i(\underset{\sim}{\theta}^*_i;\ \underset{\sim}{z}_i)\end{aligned}$$

and hence that the Bayesian-i approach to inference about $\theta_{i+1}, \ldots, \theta_{k+1}$ would be nearly equivalent to the Bayesian-(i-1) approach. Conversely, if $d_i(\theta_i)(2\pi\theta_i)^{-r_i/2}$ were not relatively small for small values of θ_i, then approximation (8.4) could be inadequate, and the Bayesian-i and Bayesian-(i-1) approaches could differ significantly.

9. CONFIDENCE SETS

We now consider the construction of a confidence set for the ith variance component $\theta_i (i = 1, \ldots, k+1)$. Except for the last component θ_{k+1}, we must, in general, settle for an approximate confidence set (e.g., Graybill, 1976, chap. 15). We begin by describing a general procedure for obtaining an exact or approximate confidence set — one that is closely related to the parametric version of Efron's (1982, 1985) bootstrap procedure.

Suppose that the distribution of an $n \times 1$ observable random vector $\underset{\sim}{y}$ (an observable random vector that follows Model Ia or IIa in our application) depends on one or more unknown parameters ($\theta_1, \ldots, \theta_{k+1}$ and the elements of $\underset{\sim}{\beta}_0$ in our application) and that, for some real number α ($0 < \alpha \leq 1$), we seek an exact or approximate $100(1-\alpha)\%$ confidence set for some function ϕ of the unknown parameters (the "function" θ_i in our application).

Let $t(\underset{\sim}{y}, \phi)$ represent a function of $\underset{\sim}{y}$ and ϕ whose distribution function $G(\cdot; \phi, \underset{\sim}{\eta})$ depends on the unknown parameters (to a "moderate" extent, if at all) through the value of the function ϕ and possibly through the value of some vector $\underset{\sim}{\eta}$ of additional functions of the unknown parameters. If $G(\cdot; \phi, \underset{\sim}{\eta})$ does not depend on ϕ and $\underset{\sim}{\eta}$, then $t(\underset{\sim}{y}, \phi)$ is called a pivot or pivotal quantity (e.g., Cox and Hinkley, 1974, p. 211). If $G(\cdot; \phi, \underset{\sim}{\eta})$ varies with ϕ and $\underset{\sim}{\eta}$ to a moderate extent, we refer to $t(\underset{\sim}{y}, \phi)$ as an approximate pivot.

Assuming that the distribution function $G(\cdot; \phi, \underset{\sim}{\eta})$ is continuous, take $t_1(\phi, \underset{\sim}{\eta})$ and $t_2(\phi, \underset{\sim}{\eta})$ to be real numbers (or $\pm\infty$) such that

$$\Pr[t_1(\phi, \underset{\sim}{\eta}) \leq t(\underset{\sim}{y}; \phi) \leq t_2(\phi, \underset{\sim}{\eta})] = 1-\alpha \ .$$

Define $S(\underset{\sim}{y}; \underset{\sim}{\eta})$ to be the set whose elements consist of those values of ϕ that satisfy the inequality

$$t_1(\phi, \underset{\sim}{\eta}) \leq t(\underset{\sim}{y}, \phi) \leq t_2(\phi, \underset{\sim}{\eta}) \ .$$

Then, $\Pr[\phi \in S(\underset{\sim}{y}; \underset{\sim}{\eta})] = 1-\alpha$.

If $t(\underset{\sim}{y}, \phi)$ is an exact pivot or, more generally, if $G(\cdot; \phi, \underset{\sim}{\eta})$ does not depend on $\underset{\sim}{\eta}$ (but possibly depends on ϕ), then $t_1(\phi, \underset{\sim}{\eta})$ and $t_2(\phi, \underset{\sim}{\eta})$ can be chosen so as not to depend on $\underset{\sim}{\eta}$, in which case the set $S(\underset{\sim}{y}; \underset{\sim}{\eta})$ does not vary with $\underset{\sim}{\eta}$ and can be regarded as an exact $100(1-\alpha)\%$ confidence set for ϕ.

Let $\hat{\underset{\sim}{\eta}}$ represent an estimator of $\underset{\sim}{\eta}$. In the most general case, where $G(\cdot; \phi, \underset{\sim}{\eta})$ may depend on $\underset{\sim}{\eta}$, we can regard the set $S'(\underset{\sim}{y}) = S(\underset{\sim}{y}; \hat{\underset{\sim}{\eta}})$ as an approximate $100(1-\alpha)\%$ confidence set for ϕ.

To determine the set $S'(\underset{\sim}{y})$, we may have to compute $t_1(\phi, \hat{\eta})$ and $t_2(\phi, \hat{\eta})$ for each of a number of different values of ϕ. To do so, we require knowledge of the distribution function $G(\cdot; \phi, \hat{\underset{\sim}{\eta}})$. If this knowledge is not otherwise available, it can be accumulateed by simulation, that is, by repeatedly sampling from the distribution function $G(\cdot; \phi, \hat{\underset{\sim}{\eta}})$ and by then regarding the empirical distribution function $\tilde{G}(\cdot; \phi, \hat{\underset{\sim}{\eta}})$ as an approximation to $G(\cdot; \phi, \hat{\underset{\sim}{\eta}})$. The approximation $\tilde{G}(\cdot; \phi, \hat{\underset{\sim}{\eta}})$ can, for purposes of computing $t_1(\phi, \hat{\underset{\sim}{\eta}})$ and $t_2(\phi, \hat{\underset{\sim}{\eta}})$, be used in place of $G(\cdot; \phi, \hat{\underset{\sim}{\eta}})$.

Note that, if we must resort to simulation and if $G(\cdot; \phi, \hat{\underset{\sim}{\eta}})$ depends (nontrivially) on ϕ, then the simulation must be repeated for each value of ϕ. Repeated simulation can be avoided by replacing the approximate $100(1-\alpha)\%$ confidence set $S'(\underset{\sim}{y})$ with the set $S^*(\underset{\sim}{y})$ whose elements consist of those values of ϕ that satisfy the inequality

$$t_1(\hat{\phi}, \hat{\underset{\sim}{\eta}}) \leq t(\underset{\sim}{y}, \phi) \leq t_2(\hat{\phi}, \hat{\underset{\sim}{\eta}})$$

(where $\hat{\phi}$ is an estimator of ϕ). To determine the alternative approximate $100(1-\alpha)\%$ confidence set $S^*(\underset{\sim}{y})$, we require knowledge of $G(\cdot; \phi, \underset{\sim}{\eta})$ for only a single value of ϕ (namely, the estimated value $\hat{\phi}$).

To insure that the approximate confidence set $S'(y)$ or $S^*(y)$ is satisfactory, it would seem that the approximate pivot $t(y, \phi)$ should be chosen so that the dependence of its distribution function $G(\cdot;\ \phi,\ \eta)$ on η and [in the case of $S^*(y)$] on ϕ is as slight as possible.

Suppose now that the observable random vector y follows Model Ia or IIa and that the function ϕ for which an exact or approximate $100(1-\alpha)\%$ confidence set is sought is $\phi = \theta_i$. We discuss five possible choices for the exact or approximate pivotal quantity $t(y,\phi)$.

(1) One choice for $t(y,\phi)$ is

$$\begin{aligned} t(y,\ \phi) &= Q_{k+1}/\theta_{k+1}\ , \text{ if } i = k+1\ , \\ &= Q_i/[\hat{\theta}_{k+1} + r_i^{-1}(\textstyle\sum_{j=i+1}^{k} \lambda_{ji}\hat{\theta}_j + \lambda_{ii}\theta_i)]\ , \text{ if } 1 \le i \le k\ , \end{aligned} \tag{9.1}$$

where, as in Section 5, $Q_1, \ldots, Q_{k+1}$ are the quadratic forms employed in the method of fitting constants. The quantities $\hat{\theta}_1, \ldots, \hat{\theta}_{k+1}$ are the method-of-fitting-constants estimators of $\theta_1, \ldots, \theta_{k+1}$, respectively, suitably modified to insure that $(\hat{\theta}_1, \ldots, \hat{\theta}_{k+1})'$ is contained in Ω^+ or Ω. Except in special cases, the distribution of (9.1) depends on the value of $\phi = \theta_i$ and on the value of the vector $\eta = (\theta_{i+1}, \ldots, \theta_{k+1})'$. The natural choices for $\hat{\eta}$ and $\hat{\phi}$ would be $\hat{\eta} = (\hat{\theta}_{i+1}, \ldots, \hat{\theta}_{k+1})'$ and $\hat{\phi} = \hat{\theta}_i$. When $\hat{\eta}$ and $\hat{\phi}$ are chosen in this way, the confidence set $S'(y)$ or $S^*(y)$ for $\phi = \theta_i$ would be X_{i-1}^*-invariant.

(2) A second choice for $t(y,\phi)$ is the log-likelihood ratio associated with z_0. Taking

$$L_0(\theta;z_0) = \log[\ell_0(\theta;z_0)]$$

to be the log-likelihood function associated with z_0 and letting (as in Section 6) $\hat{\theta} = (\hat{\theta}_1, \ldots, \hat{\theta}_{k+1})'$ represent the REML estimator of θ, this choice is (in the case of Model IIa)

$$t(y,\ \phi) = L_0(\hat{\theta};\ z_0) - \underset{\{\theta_1, \ldots, \theta_{i-1}, \theta_{i+1}, \ldots, \theta_{k+1}:\ \theta \in \Omega\}}{\text{maximum}} L_0(\theta;\ z_0)$$

In general, the distribution of the log-likelihood ratio associated with z_0 depends on the value of $\phi = \theta_i$ and on the value of the vector $\eta = (\theta_1, \ldots, \theta_{i-1}, \theta_{i+1}, \ldots, \theta_{k+1})'$. The natural choices for η and ϕ would be the REML estimators $\hat{\eta} = (\hat{\theta}_1, \ldots, \hat{\theta}_{i-1}, \hat{\theta}_{i+1}, \ldots, \hat{\theta}_{k+1})'$ and $\hat{\phi} = \hat{\theta}_i$. With these choices for $\hat{\eta}$ and $\hat{\phi}$, the confidence set $S'(y)$ or $S^*(y)$ for $\phi = \theta_i$ would be X_0-invariant but not, in general, X_i^*-invariant.

(3) A third choice for $t(y,\phi)$ is the log-likelihood ratio associated with z_{i-1}. Taking

$$L_{i-1}(\theta_{i-1}^*;\ z_{i-1}) = \log[\ell_{i-1}(\theta_{i-1}^*;\ z_{i-1})]$$

to be the log-likelihood function associated with z_{i-1} and letting $\tilde{\theta}_{i-1}^* = (\tilde{\theta}_i, \ldots, \tilde{\theta}_{k+1})'$ represent the REML-(i-1) estimator of θ_{i-1}^*, this choice is (in the case of Model IIa)

$$t(\underset{\sim}{y}, \phi) = L_{i-1}(\tilde{\underset{\sim}{\theta}}^*_{i-1}; \underset{\sim}{z}_{i-1}) - \underset{\{\theta_{i+1},\ldots,\theta_{k+1}:\ \underset{\sim}{\theta}^*_{i-1} \in \Omega_{i-1}\}}{\text{maximum}} L_{i-1}(\underset{\sim}{\theta}^*_{i-1}; \underset{\sim}{z}_{i-1}).$$

The natural choices for $\hat{\underset{\sim}{\eta}}$ and $\hat{\phi}$ would be the REML-i estimators $\hat{\underset{\sim}{\eta}} = (\tilde{\theta}_{i=1}, \ldots, \tilde{\theta}_{k+1})'$ and $\hat{\phi} = \tilde{\theta}_i$. With these choices for $\hat{\underset{\sim}{\eta}}$ and $\hat{\phi}$, the confidence set $S'(\underset{\sim}{y})$ or $S^*(\underset{\sim}{y})$ would be $\underset{\sim}{X}^*_{i-1}$-invariant.

(4) The third choice for $t(\underset{\sim}{y},\phi)$ imposes rather severe computational requirements, which can be reduced by choosing the modification

$$t(\underset{\sim}{y}, \phi) = L_{i-1}(\bar{\underset{\sim}{\theta}}^*_{i-1}; \underset{\sim}{z}_{i-1}) - L_{i-1}[(\theta_i, \bar{\theta}_{i+1}, \ldots, \bar{\theta}_{k+1})'; \underset{\sim}{z}_{i-1}] ,$$

where, as in Section 6, $\bar{\underset{\sim}{\theta}}^*_{i-1} = (\bar{\theta}_i, \ldots, \bar{\theta}_{k+1})'$ is the vector of sequential REML estimators of $\theta_i, \ldots, \theta_{k+1}$. The distribution of this choice for $t(\underset{\sim}{y},\phi)$, like that of the third choice, depends (in general) on the value of $\phi = \theta_i$ and on the value of the vector $\underset{\sim}{\eta} = (\theta_{i+1}, \ldots, \theta_{k+1})'$. The natural choices for $\hat{\underset{\sim}{\eta}}$ and $\hat{\phi}$ would be the sequential REML estimators $\hat{\underset{\sim}{\eta}} = (\bar{\theta}_{i+1}, \ldots, \bar{\theta}_{k+1})'$ and $\hat{\phi} = \bar{\theta}_i$. These choices for $\hat{\underset{\sim}{\eta}}$ and $\hat{\phi}$ are such that the confidence set $S'(\underset{\sim}{y})$ or $S^*(\underset{\sim}{y})$ would be $\underset{\sim}{X}^*_{i-1}$-invariant.

(5) Another choice for $t(\underset{\sim}{y}, \phi)$ is suggested by Fenech and Harville's (1986) procedure for constructing an exact confidence region for the vector $\underset{\sim}{\theta}$. Their procedure uses quadratic forms

$$q_i(\theta_i, \ldots, \theta_{k+1}; \underset{\sim}{y}) = \underset{\sim}{y}'\underset{\sim}{R}_i(\underset{\sim}{R}_i'\underset{\sim}{\Sigma}_{i-1}\underset{\sim}{R}_i)^{-1}\underset{\sim}{R}_i'\underset{\sim}{y} \qquad (i = 1, \ldots, k) ,$$

$$q_{k+1}(\theta_{k+1}; \underset{\sim}{y}) = \underset{\sim}{y}'(\underset{\sim}{I} - \underset{\sim}{P}_k)\underset{\sim}{y}/\theta_{k+1} .$$

Here, $\underset{\sim}{R}_i = \underset{\sim}{S}_i\underset{\sim}{X}_i\underset{\sim}{L}_i$, where

$$\underset{\sim}{S}_i = \underset{\sim}{\Sigma}_i^{-1} - \underset{\sim}{\Sigma}_i^{-1}\underset{\sim}{X}^*_{i-1}(\underset{\sim}{X}^{*\prime}_{i-1}\underset{\sim}{\Sigma}_i^{-1}\underset{\sim}{X}^*_{i-1})^{-}\underset{\sim}{X}^{*\prime}_{i-1}\underset{\sim}{\Sigma}_i^{-1}$$

and $\underset{\sim}{L}_i$ is any $m_i \times r_i$ matrix such that $\text{rank}(\underset{\sim}{L}_i'\underset{\sim}{X}_i'\underset{\sim}{S}_i\underset{\sim}{X}_i) = r_i$. Fenech and Harville showed that the k+1 quadratic forms $q_i(\theta_i, \ldots, \theta_{k+1};\underset{\sim}{y})$ $(i = 1, \ldots, k+1)$ are distributed independently as chi-square random variables with degrees of freedom $r_1, \ldots, r_{k+1}$, respectively. Estimators $\hat{\theta}_1, \ldots, \hat{\theta}_{k+1}$ of $\theta_1, \ldots, \theta_{k+1}$ can be obtained from these quadratic forms by equating them to the means $r_1, \ldots, r_{k+1}$ or the medians of their respective chi-square distributions. We could choose

$$t(\underset{\sim}{y}, \phi) = q_i(\theta_i, \hat{\theta}_{i+1}, \ldots, \hat{\theta}_{k+1}; \underset{\sim}{y}) .$$

[The estimators $\hat{\theta}_{i+1}, \ldots, \hat{\theta}_{k+1}$ could be suitably modified to insure that $(\hat{\theta}_{i+1}, \ldots, \hat{\theta}_{k+1})' \in \Omega_i$.] In general, the distribution of $t(\underset{\sim}{y},\phi)$ would depend on the value of $\phi = \theta_i$ and on the value of the vector $\underset{\sim}{\eta} = (\theta_{i+1}, \ldots, \theta_{k+1})'$. The natural choices for $\hat{\underset{\sim}{\eta}}$ and $\hat{\phi}$ would be $\hat{\underset{\sim}{\eta}} = (\hat{\theta}_{i+1}, \ldots, \hat{\theta}_{k+1})'$ and $\hat{\phi} = \hat{\theta}_i$. These choices for $\hat{\underset{\sim}{\eta}}$ and $\hat{\phi}$ are such that the confidence set $S'(\underset{\sim}{y})$ or $S^*(\underset{\sim}{y})$ would be $\underset{\sim}{X}^*_{i-1}$-invariant.

The general procedure for obtaining exact or approximate confidence intervals could also be used to obtain a confidence interval for a ratio $\gamma_{i+1} = \theta_{i+1}/\theta_{k+1}$ or a more complicated function of $\theta_{i+1}, \ldots, \theta_{k+1}$.

ACKNOWLEDGEMENT

The author's work was supported in part by Office of Naval Research Contract N0014-85-K-0418.

REFERENCES

[1] Callanan, T. P., Restricted maximum likelihood estimation of variance components: computational aspects, Ph.D. Dissertation, Department of Statistics, Iowa State University, Ames, Iowa (December 1985).

[2] Cox, D. R. and Hinkley, D. V., Theoretical Statistics (Chapman and Hall, London, 1974).

[3] Dempster, A. P., Rubin, D. B., and Tsutakawa, R. K., Estimation in covariance components models, J. Amer. Statist. Assoc. 76 (1981) 341-353.

[4] Dempster, A. P., Selwyn, M. R., Patel, C. M., and Roth, A. J., Statistical and computational aspects of mixed model analysis, Applied Statistics 33 (1984) 203-214.

[5] Efron, B., The Jacknife, the Bootstrap, and Other Resampling Plans, Number 38 in CBMS-NSF Regional Conference Series in Applied Mathematics (Society for Industrial and Applied Mathematics, Philadelphia, Pennsylvania, 1982).

[6] Efron, B., Bootstrap confidence intervals for a class of parametric problems, Biometrika 72 (1985) 45-58.

[7] Fenech, A. P. and Harville, D. A., Confidence regions for variance components in unbalanced mixed linear models, Preprint No. 86-12, Department of Statistics, Iowa State University, Ames, Iowa (January 1986).

[8] Ferguson, T. S., Mathematical Statistics: A Decision Theoretic Approach (Academic Press, New York, 1967).

[9] Graybill, F. A., Theory and Application of the Linear Model (Duxbury Press, North Scituate, Massachusetts, 1976).

[10] Harville, D. A., Quadratic unbiased estimation of variance components for the one-way classification, Biometrika 56 (1969) 313-326.

[11] Harville, D. A., Bayesian inference for variance components using only error contrasts, Biometrika 61 (1974) 383-385.

[12] Harville, D. A., Maximum likelihood approaches to variance component estimation and to related problems, J. Amer. Statist. Assoc. 72 (1977) 320-338.

[13] Henderson, C. R., Estimation of variance and covariance components, Biometrics 9 (1953) 226-252.

[14] LaMotte, L. R., Quadratic estimation of variance components, Biometrics 29 (1973) 311-330.

[15] Patterson, H. D. and Thompson, R., Recovery of interblock information when block sizes are unequal, Biometrika 58 (1971) 545-554.

[16] Rao, C. R., Minimum variance quadratic unbiased estimation of variance components, Journal of Multivariate Analysis 1 (1971) 445-456.

[17] Sallas, W. M., Recursive mixed model estimation, Ph.D. Dissertation, Department of Statistics, Iowa State University, Ames, Iowa (November 1979).

[18] Searle, S. R., Linear Models (John Wiley and Sons, New York, 1971).

[19] Thompson, W. A., Jr., The problem of negative estimates of variance components. Ann. Math. Statist. 33 (1962) 273-289.

Original paper received: 04.03.85
Final paper received: 02.07.87

Paper recommended by J.S. Williams

PROBABILITY AND STATISTICS
Essays in Honor of Franklin A. Graybill
J.N. Srivastava (Editor)

AN ARGUMENT FOR ADAPTIVE ROBUST ESTIMATION

Robert V. HOGG, University of Iowa, Iowa City, IA 52242 U.S.A.
Gordon K. BRIL, Luther College, Decorah, IA 52101 U.S.A.
Sang M. HAN, Seoul City University, Seoul 131, Korea
Lianng YUH, Merrell Dow Pharmaceuticals, Cincinnati, OH 45215 U.S.A.

ABSTRACT: Consideration is first given to developing an easy M-estimator. To To choose the tuning constants of M-estimators better, realistic measures of skewness, peakedness, and tail thickness are given. A new adaptive L-estimator is presented; and its performance is compared, in estimating a location parameter, to standard L-estimators as well as to adaptive and regular M-estimators when the underlying distribution is symmetric. A modification of this adaptive L-estimator makes it suitable for estimating regression parameters. In addition, adaptive M- and R-estimators are introduced for the regression situations in which the underlying error variable could be symmetric or skewed. A Monte Carlo study clearly indicates that least squares must not be used blindly, and adaptive M-estimation is the best procedure at the present time. However, there is also great promise for modified adaptive L-estimation in the future.

AMS Subject Classification: 62F35

Key words: Robust and adaptive estimators, selector statistics, M-, R-, and L-estimators, regression estimators.

1. Introduction to M-estimation

In making an argument for adaptive robust estimation, we must first convince the reader of the need of using some type of robust procedure, along with the usual least squares method. We think that the easiest way to do that is as follows.

Most statisticians recognize that many data sets have seemingly come from distributions that are skewed or have heavier tails than those of a normal distributions. For the moment, let us not worry about the skewed case, but accept the fact that sometimes the tails are heavier than those of the normal. Suppose we are willing to assume that the underlying distribution is something like Student's $t(r)$-distribution with median θ and spread parameter ρ. Our primary interest is the center θ; that is, we assume we know ρ or it can be approximated easily. With a random sample from this distribution the likelihood function is

$$L(\theta) \propto \prod_{i=1}^{n} \frac{1}{\left[1+\left(\frac{x_i-\theta}{\rho}\right)^2 \Big/ r\right]^{(r+1)/2}}.$$

To find the maximum likelihood estimator of θ, equate the first derivative of $\ln L(\theta)$ to zero. This results in the equation

$$\sum_{i=1}^{n} \frac{1}{\left[1+\left(\frac{x_i-\theta}{\rho}\right)^2 \Big/ r\right]} (x_i-\theta) = 0,$$

which must be solved for θ by some iterative method. One easy way of doing this replaces θ and ρ in the first factor of the summand by the preliminary estimates, $\hat{\theta}_0$ and $\hat{\rho}_0$. This gives

$$\sum_{i=1}^{n} (w_i)(x_i-\theta) = 0 \quad \text{and} \quad \hat{\theta}_1 = \frac{\Sigma w_i x_i}{\Sigma w_i},$$

where

$$w_i = \frac{1}{1+\left[\frac{x_i-\hat{\theta}_0}{\hat{\rho}_0}\right]^2 \Big/ r}, \quad i = 1,2,\ldots,n.$$

We call $\hat{\theta}_1$ a one-step estimate. Of course, the process can continue using $\hat{\theta}_1$ as the preliminary estimate resulting in $\hat{\theta}_2$, and so on. However, many times, with a good preliminary robust estimate of θ, a one-step estimate is quite adequate. "Robustniks" frequently use

$$\hat{\theta}_0 = \text{median of the sample} = M$$

and

$$\hat{\rho}_0 = \frac{\text{median}\,|x_i-\hat{\theta}_0|}{0.6745} = \frac{\text{MAD}}{0.6745}.$$

The reason that the value 0.6745 is in the denominator of $\hat{\rho}_0$ is that then $\hat{\rho}_0$ is a robust and consistent estimate of the standard deviation of the population in case the sample actually arises from a normal distribution.

As an illustration, if we use $r = 6$ degrees of freedom in the preceding, the resulting estimator is a very satisfactory robust estimator of θ. That is, one that works quite well for normally distributed data but also gives protection with Cauchy-type data. However, if one knows absolutely that the data did not arise from an extremely heavy-tailed distribution, like the Cauchy, then $r = 8$ or 10 can be used giving better performance with normally distributed data but still adequate protection from many heavier-tailed situations. However, if more protection is desired against possible outliers, then use $r = 4$ or 2; of course, then the performance is not as good in the normal case. Once we accept the idea that r can be changed to fit the data set under consideration, we have taken the first important step towards *adapting*.

A few comments should be made, however, before we leave this introductory exposition of M-estimation; here M stands for maximum likelihood.

1. Using the t-weights as developed above, we can determine an approximate error structure of our estimate $\hat{\theta}$ from the usual asymptotic results associated with maximum likelihood estimation. See Hogg (1979) and Huber (1981) for more details about this and other aspects of robust estimation.

2. There are many good weight functions used in practice. With $z_i = (x_i-\theta)/\rho$,

two of these are:

a. Tukey's *biweight* , B(k), is

$$w_i = (1-z_i^2/k^2)^2, \quad |z_i| \le k,$$

equals zero otherwise.

b. Huber's *weight*, H(k), is

$$\begin{aligned} w_i &= 1, & |z_i| &\le k, \\ &= k/|z_i|, & |z_i| &> k. \end{aligned}$$

The B(4.82) and H(1.25) correspond fairly well to the t-weights with $r = 6$, denoted by T(6), but H(1.25) is much more conservative in the treatment of outliers than are the others; thus it is not as good as T(6) and B(4.82) if the sample actually arises from a distribution with tails as heavy as those of t(2). So as not to complicate the situation, we use t-weights throughout our exposition, but we recognize others might want to use their favorite robust weight functions.

3. M-estimation can be easily extended to regression. For example, if Y_i is a random variable with middle $\alpha + \beta x_i$, use

$$L(\alpha,\beta) \propto \prod_{i=1}^{n} \frac{1}{\left[1+\left(\frac{y_i-\alpha-\beta x_i}{\rho}\right)^2\Big/r\right]^{(r+1)/2}}.$$

In the two equations

$$\frac{\partial \ln L(\alpha,\beta)}{\partial\alpha} = 0, \qquad \frac{\partial \ln L(\alpha,\beta)}{\partial\beta} = 0$$

replace α, β, and ρ in the weight factor by preliminary robust estimates, $\hat{\alpha}_0$, $\hat{\beta}_0$, and $\hat{\rho}_0$. Say $\hat{\alpha}_0$ and $\hat{\beta}_0$ are L_1 estimates (least absolute values) and

$$\hat{\rho}_0 = \frac{\text{median}\,|y_i-\hat{\alpha}_0-\hat{\beta}_0 x_i|}{0.6745}.$$

This actually is equivalent to the weighted least squares solutions, $\hat{\alpha}_1$ and $\hat{\beta}_1$, for α and β using

$$\sum_{i=1}^{n} (w_i)(y_i-\alpha-\beta x_i)^2,$$

where

$$w_i = \frac{1}{\left[1+\left(\frac{y_i-\hat{\alpha}_0-\hat{\beta}_0 x_i}{\hat{\rho}_0}\right)^2\Big/r\right]}.$$

Again we can continue to iterate using $\hat{\alpha}_1$ and $\hat{\beta}_1$ as the preliminary estimates obtaining $\hat{\alpha}_2$ and $\hat{\beta}_2$, and so on, if we so desire.

4. No robustnik, as far as we know, is suggesting that we give up least squares methods; most recommend the folloiwng.

a. Use least squares.

b. Use one robust procedure, say at least a one-step estimate with $r = 6$.

c. If (a) and (b) agree, report that fact and the usual statistical summary associated with the least squares estimators.

d. If (a) and (b) do not agree, take another hard look at the data. It is helpful to print out the weights, $w_1, w_2, \ldots, w_n$, on the last iteration. Low weights can help us spot "outliers" so that we can ask appropriate questions and have a better understanding of what the data set is actually telling us.

2. The selector statistics.

To use this adaptive approach [or what Bickel (1982) calls *partially* adaptive methods], we need to have some rough idea about the nature of the underlying distribution. One of us, Yuh (1984), has developed good measures of skewness, peakedness, and tail thickness. To describe these selector statistics, we need the folloiwng notation. In our exposition we use the order statistics of a simple random sample, but these can be replaced by the ordered residuals from a preliminary and robust estimate of the middle in a regression situation.

Say $\alpha+\beta+\gamma = 0.5$, where α, β, and γ are nonnegative fractions. Let L_α, B_β, C_γ, D_γ, E_β, and U_α be the respective means of the smallest $n\alpha$ order statistics, of the next smallest $n\beta$ order statistics, of the next smallest $n\gamma$ order statistics, and so on, until U_α is taken to be the mean of the largest $n\alpha$ order statistics. If $n\alpha$, $n\beta$, and $n\gamma$ are not integers, we use fractional observations in computing these means. Also we let the trimmed mean, $M_{1-2\alpha} = \overline{X}_\alpha$, be the mean of the middle $n(1-2\alpha)$ order statistics, again using fractional observations if necessary. Finally, we call the distribution for which the Huber weight, $H(k)$, gives the maximum likelihood estimate of θ the *Huber distribution* with parameter k. The Huber density is normal in the middle, $(-k \le x \le k)$, and double exponential in the tails, where these curves are joined together continuously at $x = -k$ and $x = k$. Thus the Huber distribution is not peaked but has thicker tails than those of the normal. The two limiting cases, $k = 0$ and $k = \infty$, give the double exponential (DE) and the normal (NOR) distributions, respectively.

The test of the peaked distribution, DE, against the nonpeaked distribution, Huber (k) (each having the same type of tails), led Yuh to a test statistic like

$$H_2 = \frac{E_\beta - B_\beta}{D_\gamma - C_\gamma}.$$

This agrees with our intuition because $D_\gamma - C_\gamma$ tends to become small with peaked distributions and thus we say the distribution is peaked when H_2 is large.

In addition, to test the heavy tailed distribution, Huber (k), against the light tailed NOR (each with the same type of middles), Yuh used

$$H_3 = \frac{U_\alpha - L_\alpha}{D_{\beta+\gamma} - C_{\beta+\gamma}} \quad \text{or} \quad H_3 = \frac{U_\alpha - L_\alpha}{D_\delta - C_\delta}, \quad \text{where} \quad \delta = \beta+\gamma.$$

Again this makes sense since $U_\alpha - L_\alpha$ tends to be large with heavier tails and thus large values of H_3 lead to the rejection of lighter tails.

Finally, Hogg, Fisher, and Randles (1975) had used

$$H_1 = \frac{U_\eta - M_{1-2\tau}}{M_{1-2\tau} - L_\eta}$$

to test for symmetry of a distribution. Again this agrees with our intuition as large values of H_1 indicate that the distribution is skewed to the right while very small values suggest left skewed distributions. Values of H_1 around one suggest symmetry.

The use of H_1, H_2, and H_3 is better than the usual standardized third and fourth moments because we obtained faster convergence properties. In particular, we need only the existence of the first two moments of the underlying distribution to obtain the asymptotic distributions of H_1, H_2, and H_3. It is these distributions that help us determine the best values of α, β, γ, $\delta = \beta+\gamma$, η, and τ.

Under the assumption that the underlying distribution is symmetric and has two moments, H_2 and H_3 have an asymptotic bivariate normal distribution. However, this bivariate distribution is different for different underlying distributions. For illustrations, Yuh found three bivariate normal distributions using the uniform, the NOR, and the DE as representative underlying distributions. We denote these as B_U, B_N, and B_D, respectively. Moreover, as we move to the more peaked and thicker tailed underlying distributions, the means and variances of H_2 and H_3 increase. If we wish to test the uniform distribution against the NOR using (H_2,H_3), the asymptotic theory leads to a quadratic discriminant function. We obtain another quadratic function in testing the NOR against the DE.

To help us discriminate better between the uniform, NOR, and DE, Yuh selected α, β, and γ to maximize the Mahalanobis distances between B_U and B_N and between B_N and B_D. Here the Mahalanobis distance is defined by averaging the two covariance matrices under consideration. Of course, those two distances, from B_U to B_N and from B_N to B_D, cannot be maximized simultaneously; but $\alpha = 0.02$, $\beta = 0.10$, $\gamma = 0.38$, and $\delta = \beta+\gamma = 0.48$ provided an extremely good compromise solution.

In addition to these asymptotic results, Yuh actually found the empirical distributions of (H_2,H_3) with these three underlying distributions for sample sizes of $n = 20, 40,$ and 80. It was interesting to observe that the quadratic discriminant functions could be approximated, in both the asymptotic and empiric cases, by a simple linear function of the form H_2+H_3. Thus we use the test statistic $H = H_2+H_3$ to discriminate between light tailed and nonpeaked distributions and heavy tailed and peaked distributions.

Using H and H_1, we find rules classifying skewed distributions as well as those with light or heavy tails. By considering certain Mahalanobis distances, Yuh determined that good values for η and τ are $\eta = 0.05$ and $\tau = 0.3$. Thus the following selector statistics are recommended:

$$H_1 = \frac{U_{0.05}-M_{0.4}}{M_{0.4}-L_{0.05}},$$

$$H_2 = \frac{E_{0.10}-B_{0.10}}{D_{0.38}-C_{0.38}},$$

$$H_3 = \frac{U_{0.02}-L_{0.02}}{D_{0.48}-C_{0.48}},$$

and $H = H_2+H_3$.

3. An adaptive L-estimator

We want to compare some nonadaptive estimators, like the trimmed means, and regular M-estimators to some adaptive M-estimators. But before doing this, we introduce one other kind of adaptive estimator, namely an adaptive L-estimator (here L stands for a linear function of order statistics). It has much the flavor of an estimator of Johns (1974).

By considering rough approximations to f'/f to help find a score function, Han (1985) developed an asymptotic estimator of the following form. Let p_0 be a small positive fraction. Divide the interval $(p_0, 1-p_0)$ into $2k$ parts. For convenience, make these parts of equal length, namely of length $q = (1-2p_0)/2k$. Consider the order statistics $X_{(1)} < X_{(2)} < \cdots < X_{(n)}$ and define the pth sample quantile, $\hat{\xi}_p$, as the $(n+1)p^{th}$ order statistic, using weighted averages if $(n+1)p$ is not an integer. In particular, we need $\hat{\xi}_{p_0+iq}$, $i = 0,1,2, \ldots, 2k$. Define

$$d_i = (0.5)[\hat{\xi}_{p_0+iq} - \hat{\xi}_{p_0+(i-1)q} + \hat{\xi}_{p_0+(2k+1-i)q} - \hat{\xi}_{p_0+(2k-i)q}],$$

$i = 1,2,\ldots,k$. In addition, let $\overline{S}_i$ be the average of the order statistics in one of the two intervals, $(\hat{\xi}_{p_0+(i-1)q}, \hat{\xi}_{p_0+iq}]$ and $[\hat{\xi}_{p_0+(2k-i)q}, \hat{\xi}_{p_0+(2k+1-i)})$, $i = 1,2,\ldots,k$. Han's estimator is of the form

$$HH = \sum_{i=1}^{k} w_i \overline{S}_i \Big/ \sum_{i=1}^{k} w_i,$$

where $w_1, w_2, \ldots, w_k$ are functions of $d_1, d_2, \ldots, d_k$. These weight functions are selected so the HH has the optimal asymptotic properties as $n \to \infty$ and $k \to \infty$ for every fixed p_0.

It was discovered, by trying different values of k, for sample sizes roughly between 20 and 100, that $k = 2$ was the best choice. By using $k > 2$ with sample sizes less than 100, there is a tendency to overadapt. That is, those variable weights $w_1, w_2, \ldots, w_k$ try to respond to the data too much, creating more variability in the estimator than desired.

So that the reader gets a good idea about the construction of the HH estimator, we spell it out in some detail in the case $k = 2$ and $n = 40$, with $p_0 = 0.05$, as follows.

$$\hat{\xi}_{0.05} = (0.95)X_{(2)} + (0.05)X_{(3)},$$

$$\hat{\xi}_{0.275} = (0.725)X_{(11)} + (0.275)X_{(12)},$$

$$\hat{\xi}_{0.5} = (0.5)X_{(20)} + (0.5)X_{(21)},$$

$$\hat{\xi}_{0.725} = (0.275)X_{(29)} + (0.725)X_{(30)},$$

$$\hat{\xi}_{0.95} = (0.05)X_{(38)} + (0.95)X_{(39)},$$

$$d_1 = (0.5)(\hat{\xi}_{0.275} - \hat{\xi}_{0.05} + \hat{\xi}_{0.95} - \hat{\xi}_{0.725}),$$

$$d_2 = (0.05)(\hat{\xi}_{0.725} - \hat{\xi}_{0.275}),$$

$$\bar{S}_1 = \frac{[X_{(3)}+\cdots+X_{(11)}]+[X_{(30)}+\cdots+X_{(38)}]}{18},$$

$$\bar{S}_2 = \frac{X_{(12)}+\cdots+X_{(29)}}{18}.$$

To obtain w_1 and w_2 as functions d_1 and d_2, Han found that

$$w_1 = \frac{0.225}{(0.05+0.1125m)d_1^2} - \frac{2(d_1-d_2)}{(d_1^2+d_2^2)d_1}$$

and

$$w_2 = \frac{d_1-d_2}{(d_1^2+d_2^2)d_2}.$$

There is the factor m in w_1, which is selected adaptively, depending upon peakedness and tail thickness, as follows:

$$m = \begin{cases} 0.25, & H \le 7.4-28/n, \\ 0.5, & 7.4-28/n < H \le 10-40/n, \\ 1.0, & 10-40/n < H \le 12.5-40/n, \\ 1.5, & 12.5-40/n \le H, \end{cases}$$

where, in this case, $n = 40$. This adaptation, which was determined empirically, seems to adjust fairly well for different values of tail thickness. The Han statistics is then

$$HH = (w_1\bar{S}_1+w_2\bar{S}_2)/(w_1+w_2).$$

In the asymptotic situation ($n \to \infty$ and $k \to \infty$) the finite m value is unimportant to proving certain properties of HH, but it does make a difference with these smaller samples.

The three adaptive esimtators, besides HH, that we consider are:
a) the adaptive t-weight

$$AT = \begin{cases} T(11), & H \le 7.4-28/n, \\ T(3), & H > 7.4-28/n; \end{cases}$$

b) the adaptive biweight

$$AB = \begin{cases} B(5.5), & H \le 7.4-28/n, \\ B(4.1), & H > 7.4-28/n; \end{cases}$$

c) the adaptive HH

$$AHH = \begin{cases} \bar{X}_{0.05}, & H \le 7.4-28/n, \\ HH, & H > 7.4-28/n. \end{cases}$$

Here $\bar{X}_{0.05}$ represents Han's statistic when $k = 1$.

Of course, when $H \le 7.4-28/n$ (incidentally, this cutoff point was determined empirically), the underlying distribution seems to have fairly light tails and we use estimators, T(11), B(4.1), and $\bar{X}_{0.05}$, which are more favorable to distributions close to the normal. On the other hand, heavy tailed, peaked distributions are suggested by $H > 7.4-28/n$, and thus estimators more suitable to these distributions are used.

4. Comparison of location estimators

Our Monte Carlo studies have been very exhaustive, considering a large number of statistics. Also we have used many underlying distributions, primarily those with which we can use the swindle technique; see Gross (1973). In addition we have used many different sample sizes, most with n under 100 but we have actually taken some larger than that. So as not to complicate the exposition here, we restrict our report to one table which gives the efficiencies of 13 estimators for 6 underlying distributions when $n = 40$. We selected these six distributions because there are two light tailed ones (NOR and TE) for which $\bar{X}$ is an excellent estimator, two with little heavier tails (CU and CON) for which $\bar{X}_{0.15}$ is a very good estimator, and two that are peaked with fairly heavy tails (DE and CA) for which $\bar{X}_{0.35}$ is quite good as an estimator. All these distributions can be generated through $X = Y/Z$, where Y and Z are independent and Y is the standardized normal random variable, $N(0,1)$.

Normal (NOR): $Z \equiv 1$.

Slate (TE): $Z = U^{1/10}$, where U is uniform $(0,1)$.

Slacu (CU): $Z = U^{1/3}$.

Contaminated (CON): $Z = \begin{cases} 1 & \text{with probability } 0.9 \\ 1/3 & \text{with probability } 0.1. \end{cases}$

Double Exponential (DE): $Z = 1/\sqrt{W}$, where W is $\chi^2(2)$.

Cauchy (CA): $Z = |V|$, where V is $N(0,1)$.

All of the efficiencies reported in Table 1 are with respect to the best estimator in the study (sometimes not inlcuded in this table and thus a few columns have all efficiencies under 1.000). Moreover the Monte Carlo efficiencies are usually within 0.01 or 0.02 of the correct value, except for $\overline{X}$ in the Cauchy case where we report the exact theoretical value.

TABLE 1
Efficiencies of Estimators (n = 40)

Estimators	Distributions					
	NOR	TE	CU	CON	DE	CA
$\overline{x}$	1.000	1.000	0.734	0.721	0.582	0.000
$\overline{x}_{.05}$	0.997	0.990	0.962	0.950	0.692	0.157
$\overline{x}_{.15}$	0.916	0.937	0.994*	0.991	0.827	0.565
$\overline{x}_{.25}$	0.851	0.874	0.960	0.946	0.931	0.813
$\overline{x}_{.35}$	0.781	0.804	0.901	0.875	1.000	0.920
M	0.686	0.705	0.796	0.763	0.971	0.869
H(1.25)	0.919	0.938	0.993	0.985	0.885	0.683
B(4.82)	0.924	0.941	0.991	0.995*	0.883	0.811
T(6)	0.919	0.938	0.991	0.980	0.904	0.776
HH	0.943	0.954	0.965	0.959	0.891	0.970*
AB	0.932	0.946	0.980	0.984	0.901	0.876
AT	0.940	0.954	0.980	0.967	0.938	0.888
AHH	0.960	0.971	0.972	0.962	0.882	0.970*

*The best estimator is not included among our 13 estimators. The error in each of these efficiencies recorded in this table is usually less than 0.02.

It is interesting to observe that if there is no concern about the peaked, heavy tailed distributions, like DE or CA, the trimmed means, $\overline{X}_{0.05}$ and $\overline{X}_{0.15}$, are very good. Certainly the trimmed mean $\overline{X}_{0.10}$ would be very acceptable to most statisticians. If we also have concern about those heavier tailed distributions, the mid-mean $\overline{X}_{0.25}$ is really quite good.

The so-called "redescending M-estimators," B(4.82) and T(6), are roughly equivalent to Huber's H(1.25) through the first five distributions included in the table, but clearly they are much better than H(1.25) with Cauchy errors. However, Han's HH is better than all of them (and is also better than Johns' estimator, although the latter is extremely good). This suggests that a quasi-linear combination of averages, like $\overline{S}_1$ and $\overline{S}_2$, which are least squares estimators of "blocks" of observations, is extremely promising.

The adaptive estimators, AB, AT, and AHH, are better than their regular counterparts. Surprisingly, this even holds when $n = 20$, and it is extremely clear that we must consider adapting when n is as small as 25. Our suggestion is to use B(4.82) or T(6) for $n \leq 25$. but shift to AT—or AHH—for larger sample sizes. We have analyzed all of our results many different ways and this seems to be the conclusion by all measures. Statisticians, at this moment, might prefer the adaptive M-estimators to AHH because they are easier to compute and have simpler error structures. However, we envision a bright future for adaptive L-estimators, like Han's; but as we see in the next two sections, the regression case clearly indicates that more work is needed with this type of estimator.

5. M-, R-, and L-regression estimators

While the location case can act as a guide to the value of these estimators, the real test is in a more complicated situation. Thus we consider a simple linear regression model with one independent variable x and one slope β. Our objective is to estimate β considering a number of different underlying distributions, some symmetric and some skewed. We recognize that most statisticians would try to transform the variables in the skewed case. We applaud this; but note that a transformation, like the square root or the logarithm, would destroy a linear structure, if one existed. However, in practice, we might get a linear structure after the transformation and this would be fine; but, under the assumptions we use here, such a transformation would not keep the linear relationship.

In particular, we assume that $Y_1, Y_2, \ldots, Y_n$ are independent with respective densities

$$\frac{1}{\rho} f\left[\frac{y_i - \alpha - \beta x_i}{\rho}\right], \quad i = 1,2,\ldots,n,$$

where $f(z)$ is a density function and $x_1, x_2, \ldots, x_n$ are values of the explanatory variable. For our exposition, we assume the density $f(z)$ is symmetric or skewed to the right. From our discussion, it will be clear how we handle the left skewed case in practice.

Of course, we know how to find least squares (LS) and least absolute values (L_1) estimates of β. Moreover, M-estimators, as indicated earlier, are easy to extend to the regression case. But we do want to introduce a few more estimators, including the quantile estimators as we need these to extend the Han estimator to the regression case.

If we let, with $0 < p < 1$,

$$\rho(z) = \begin{cases} -(1-p)z, & z < 0, \\ pz, & 0 \leq z, \end{cases}$$

then the minimizing solution to

$$\min_{\alpha,\beta} \sum_{i=1}^{n} \rho(y_i-\alpha-\beta x_i)$$

produces the pth quantile linear estimator. We denote this line by K_p and the corresponding β estimator by $\hat{\beta}_p$. Roughly, K_p is the 100p percentile estimator and about 100p percent of the points are underneath this line. In particular, if $p = 1/2$, then $K_{1/2}$ is the L_1 estimator.

Without explaining all of the details, in the symmmetric case Han obtains his weights, w_1 and w_2, using the order residuals from a preliminary L_1 fit. He applies w_1 to the least squares estimate of β that he finds using all the points either between $K_{0.05}$ and $K_{0.275}$ or between $K_{0.725}$ and $K_{0.95}$. The weight w_2 is assigned to the least squares estimate of β found by using the points between $K_{0.275}$ and $K_{0.725}$. This weighted least squares estimate of β is denoted by HH.

In case the distribution is skewed to the right, Han gets two least squares estimates of β: that with all the points between $K_{0.05}$ and $K_{0.275}$ and that with all the points between $K_{0.275}$ and $K_{0.95}$. Using residuals from the $K_{0.275}$ fit, he determines weights to apply to each of these least square estimates. The resulting weighted estimate is denoted by SHH; for details, see Han (1985).

Finally, an adaptive Han estimator is achieved as follows. Note the three regions in the (H,H_1) plane as depicted in Figure 1 by

Light: symmetric, light tailed distributions

Heavy: somewhat symmetric, heavy tailed and peaked distributions

Skewed: distributions very skewed to the right

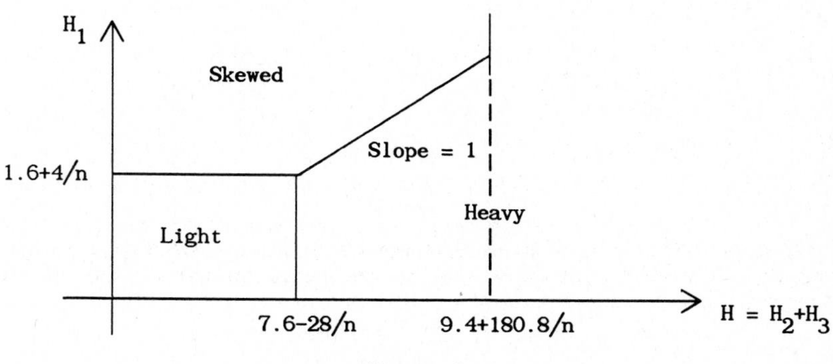

Figure 1

Here H_1, H_2, and H_3 are computed using the ordered residuals from an L_1 fit. The adaptive Han estimator is

$$\text{AHH} = \begin{cases} \text{LS} , & (H,H_1) \in \text{Light}, \\ \text{HH} , & (H,H_1) \in \text{Heavy}, \\ \text{SHH} , & (H,H_1) \in \text{Skewed}. \end{cases}$$

Another estimator is the R-estimator, one determined using the ranks, $R(y_i-\alpha-\beta x_i)$, of $y_i-\alpha-\beta x_i$, $i = 1,2,\ldots,n$. Bril (1984) considered many estimators that were found by minimizing

$$D(\beta) = \sum_{i=1}^{n} (y_i-\alpha-\beta x_i)a[R(y_i-\alpha-\beta x_i)]$$

for different score functions, $a(\cdot)$. Since he chooses each score function so that $\Sigma a(i) = 0$ and since α does not influence the rank of $y_i-\alpha-\beta x_i$, the α can be dropped in this expression; that is,

$$D(\beta) = \sum_{i=1}^{n} (y_i-\beta x_i)a[R(y_i-\beta x_i)]$$

is truly a function of β only. Bril has an algorithm for minimizing $D(\beta)$ for different score functions, $a(\cdot)$.

Three of the score functions he uses are the Wilcoxon scores

$$a_W(i) = i-(n+1)/2, \quad i = 1,2,\ldots,n,$$

for moderate tailed distributions;

$$a_H(i) = \begin{cases} -(n+1)/4 , & i < (n+1)/4, \\ i-(n+1)/2 , & (n+1)/4 \leq i \leq 3(n+1)/4, \\ 3(n+1)/4 , & 3(n+1)/4 < i, \end{cases}$$

with heavy tailed distributions; and

$$a_S(i) = \begin{cases} i-(n+1)/2 + c , & i \leq (n+1)/2, \\ c , & (n+1)/2 < i, \end{cases}$$

with right skewed distributions, where the constant c is selected so that $\Sigma a_S(i) = 0$.

If we denote the three corresponding rank estiamtes of β by β_W, β_H, and β_S, Bril's adaptive rank estimator is

$$\text{AR} = \begin{cases} \beta_W , & (H,H_1) \in \text{Light}, \\ \beta_H , & (H,H_1) \in \text{Heavy}, \\ \beta_S , & (H,H_1) \in \text{Skewed}. \end{cases}$$

He uses L_1 starts in the first two situations and a $K_{0.25}$ start in the skewed case.

Finally, we create the skewed branch of the adaptive t-weight in the skewed case. It is ST given by

$$w_i = \frac{1}{1+5(z_i+0.5)^2},$$

where $z_i = (y_i-\hat{\alpha}_0-\hat{\beta}_0 x_i)/\hat{\rho}_0$, $i = 1,2,\ldots,n$. Here, in the skewed case, we usually find $\hat{\beta}_0$ from $K_{0.25}$ or $K_{0.275}$. Note that w_i has been translated to center around -0.5 and the weights fall of *much faster* than with T(3), T(6), or T(11). In effect, this gives essentially no weight to points above $K_{0.5}$, with points around $K_{0.2}$ and $K_{0.3}$ getting the most weight. The adaptive t-weight estimator becomes

$$AT = \begin{cases} T(11), & (H,H_1) \in \text{Light}, \\ T(3), & (H,H_1) \in \text{Heavy}, \\ ST, & (H,H_1) \in \text{Skewed}. \end{cases}$$

6. <u>Comparison of regression estimators</u>

Again we have studied a number of different situations in all of our Monte Carlos. To keep it simple for this exposition, we give only one table, when n = 40. Note that we now have four skewed distributions, namely chi-square distributions with one, two, four, and eight degrees of freedom, respectively. The efficiencies of the estimators of β are reported in Table 2.

From our Monte Carlos, it is extremely clear that least squares should not be used if the tails are very heavy or if the distribution is skewed. While robust, the L_1 estimator is not good in extremely skewed cases. Of course, this was the reason we used $K_{0.25}$ or $K_{0.275}$ as a preliminary estimator in the right skewed cases. But, it seems, without question, that statisticians must use

TABLE 2
Efficiencies of β Estimators (n = 40)

Estimators	NOR	TE	CU	CON	DE	CA	$\chi^2(1)$	$\chi^2(2)$	$\chi^2(4)$	$\chi^2(8)$
	Distributions									
LS	1.000	1.000	0.736	0.748	0.725	0.000	0.048	0.225	0.538	0.859
L_1	0.646	0.632	0.727	0.708	0.931	0.882	0.088	0.231	0.430	0.603
B(4.82)	0.907	0.921	1.000	1.000	0.999	0.956	0.139	0.360	0.660	0.900
AR	0.906	0.916	0.952	0.933	1.000	0.900	0.500	0.655	0.863	0.951
AT	0.939	0.946	0.966	0.952	0.982	1.000	1.000	1.000	1.000	0.976*
AHH	0.946	0.942	0.930	0.900	0.957	0.899	0.784	0.858	0.955	0.952

*The best estimator is not included among the six estimators in this study. The error in each of these efficiencies recorded in this table is usually less than 0.04.

adaptive estimators (or transform the data, which is also adaptive) when the sample size is greater than 25. The adaptive L-type estimators do show some promise and have great appeal as all of us like the idea of using least squares on blocks of points. However, at this moment we must recommend something like the adaptive t-weights. And, while we make this recommendation based on regression with only one independent variable, a small study has been made with two independent vari-ables and the results tended to be similar. These t-weights (or Biweights)are easy to use. Accordingly, we urge that, along with least squares, adaptive t-weights estimators be found, at least a one-step version. If L_1 or $K_{0.25}$ estimates are too difficult to determine, there are nonparametric estimates that could be substituted for them; see Hogg (1975). We simply must not use least squares in some of these heavy tailed and skewed cases.

In summary, our theoretical and Monte Carlo results clearly indicate that we must adapt: (a) either the data should be transformed to make the underlying distribution more symmetric with lighter tails; or (b) adaptive estimators must be used on the raw data. In those cases in which transformations do not make sense, we should explore adaptive estimation, at least as a partner to the usual least squares technique.

References

Bickel, P.J. (1982). On adaptive estimation. *Ann. Statist.* **10**, 647-671.

Bril, G.K. (1984). *Adaptive Rank Estimates for Linear Regression*. Unpublished thesis, University of Iowa, Iowa City, Iowa.

Gross, A.M. (1973). A Monte Carlo swindle for estimates of location. *Appl. Statist.* **22**, 347-353.

Han, S.M. (1985). *Adaptive L-Estimation*. Unpublished thesis, University of Iowa, Iowa City, Iowa.

Hogg, R.V. (1975). Estimates of percentile regression lines using salary data. *J. Amer. Statist. Assoc.* **70**, 56-59.

Hogg, R.V., D.M. Fisher, and R.H. Randles (1975). A two-sample adaptive distribution-free test. *J. Amer. Statist. Assoc.* **70**, 656-661.

Hogg, R.V. (1979). Statistical robustness: one view on its use in applications today. *Amer. Statistician* **33**, 108-115.

Huber, P.J. (1981). *Robust Statistics*. NewYork: Wiley.

Johns, M.V. (1974). Nonparametric estimation of location. *J. Amer. Statist. Assoc.* **69**, 453-467.

Yuh, L. (1984). *Selection of Models and Adaptive Robust Estimation*. Unpublished thesis, University of Iowa, Iowa City, Iowa.

Original paper received: 08.04.85
Final paper received: 21.10.87

Paper recommended by M.M. Siddiqui

PROBABILITY AND STATISTICS
Essays in Honor of Franklin A. Graybill
J.N. Srivastava (Editor)

DISTRIBUTION FREE PROCEDURES FOR SOME RATIO PROBLEMS

Hariharan Iyer
Department of Statistics
Colorado State University
Fort Collins, Colorado.

and

Lee Kaiser
Sterling-Winthrop
Research Institute
Rensselaer, New York.

ABSTRACT

Assume that (a_i, b_i) $i = 1,2,\ldots,m$ are independent bivariate random vectors distributed symmetrically (but not necessarily identically) about (α, β) where $\beta \neq 0$. Let $\rho = -\alpha/\beta$ and suppose that we wish to test H_0: $\rho = \rho_0$ versus H_a: $\rho \neq \rho_0$. We propose two families of distribution free tests. One is a family of weighted sign (WS) tests and the other is a family of modified Wilcoxon signed rank (MWSR) tests. Confidence regions for ρ can be obtained by inverting these tests. Several applications are indicated.

1. INTRODUCTION

Assume that (a_i, b_i) $i = 1,2,\ldots,n$ are independent bivariate random vectors distributed symmetrically (but not necessarily identically) about (α, β) where $\beta \neq 0$. The notion of bivariate symmetry as required in this paper will be precisely defined in section 2. Let $\rho = -\alpha/\beta$ and suppose that we wish to test H_0: $\rho = \rho_0$ versus H_a: $\rho \neq \rho_0$. We propose two families of distribution free tests. One is a family of weighted sign (WS) tests and the other is a family of modified Wilcoxon signed rank (MWSR) tests. The MWSR tests are essentially Wilcoxon signed rank tests applied to weighted observations. Each choice of the weights yields a test and hence it is necessary to consider the problem of optimal or at least reasonable choices of the weights. The optimality criterion used in this paper is based on the local power of the tests. Confidence regions for ρ can be obtained by inverting any of the above tests. However the actual process of inverting the test is not always easy.

Many important ratio problems which arise in the general linear model framework can be reduced to the problem stated above. These are discussed in section 2. The procedure for reducing these general linear model ratio problems to the problem above involves "partitioning" of the set of observations. A special case of this "partitioning", viz. "pairing", of observations has been implicitly considered by Theil (1950) and more explicitly by Maritz (1979). Theil considers "pairing" of the observations in connection with testing the slope parameter while Maritz considers it in connection with testing the intercept parameter in the straight line regression problem. B.M. Brown of the University of Tasmania has considered "pairing" as well as "grouping" of observations to obtain exact distribution free tests for any one of several parameters in the linear model. He has also considered the "optimal grouping" problem and has obtained solutions in special cases. While we discuss the "optimal partitions" or "optimal grouping" problem in this paper, no explicit solutions are offered. Actually it turns out that in general the problem of optimal partitioning" and that of "optimal weights" need to be considered simultaneously. This appears to be a very difficult problem. However, we suggest several criteria which may be used to compare two or more such procedures and choose one which is relatively good.

The "usual" procedure for obtaining a confidence region for the ratio of the marginal means of a bivariate normal distribution is attributed to Fieller (1940, 1944, 1954) while Fieller himself attributes the basic method to Bliss (1935 a,b). Creasy (1954) studied this problem from the point of view of the fiducial probability theory. Williams (1959, chapter 6) has discussed related questions in connection with inverse estimation in regression analysis. The problem has also been treated by James, Wilkinson and Venables (1974). Zerbe (1978) has investigated the problem for the ratio of linear combinations of the parameters in the general linear model using Fieller's approach. Buonaccorsi (1982) has studied the performance of a large sample procedure based on the asymptotic normality of $\hat{\rho}$, the maximum likelihood estimator of ρ under normal errors, under various conditions and has recommended some caution in using this procedure.

A distribution free procedure for testing H_0: $\rho = \rho_0$ based on the Wilcoxon signed rank test was suggested by Bennett (1965) who discussed the possibility of inverting this test to obtain confidence intervals for ρ. Recently Bennett and Mease (1980) have used the Wilcoxon signed rank procedure to obtain a distribution free method for combining estimates of a ratio of means.

It should be pointed out that while we restrict ourselves to the WS tests and MWSR tests, any one sample symmetric location difference test procedure can be adapted to the situation at hand.

2. RATIO PROBLEMS IN THE GENERAL LINEAR MODEL.

2.1 A Definition of Symmetry.

We need a notion of symmetry for the distribution of a random vector in R^n. The appropriate notion of symmetry for our purposes is the following.

2.1.1 Definition: A random vector $\underline{X}$ in R^n is said to be distributed symmetrically about the point $\underline{\mu} = (\mu_1, \mu_2, \ldots, \mu_n)$ in R^n if and only if the distribution of the scalar random variable $\underline{\ell}^t\underline{X}$ is symmetric about $\underline{\ell}^t\underline{\mu}$ for every vector $\underline{\ell}$ in R^n.

Obviously many meaningful notions of higher dimensional symmetry exist but we do not pursue this matter here. The following lemma which is easily proved is of some interest.

2.1.2 Lemma: Let (X,Y) be a random vector with distribution function F, characteristic function ϕ, and probability density function f. Then the following statements are equivalent.

(a) The random vector (X,Y) is symmetric about (μ, υ).
(b) $X + bY$ is distributed symmetrically about $\mu + b\upsilon$ for all b in R.
(c) $aX + Y$ is distributed symmetrically about $a\mu + \upsilon$ for all a in R.
(d) There exists a real characteristic function $\phi_0(t_1, t_2)$ which may or may not involve μ or υ, such that

$$\phi(t_1, t_2) = e^{i(\mu t_1 + \upsilon t_2)} \phi_0(t_1, t_2)$$

for all t_1, t_2 in R.

(e) There exists a probability distribution function $F_0(x,y)$ which may or may not involve μ or υ with the property that
$F_0(x,y)-F_0(x,0)-F_0(0,y)+F_0(0,0)=F_0(-x,-y)-F_0(-x,0)-F_0(0,-y)+F_0(0,0)$
for all x and y, such that the distribution function F of (X,Y) is given by $F(x,y) = F_0(x-\mu, y-\upsilon)$ for all x,y in R.

(f) There exists a probability density function f_0 with the property that

$$f_0(x,y) = f_0(-x,-y)$$

for all x and y, such that the probability density function f of (X,Y) is given by

$$f(x,y) = f_0(x-\mu,\ y-\upsilon)$$

for all x,y in R.

2.2 <u>Ratio Problems in the Linear Model set-up</u>. Consider the usual general linear model $\underline{Y} = \underline{X}\,\underline{\gamma} + \underline{\epsilon}$, where the errors are distributed independently and symmetrically about zero. Let $\underline{\ell}_1$ and $\underline{\ell}_2$ be any two vectors in R_p such that $\underline{\ell}_2^t\,\underline{\gamma} \neq 0$. Let $\rho = -\alpha/\beta$ where $\alpha = \underline{\ell}_1^t\,\underline{\gamma}$ and $\beta = \underline{\ell}_2^t\,\underline{\gamma}$. As subsequently shown, many important problems reduce to one of testing a hypothesis of the form H_0: $\rho = \rho_0$.

Let $\underline{x}_1, \underline{x}_2, \ldots, \underline{x}_m$ be the "design" points not necessarily all distinct. Let y_i be the response variable observed at $\underline{x}_i$. We will call the pair $(\underline{x}_i, y_i)$ an observation. Let $\Delta = \{(\underline{x}_i, y_i) | i=1,2,\ldots,m\}$ be the set of all observations. Let π denote a partition of Δ into $n(\pi)$ disjoint subsets $\Delta_1, \Delta_2, \ldots, \Delta_{n(\pi)}$ where $n = n(\pi)$ may depend on the partition, i.e., we may consider partitions of Δ into different numbers of subsets. In this paper we will only consider partitions for which both α and β are estimable using the observations of Δ_i for each $i = 1,2,\ldots,n(\pi)$.

A difficulty arises if the $\underline{x}_i$'s are not all distinct. For instance, if $\underline{x}_r$ and $\underline{x}_s$ are equal, unless both of these points are assigned to the same subset of Δ, one is confronted with the problem of how to split these up to be assigned to different subsets of Δ. One way of dealing with this situation is to assign all the observations with the same x value to the same subset of Δ.

Let (a_i, b_i) be the usual least squares estimators of (α,β) obtained using the subset of observations in Δ_i. It is easy to see that $(a_i-\alpha, b_i-\beta)$ is distributed symmetrically about (0,0) and that its distribution is free of α and β. Let $X_i = a_i + \rho_0 b_i$, $i = 1,2,\ldots,n$. Then $X_1, X_2, \ldots, X_n$ are independent and distributed symmetrically about $\alpha + \rho_0\beta$. In fact, $X_i = a_i - \alpha + \rho_0(b_i-\beta) + \beta(\rho_0-\rho)$ and if H_0: $\rho = \rho_0$ is true then $X_1, \ldots, X_n$ are distributed independently and symmetrically about zero. If the null hypothesis is not true then the distribution of X_i is just a shifted version of its null distribution. Hence any one sample location test procedure may be used to test H_0: $\rho = \rho_0$ Some examples of commonly occurring ratio problems are presented below.

{A} <u>x-intercept in straight line regression</u>: Consider the straight line regression model $Y = \alpha + \beta x + \epsilon$. The x-intercept is given by $\rho = -\alpha/\beta$. Suppose $x_1 < x_2 < \ldots < x_m$ are the distinct x-values at which Y is observed. The following partitions have been considered by Brown (1981) when m is even, say 2n.

(i) <u>Rainbow</u>: The set $\Lambda = \{(x_1,y_1),\ldots,(x_m,y_m)\}$ is partitioned into $\Lambda_1 = \{(x_1,y_1),\ (x_m,y_m)\}$, $\Lambda_2 = \{(x_2,y_2),\ (x_{m-1},y_{m-1})\}$ ...., $\Lambda_n = \{(x_n,y_n),\ (x_{n+1},Y_{n+1})\}$.

(ii) <u>Tunnel</u>: Λ is partitioned into $\Lambda_1 = \{(x_1,y_1),\ (x_{n+1},y_{n+1})\}$, $\Lambda_2 = \{(x_2,y_2),\ (x_{n+2},y_{n+2})\}$ ...., $\Lambda_n = \{(x_n,y_n),(x_m,y_m)\}$.

These partitions have the following interpretations. Optimal design theory under least squares estimation indicates that for efficient estimation of the parameters of a straight line regression, observations should be taken at two points as far away from each other as is practical. The "tunnel" partition attempts to achieve this arrangement with respect to all the pairs of observations. The "rainbow" partition, on the other hand, attempts to maximize the efficiency of estimation in some of the pairs while suffering a loss of precision in others. It remains to be seen which of these partitions is "better". In section 4 we discuss methods of comparison of partitions and in section 5 we look at the performance of the "rainbow" and "tunnel" partitions with respect to a specific example.

{B} <u>Intersection of two straight lines</u>: Let $Y = \gamma_1+\delta_1 x+\epsilon$ and $Y = \gamma_2+\delta_2 x+\epsilon$ denote two straight line regresions. The abscissa of the point of intersection of the two lines is given by $\rho = -\alpha/\beta$ where $\alpha = \gamma_1 - \gamma_2$ and $\beta = \delta_1 - \delta_2$.

When there are m = 2n observations from each straight line model, one method of generating partitions is to combine the "rainbow" and/or "tunnel" partitions for each straight line. An obvious partition obtained by combining the "tunnel" partition for one straight line regression with the "tunnel" partition for the other is given by

$$\Lambda_1 = \{(x_{1,1},y_{1,1}),\ (x_{1,n+1},y_{1,n+1}),\ (x_{2,1},y_{2,1}),\ (x_{2,n+1},y_{2,n+1})\},$$
$$\Lambda_2 = \{(x_{1,2},y_{1,2}),\ (x_{1,n+2},y_{1,n+2}),\ (x_{2,2},y_{2,2}),\ (x_{2,n+2},y_{2,n+2})\},$$
$$\vdots$$
$$\Lambda_n = \{(x_{1,n},y_{1,n}),\ (x_{1,2n},y_{1,2n}),\ (x_{2,n},y_{2,n}),\ (x_{2,2n},y_{2,2n})\}$$

where $x_{1,1} < x_{1,2} < \ldots < x_{1,2n}$ and $x_{2,1} < x_{2,2} < \ldots < x_{2,2n}$ are the 2n observations from each of the two straight line regressions respectively.

{C} <u>Extremum of a Quadratic</u>: Consider the quadratic regression model $Y = \gamma_0 + \gamma_1 x + \gamma_2 x^2 + \epsilon$. The abscissa at which the extremum of the quadratic occurs is given by $\rho = -\alpha/\beta$ where $\alpha = \gamma_1$ and $\beta = 2\ \gamma_2$.

When the number of observations m is a multiple of 3, say 3n, an obvious partition is obtained by splitting Λ into $\Lambda_1 = \{(x_1,y_1),\ (x_{n+1},y_{n+1}),\ (x_{2n+1},y_{2n+1})\}$, $\Lambda_2 = \{(x_2,y_2),\ (x_{n+2},y_{n+2}),\ (x_{2n+2},y_{2n+2})\}$ ...., $\Lambda_n = \{(x_n,y_n),$

(x_{2n}, y_{2n}), $(x_{3n}, y_{3n})\}$ where $x_1 < x_2 < \ldots < x_m$ are the design points. This is based on the optimal design theory for least squares estimation according to which, for efficient estimation of parameters of a quadratic regression on $[a,b]$, the observations should be divided equally between three points viz., a, $(a+b)/2$ and b.

3. WEIGHTED SIGN TESTS AND MODIFIED WILCOXON SIGNED RANK TESTS.

3.1. Weighted Sign Tests. We shall introduce the WS tests in the single location parameter case. However for applications to the ratio problem, we will have to consider the presence of nuisance parameters as well. It turns out that only minor modifications are necessary in that case.

Let $X_1, X_2, \ldots, X_n$ be distributed independently and symmetrically about the location parameter. Let $f_i(x)$ be the density function of $X_i - \theta$, where f_i is independent of θ. All probability densities considered in this paper are assumed to be positive everywhere and possess continuous derivatives. Let $d_1, d_2, \ldots, d_n$ be any set of real numbers and $\underline{d} = (d_1, d_2, \ldots, d_n)^t$. Let $c(x) = 1$ for $x \geq 0$ and 0 otherwise. Consider testing the hypothesis H_0: $\theta = 0$ versus H_a: $\theta \neq 0$. Define $S = S_n = S_n(\underline{d}) = \sum_{i=1}^{n} d_i c(x_i)$. For each choice of $\underline{d}$ we get a statistic which may be used to test H_0 versus H_a. We call $S(\underline{d})$ the WS test statistic with weights $\underline{d}$. Clearly when the d_i's are all 1 we get the usual sign test statistic.

Under the null hypothesis, S is a linear combination of i.i.d. Bernoulli random variables with success probability 1/2 and therefore the exact null distribution of S could be worked out. For large sample sizes this is a tedious task and a normal approximation may be used instead. The normal approximation is justified by the following result.

3.1.1 Theorem: Let $d_1, d_2, \ldots, d_n, \ldots$ be a sequence of constants such that

$$\lim_{n\to\infty} \max_{\{1,2,\ldots,n\}} d_i^2 / \sum_1^n d_i^2 \to \infty \text{ as } n \to \infty.$$

Let $U_1, U_2, \ldots U_n, \ldots$ be a sequence of independent random variables with U_i distributed as a Bernoulli random variable with success probability p_i. Assume that there exists $\epsilon_0 > 0$ such that $p_i(1-p_i) > \epsilon_0$ for all i. Let $S_n = \sum_1^n d_i U_i$. Then,

$$\frac{S_n - E(S_n)}{\sqrt{\text{var}(S_n)}} \longrightarrow N(0,1)$$

in distribution.

The proof follows directly from the Lindeberg-Feller central limit theorem.

The weights d_i may be chosen so as to optimize various criterion functions. We propose the criterion of maximizing local power. However, since exact expressions for power are very difficult to obtain we in fact maximize only an approximate expression for the local power.

Consider the WS test with weights $\underline{d}$. The following notation will be found useful in our discussion. We define $\mu_\theta = \mu_\theta(\underline{d}) = E(S(\underline{d})|\theta) = \sum_1^n d_i\, p_i(\theta)$ and $v_\theta = \mathrm{var}(S(\underline{d})|\theta) = \sum_1^n d_i p_i(\theta)(1-p_i(\theta))$, where $p_i(\theta) = \mathrm{Prob}(X_i > 0|\theta)$. We also let $\sigma_\theta = \sigma_\theta(\underline{d}) = v_\theta(\underline{d})^{1/2}$. A subscript of zero in place of θ in any of these quantities refers to their values under the null hypothesis. Then $\mu_0(\underline{d}) = \sum_1^n d_i/2$ and $v_0(\underline{d}) = \sum_1^n d_i^2/4$. For an α-level test, the procedure using the normal approximation for the distribution of $S(\underline{d})$ is to reject H_0 whenever $|S(\underline{d})-\mu_0(\underline{d})|/\sigma_o(\underline{d}) > z_{\alpha/2}$ where $z_{\alpha/2}$ is the $1-(\alpha/2)$ quantile of the standard normal distribution. Then, using the symbol $\approx$ to mean "approximately equal to" and Φ and ϕ for the standard normal c.d.f and p.d.f respectively, we have

$$
\begin{aligned}
&P[H_0 \text{ is rejected}|\theta] \\
&= 1 - P[\mu_0(\underline{d}) - z_{\alpha/2}\sigma_0(\underline{d}) < S(\underline{d}) < \mu_0(\underline{d}) + z_{\alpha/2}\sigma_0(\underline{d})] \\
&\approx 1 - \Phi((\mu_0 + z_{\alpha/2}\sigma_0 - \mu_\theta)/\sigma_\theta) + \Phi((\mu_0 - z_{\alpha/2}\sigma_0 - \mu_\theta)/\sigma_\theta)
\end{aligned}
$$

and on expanding this last expression about $\theta = 0$ and neglecting terms involving θ^r for $r \geq 3$ we get,

$$
\begin{aligned}
&P[H_0 \text{ is rejected}|\theta] \\
&\approx \alpha + 4\theta^2 z_{\alpha/2}\, \phi(z_{\alpha/2})\{((\Sigma d_i p_i{}'(0))^2 - \Sigma d_i^2 p_i{}'(0)^2)/(\Sigma d_i^2)\} \\
&= \alpha + 4\theta^2 z_{\alpha/2}\, \phi(z_{\alpha/2})(\underline{d}^t \underline{Q}\, \underline{d}/\underline{d}^t\underline{d}) \text{ say,}
\end{aligned}
$$

where,

$$
\begin{aligned}
&\underline{Q} = \underline{P}\, \underline{P}^t - \mathrm{diag}^2\,(\underline{P}), \text{ with,} \\
&\underline{P} = (p_1'(0),\ldots,p_n'(0))^t.
\end{aligned}
$$

The maximum value of $(\underline{d}^t\underline{Q}\,\underline{d})$ equals the largest characteristic root, λ_{max} say, of $\underline{Q}$ and the maximum is attained when $\underline{d}$ is any characteristic vector of $\underline{Q}$ corresponding to λ_{max} (Rao, 1973).

Since $f_i(x-\theta)$ is the density of X_i, $p_i(\theta) = P(X_i \geq 0|\theta) = \int_0^\infty f_i(x-\theta)\,dx$ and so $p_i{}'(0) = f_i(0)$. Furthermore, the elements of $\underline{Q}$ are all nonnegative and it follows from the theory of nonnegative matrices (see Bellman, 1970, chapter 16) that there exists a maximizing vector $\underline{d}$ whose elements are all positive. In any case, the set of weights $d_1,\ldots,d_n$ which maximizes the (approximate) local power can be calculated if the $f_i(0)$'s are known up to a scalar multiple.

Let $d_i^* = p_i'(0)$. The following observation suggests that $\underline{d}^* = (d_1^*,\ldots,d_n^*)^t$ is a reasonable vector of weights to use. Now,

$$(\underline{d}^{*t} \underline{Q}\, \underline{d}^*)/(\underline{d}^{*t} \underline{d}^*) = \underline{P}^t\,(\underline{PP}^t - \text{diag}^2\,(\underline{P}))\underline{P}/\underline{P}^t\underline{P}$$

$$= [(\Sigma p_i'(0)^2)^2 - \Sigma p_i'(0)^4]/\Sigma p_i'(0)^2.$$

Furthermore, $\lambda_{max} \leq \underline{P}^t\ \underline{P} = \Sigma p_i'(0)^2$. Suppose that $0 < A \leq p_i'(0) \leq B$ for all i. Then,

$$1 \geq (d^{*t}\ \underline{Q}\ d^*)/(d^{*t}\ d^*\ \lambda_{max}) \geq 1 - [(\Sigma p_i'(0)^4)/(\Sigma p_i'(0)^2)^2]$$

$$\geq 1 - (nB^4/(nA^2)^2) \sim 1 \text{ when n is large.}$$

This leads to the conclusion that for densities bounded at 0, for large n, the difference between the local power of the test using the optimum $\underline{d}$ and that using $\underline{d}^*$ is small. In the special case when the X_i are normally distributed with mean θ and variance σ_i^2 it follows that d_i^* is proportional to σ_i^{-1}.

3.2 Modified Wilcoxon Signed Rank Test. Again let X_i, $i = 1,\ldots,n$, be distributed independently and symmetrically about the location parameter θ in R. We wish to test the hypothesis H_0: $\theta = 0$ versus H_a: $\theta \neq 0$. Let $Z_i = d_iX_i$, where $d_1,\ldots,d_n$ are real constants. If $\underline{d}^t = (d_1,\ldots,d_n)$ the MWSR statistic is given by

$$W = W_n = W_n(\underline{d}) = \Sigma\ iS_i\ ,$$

where S_i is 1 if the ith ranked $|Z_i|$ corresponds to a positive Z_i and is 0 otherwise. As shown in Lehmann (1975),

$$W = \sum_{i \leq j} c(Z_i + Z_j).$$

When $d_1 = \ldots. = d_n = 1$, W is the usual Wilcoxon signed rank statistic. The null distribution of W does not depend on the distributions of the Z_i's. This follows since under H_0 given the ranks of the $|Z_i|$'s, the signs of the Z_i's are independently + or - with probability 1/2. Thus the usual tables of the Wilcoxon signed rank statistic may be used for this case. For large n one may use the following theorem, which also gives the asymptotic non-null distribution of W, in order to find the approximate critical points of the test statistic.

3.2.1 Theorem: Under the above distributional assumptions

$$\frac{W_n - E(W_n)}{\sqrt{\mathrm{Var}\ (W_n)}} \longrightarrow N(0,1)$$

in distribution provided var $(W_n)\ /\ (n(n+1)^2)$ is bounded away from zero.

Proof: This is essentially the theorem of Huskova (1970). See also Sen (1970).

Remark: Let μ_0 and σ_0^2 be respectively the expectation and the variance of W_n under H_0: $\theta = 0$. It can be shown (Lehmann, 1975) that $\mu_0 = n(n+1)/4$ and $\sigma_0^2 = n(n+1)(2n+1)/24$. Thus the asymptotic normality of the null distribution of W_n follows immediately.

Note that the first two moments of W are given by

$$E(W) = \sum_{i \leq j} p_{ij}\ ,$$

where $p_{ij} = P(Z_i + Z_j > 0)$, and

$$\begin{aligned} \mathrm{Var}(W) = & \sum_{i \leq j} p_{ij}(1-p_{ij}) + 2 \sum_{i \neq j} (p^i_{ij} - p_{ii}\, p_{ij}) \\ & + \sum_{i \neq j} \sum_{k \neq i,j} (p^i_{jk} - p_{ij}p_{ik}) \end{aligned}$$

where $p^i_{jk} = P(Z_i + Z_j > 0,\ Z_i + Z_k > 0)$. E(W) and Var(W) and their first two derivatives are needed in the evaluation of the asymptotic local power of the test. Using the critical points for the test given by the large sample distribution we have,

$$\begin{aligned} P(\text{reject } H_0|\theta) &= 1 - P(\mu_0 - z_{\alpha/2}\, \sigma_0 < W_n(\underline{d}) < \mu_0 + z_{\alpha/2}\, \sigma_0) \\ &\approx 1 - \Phi((\mu_0 + z_{\alpha/2}\, \sigma_0 - E(W_n))/(\mathrm{Var}(W_n))^{1/2}) \\ &+ \Phi\, ((\mu_0 - z_{\alpha/2}\, \sigma_0 - E(W_n))/(\mathrm{Var}(W_n))^{1/2}). \end{aligned}$$

Following a detailed but straightforward argument, it can be shown that, ignoring terms of order θ^r for $r \geq 3$,

$$\begin{aligned} P(\text{reject } H_0|\theta) \approx \alpha + & \frac{24\theta^2\, z_{\alpha/2}\, \phi(z_{\alpha/2})}{n(n+1)(2n+1)} [\{\sum_i d_i f_i(0) + \sum_{i<j} \int_{-\infty}^{\infty} (d_i+d_j)f_i f_j dx\}^2 \\ & - \sum_i d_i^2 f_i^2(0) - \sum_{i<j} \{\int_{-\infty}^{\infty}(d_i+d_j)f_i f_j dx\}^2 + \sum_{i \neq j} \{d_i d_j f_i(0) f_j(0) - \\ & 2d_i f_i(0) \int_{-\infty}^{\infty}(d_i+d_j)f_i f_j dx\} + \sum_{i \neq j} \sum_{k \neq i,j} \{(2/3)(d_i d_j + d_i d_k)\int_{-\infty}^{\infty} f_i f_j f_k dx \\ & - (d_i+d_j)\ (d_i+d_k) \int_{-\infty}^{\infty} f_i f_j dx \int_{-\infty}^{\infty} f_i f_j dx\}], \end{aligned}$$

where $f_i = f_i(x)$ is the density of $X_i - \theta$.

There seems to be no closed form solution to the problem of maximizing the above expression by appropriate choice of d_i's. However, if the densities f_i are known, then the optimum d_i's may be approximately obtained by numerical techniques. It might be noted that when $X_i \sim N(\theta, \sigma_i^2)$ the above expression for local power has a stationary point at $d_i = 1/\sigma_i^2$.

3.3 Remark. Recall that in the original problem stated in section 1, if we put $X_i = a_i + \rho_0 b_i$ and $\theta = \alpha + \rho_0 \beta$, then the hypothesis H_0: $\rho = \rho_0$ becomes equivalent to the hypothesis H_0: $\theta = 0$ and thus the statistics $S_n(\underline{d})$ and $W_n(\underline{d})$ can be used for testing H_0: $\rho = \rho_0$. There are two parameters ρ and β (or equivalently, α and β) to deal with here. However, it is easily seen that the approximations for local power derived in the single location parameter case are valid if we replace d_i's by βd_i's.

4. CHOICE OF PARTITIONS IN LINEAR MODEL APPLICATIONS

In the linear model set up $\underline{Y} = \underline{X}\ \underline{\gamma} + \underline{\epsilon}$, suppose one is interested in testing H_0: $(\underline{\ell}_1^t \underline{\gamma})/(\underline{\ell}_2^t \underline{\gamma}) = \rho_0$ where ρ_0 is a given constant. Let $\alpha = -\underline{\ell}_1^t \underline{\gamma}$, $\beta = \underline{\ell}_2^t \underline{\gamma}$, and $\rho = -\alpha/\beta$. It was indicated in section 2 that this problem can be reduced to the problem of testing H_0: $\rho = \rho_0$ based on independent random vectors (a_i, b_i), each of which is distributed symmetrically about (α, β). This process involved partitioning the set Λ of all observations into several subsets such that both α and β are estimable using only the observations in any one of the subsets. The properties of the resulting test depend on the particular partition chosen and hence it is desirable to introduce the notion of an optimal partition. With that in mind we introduce various criteria for comparison of partitions.

In order to use one of the proposed test procedures for testing H_0: $\rho = \rho_0$ one must specify a partition π of Λ, and a test statistic T with associated weights $\underline{d} = (d_1, d_2, \ldots, d_n)^t$. We shall indicate a test procedure by the triple $(T, \pi, \underline{d})$. In a neighborhood of (π_0, β_0) (where β_0 is arbitrary), the power of an α level test has an approximation of the form

$$P(\text{reject } H_0 | \rho) \approx \alpha + \beta_0^4 (\rho_0 - \rho)^2 z_{\alpha/2}\, \phi(z_{\alpha/2})\, \Psi_{T,\pi,\underline{d}}(\rho_0).$$

The explicit form of $\Psi_{T,\pi,\underline{d}}(\rho_0)$ is given in section 3.1 for WS tests and in section 3.2 for MWSR tests. Note also that for a given T, π and ρ_0 one may in principle calculate the best set of weights which will be denoted by $\underline{d}(T, \pi, \rho_0)$.

We now define various criteria for comparison of test procedures.

Let T_i, π_i, $\underline{d}_i$ (i=1,2) specify two test procedures. $(T_1, \pi_1, \underline{d}_1)$ is better that $(T_2, \pi_2, \underline{d}_2)$ (write$(T_1, \pi_1, \underline{d}_1) > (T_2, \pi_2, \underline{d}_2)$) iff

[A] $\Psi_{T_1, \pi_1, \underline{d}_1}(\rho_0) > \Psi_{T_2, \pi_2, \underline{d}_2}(\rho_0)$ for all ρ_0 in R.
(Uniformly better)

[B] $\operatorname{Min}_{\rho_0} \Psi_{T_1, \pi_1, \underline{d}_1}(\rho_0) > \operatorname{Min}_{\rho_0} \Psi_{T_2, \pi_2, \underline{d}_2}(\rho_0)$ (Minimax better)

[C] $\int \Psi_{T_1, \pi_1, \underline{d}_1}(\rho_0)\, d\mu(\rho_0) > \int \Psi_{T_2, \pi_2, \underline{d}_2}(\rho_0)\, d\mu(\rho_0)$ (Better on the average)

[D] $\Psi_{T_1, \pi_1, \underline{d}_1}(\infty) > \Psi_{T_2, \pi_2, \underline{d}_2}(\infty)$

Using any one of the criteria [A], [B], [C] or [D] one may find a "best" inference scheme $(T^*, \pi^*, \underline{d}^*)$. The criterion [D] needs some explanation. The test of $\rho = \infty$ versus $\rho \neq \infty$ actually corresponds to the test of $\beta = 0$ versus $\beta \neq 0$, ρ being equal to $-\alpha/\beta$. It turns out that a finite confidence region for ρ can be obtained by inverting the test if and only if the test for $\beta = 0$ is rejected or equivalently, the test for $\rho = \infty$ is rejected. Thus the use of criterion [D] in essence maximizes the probability of obtaining a finite confidence region for ρ when β is close to zero.

5. AN EXAMPLE

In this section we compare the confidence regions for the x-intercept of a straight line obtained by various methods. The following artificial data set is considered. The values of Y were generated using the equation $Y_i = 7 - X_i + \epsilon_i$ where ϵ_i were generated as pseudo-normal random numbers with mean 0 and standard deviation 4.

TABLE 1. Artificial Data Set

X	1	2	3	4	5	6	7	8	9	10
Y	0.9	3.8	3.7	2.5	-4.1	7.9	5.7	1.4	-2.7	-1.6
X	11	12	13	14	15	16	17	18	19	20
Y	-6.3	-2.6	-2.0	-2.5	-9.2	-10.4	-17.8	-10.1	-10.7	-9.3

The "rainbow" and "tunnel" partitions are considered. In each case the set of observations Λ is split into 10 subsets consisting of pairs of observations. Least squares computations using each subset results in 10 independent estimates (a_i, b_i) of (α, β) where $E(Y|X=x) = \alpha + \beta x$ and $\operatorname{Var}(Y|X=x) = \sigma^2$. The system of weights considered were

(1) $d_i = 1$ for all i

(2) $d_i = 1/v_i$ for MWSR tests

(3) $d_i = 1/(v_i)^{1/2}$ for WS tests

where $\operatorname{var}(a_i + \rho_0 b_i) = \sigma^2 v_i(\rho_0) = \sigma^2 v_i$, σ^2 is unknown and $v_i(\rho_0)$ is a known function of ρ_0. Confidence regions were obtained by inverting the test procedures considered. The results are exhibited in table 2. The type-I error probabilities were kept as close to 0.02 as possible. (Due to the discreteness of the distributions involved it is not always possible to attain exact coverage levels of 0.02).

TABLE 2 Confidence regions for the x-intercept of the straight line using various methods.

Test	Pairing	Weights	$100(1-\alpha)\%$ confidence region	α
Fieller	-	-	(3.108 , 9.472)	0.0214
WS	Rainbow	1	(-2.159 , 10.158)	0.0214
WS	Rainbow	$1/(v_i)^{1/2}$	(-4.894 , 6.456)	0.0214
WS	Tunnel	1	(2.250 , 9.492)	0.0214
WS	Tunnel	$1/(v_i)^{1/2}$	(-4.161 , 7.938)	0.0214
MWSR	Rainbow	1	(2.676, 10.100)	0.0195
MWSR	Rainbow	$1/v_i$	(1.019 , 10.158)	0.0195
MWSR	Tunnel	1	(3.069 , 9.701)	0.0195
MWSR	Tunnel	$1/v_i$	(3.044 , 9.788)	0.0195

We remark that one does not always get a confidence region for ρ which is an interval although in this example they all are. In general the confidence region may be a union of several disjoint intervals including possibly some infinite ones. Such behaviour of confidence regions is well known for Fieller's method. The reason for this is the non-monotonicity of the test statistic as a function of ρ_0. The test statistic proposed by Bennett (1965) is in fact a special case of MWSR statistic proposed here with $d_i = 1$ for all i. However, he erroneously claims that the test statistic is a monotone function of ρ_0 (α in his notation). Consequently his discussion on how to obtain confidence intervals is incorrect.

6. CONCLUSIONS.

Two families of distribution free procedures for testing a ratio of linear combinations of parameters in a linear model with independent and symmetric errors have been proposed. Confidence regions for such ratios can be obtained by inverting the tests. Fieller's procedure has been compared with the proposed procedures by means of an example. Clearly extensive comparisons of the proposed test procedures with Fieller's method have to be made with various underlying error distributions before any recommendations can be made. It is clear from the example of section 5 that a proper choice of the partition and weights can lead to very good confidence regions.

ACKNOWLEDGMENTS

The first author wishes to thank the Division of Mathematics and Statistics of CSIRO/Melbourne, where part of his research was carried out. The second author wishes to thank the Department of Statistics, University of California at Davis, where he did part of his research.

REFERENCES

[1] Bennett, B.M. (1965). Confidence Limits for Ratios Using Wilcoxon's Signed Rank Test. Biometrics 21 231-234.

[2] Bennett, B.M. and R. Mease (1980). On a Distribution Free Method for Combining Estimates of a Ratio of Means. Appl. Statist., 29, 39-42.

[3] Bliss, C.I. (1935a). The Calculation of the Dosage-Mortality Curve. Ann. Appl. Biol., 22, 134-167.

[4] Bliss, C.I. (1935b). The Comparison of Dosage-Mortality Data. Ann. Appl. Biol., 22, 307-333.

[5] Bellman, R. (1970). Introduction to Matrix Analysis., second ed., New York: McGraw-Hill Book Company.

[5] Brown, B.M. (1981). Robustness in Statistics, Colorado State University Tech. Report No. 101.

[7] Buonaccorsi, J.P. (1982). Inference and Design for Ratios of Linear Combinations in the General Linear Model. Ph.D. Dissertation, Colorado State University.

[8] Creasy, M.A. (1954). Limits for the Ratio of Means. J. Roy. Statist. Soc. Ser. B, 16, No. 2, 186-194.

[9] Fieller, E.C. (1940). The Biological Standardization of Insulin. J. Roy. Statist. Soc., Suppl. no. 7, 1-64.

[10] Fieller, E.C. (1944). A Fundamental Formula in the Statistics of Biological Assay, and Some Applications. Quart. J. Pharmacol., 17, 117-123.

[11] Fieller, E.C. (1954). Some Problems in Interval Estimation. J. Roy Statist. Soc. Ser. B, 16, No. 2, 175-185.

[12] Huskova, M. (1970). Asymptotic Distribution of Simple Linear Rank Statistics for Testing Symmetry. Zeit. Wahrscheinlichkeitsth., 14, 308-322.

[13] James, A.T., G.N. Wilkinson and W.N. Venables. (1974). Interval Estimates for a Ratio of Means. Sankhya. Ser. A., 36, Pt. 2, 177-183.

[14] Lehmann, E.L. (1975). Nonparametrics: Statistical Methods Based on Ranks. San Francisco: Holden-Day Inc.

[15] Maritz, J.S. (1979). On Theil's Method in Distribution-Free Regression. Austral. J. Statist., 21, (1), 30-35.

[16] Rao, C.R. (1973). Linear Statistical Inference and Its Applications. New York,: Wiley.

[17] Sen, P.K. (1970). On the Distribution of the One-Sample Rank Order Statistics. Nonparametric Techniques in Statistical Inference (ed. M.L. Puri)., 53-74.

[18] Theil, M. (1950). A Rank Method of Linear and Polynomial Regression Analysis. I. Proc. Kon. Ned. Akad. Wetench. A., 53, 386-392.

[19] Williams, E.J. (1959). Regression Analysis. New York: Wiley.

[20] Zerbe, G.O. (1978). On Fieller's Theorem and the General Linear Model. Amer. Statist., 32, 103-105.

Original paper received: 27.02.85
Final paper received: 04.03.86

Paper recommended by P.W. Mielke

PROBABILITY AND STATISTICS
Essays in Honor of Franklin A. Graybill
J.N. Srivastava (Editor)

PREDICTION INTERVALS IN BALANCED ONE-FACTOR RANDOM MODEL

S. Jeyaratnam and S. Panchapakesan

Department of Mathematics
Southern Illinois University at Carbondale
Carbondale, Illinois
U.S.A.

An exact and three approximate prediction intervals for the overall mean of a future sample are obtained under a one-factor random model. The validity of these approximate intervals are assessed by a simulation study. A sensitivity comparison is also made.

1. INTRODUCTION

Prediction intervals for the overall mean of future observations have been considered in the literature under the usual regression (fixed effects) model; see for example, Graybill [2, pp. 267-270] and Hahn [4]. Such problems for random effects models are also of interest in practice; however, this seems to have not received enough attention. The purpose of the present paper is to consider prediction intervals in a balanced one-factor random model given by

$$Y_{ij} = \mu + A_i + \varepsilon_{ij}, \quad i = 1,\dots,I; \; j = 1,\dots,J, \tag{1.1}$$

involving I levels of a factor and J observations per level. Here the ε_{ij} are i.i.d. $N(0,\sigma_\varepsilon^2)$, and the A_i are independent of the ε_{ij} and are i.i.d. $N(0,\sigma_A^2)$. The parameters μ, σ_A^2, and σ_ε^2 are all unknown. Let Y_{ij}^* denote the observations in a future experiment according to the model (1.1) where the factor is now taken at I^* levels with J^* observations per level. We are interested in obtaining two-sided prediction intervals for $\bar{\bar{Y}}^*$, the overall mean of the future observations, based on the present data. In other words, we are interested in defining an interval $(L(Y),U(Y))$ such that

$$\Pr\{L(Y) \le \bar{\bar{Y}}^* \le U(Y)\} = 1-\alpha \tag{1.2}$$

where Y denotes the set of observations Y_{ij}, and the prediction level $1-\alpha$ is specified in advance. The limits $L(Y)$ and $U(Y)$ will, of course, depend on Y through $\bar{\bar{Y}} = \sum_i\sum_j Y_{ij}/IJ$, $S_1^2 = J\sum_i(\bar{Y}_i - \bar{\bar{Y}})^2/I-1$, and $S_2^2 = \sum_i\sum_j(Y_{ij} - \bar{Y}_i)^2/I(J-1)$, since $(\bar{\bar{Y}},S_1^2,S_2^2)$

AMS 1980 Subject Classifications: Primary 62J05; Secondary 62F99.

Key Words and Phrases: Exact and approximate intervals, validity and sensitivity.

is minimal sufficient for $(\mu,\sigma_A^2,\sigma_\varepsilon^2)$.

In dealing with prediction interval for $\bar{\bar{Y}}^*$, several cases arise. When the ratio $\sigma_A^2/\sigma_\varepsilon^2$, or equivalently, $R = (\sigma_\varepsilon^2 + J\sigma_A^2)/\sigma_\varepsilon^2$, is assumed to be known, an exact prediction interval is obtained (Section (3.1). When R is unknown, an exact interval is obtained only when $J^* \leq J$ (Section 3.2). This interval based on certain linear combinations of the Y_{ij} is not unique; however, in the particular case of $J^* = J$, this method gives a unique interval based on the minimal sufficient statistics. For the case of unknown R, approximate intervals (in the sense that the prediction level is approximately $1-\alpha$) obtained by different methods, namely, (i) Plug-in Method (Section 4.1), (ii) Modified Large Sample Method (Section 4.2), and (iii) Satterthwaite Approximation Method (Section 4.3) are discussed. The relative performances of these different procedures are evaluated by a simulation study (Section 5), which finds the modified large sample method most satisfactory in terms of validity. In view of these, the approximate interval using this method is compared with the exact interval which is available when $J^* \leq J$. It is shown (Section 6) that the approximate method is superior in terms of the expected squared length unless $J^* = J$ in which case both yield the same (exact) interval.

2. NOTATIONS AND PRELIMINARY RESULTS

In this section, we introduce a set of notations that will be often used throughout the paper. We will also state a few well-known results that will be repeatedly used. Let

$$n_1 = I - 1, \qquad n_2 = I(J-1),$$

$$M_1 = \frac{1}{J}\left(\frac{1}{I^*} + \frac{1}{I}\right), \qquad M_2 = \frac{1}{I^*}\left(\frac{1}{J^*} - \frac{1}{J}\right),$$

$$\gamma_1^2 = \sigma_\varepsilon^2 + J\sigma_A^2, \qquad \gamma_2^2 = \sigma_\varepsilon^2,$$

$$\rho = \sigma_A^2/(\sigma_\varepsilon^2 + \sigma_A^2), \qquad R = (\sigma_\varepsilon^2 + J\sigma_A^2)/\sigma_\varepsilon^2 = \gamma_1^2/\gamma_2^2,$$

$$\tau^2 = \mathrm{Var}(\bar{\bar{Y}}^* - \bar{\bar{Y}}) = \gamma_2^2[M_2 + RM_1] = M_1\gamma_1^2 + M_2\gamma_2^2,$$

$$\delta_1^2(\rho) = \gamma_1^2/\tau^2 = \frac{1 + (J-1)\rho}{M_1[1 + (J-1)\rho] + M_2(1-\rho)}, \tag{2.1}$$

$$\delta_2^2(\rho) = \gamma_2^2/\tau^2 = \frac{1 - \rho}{M_1[1 + (J-1)\rho] + M_2(1-\rho)},$$

$$U_1 = \frac{(I-1)S_1^2}{\sigma_\varepsilon^2 + J\sigma_A^2} = \frac{n_1 S_1^2}{\gamma_1^2}, \qquad U_2 = \frac{I(J-1)S_2^2}{\sigma_\varepsilon^2} = \frac{n_2 S_2^2}{\gamma_2^2},$$

$$S_p^2 = \frac{n_2 S_2^2 + (n_1 S_1^2/R)}{n_1 + n_2},$$

$u^+ = u$ if $u > 0$, and $= 0$, otherwise.

Let $N(\mu,\sigma^2)$, χ^2_ν and t_ν denote the normal distribution with mean μ and variance σ^2, chi-square distribution with ν degrees of freedom (d.f.), and Student's t-distribution with ν d.f., respectively. Further, z_γ and $t_{\gamma,\nu}$ denote the upper γ quantiles of $N(0,1)$ and t_ν distributions, respectively. As usual, $E(\cdot)$ denotes the expectation of a random variable.

We conclude this section by stating the following results which are easily verified.

1. $\bar{\bar{Y}} \sim N(\mu, \gamma_1^2/IJ)$.
2. $U_1 \sim \chi^2_{n_1}$ and $U_2 \sim \chi^2_{n_2}$.
3. $\bar{\bar{Y}}$, U_1 and U_2 are independent.
4. $\bar{\bar{Y}}^* - \bar{\bar{Y}} \sim N(0, \tau^2)$.
5. $(n_1 + n_2)S_p^2/\gamma_2^2 \sim \chi^2_{n_1+n_2}$.

3. EXACT PREDICTION INTERVALS

We consider two cases: A. R known and B. R unknown. In Case B, we assume that $J^* \leq J$.

3.1 Case A: R Known. Using the fact that

$$\frac{\bar{\bar{Y}}^* - \bar{\bar{Y}}}{S_p\sqrt{M_2 + RM_1}} \sim t_{n_1+n_2},$$

we get an exact $(1-\alpha)$ level prediction interval for $\bar{\bar{Y}}^*$ given by

$$(3.1) \qquad I_{ER}: \bar{\bar{Y}} \pm t_{\alpha/2,n_1+n_2} S_p\sqrt{M_2 + RM_1}.$$

3.2 Case B: R Unknown and $J^* \leq J$. Let us consider $D_i = \sum_{j=1}^{J} \ell_{ij}Y_{ij}$, $i=1,\ldots,I$, where the coefficients ℓ_{ij} are chosen such that, for all $i=1,\ldots,I$,

$$\sum_{j=1}^{J} \ell_{ij} = 1 \text{ and } \mathrm{Var}\,(D_i) = (\sigma_\varepsilon^2 + J^*\sigma_A^2)/J^* = \sigma_D^2, \text{ say.}$$

Since $\mathrm{Var}\,(D_i) = \sigma_A^2\left(\sum_j \ell_{ij}\right)^2 + \sigma_\varepsilon^2 \sum_j \ell_{ij}^2$, one possible choice is $\ell_{ij} = \frac{1}{J^*}$, $j=1,\ldots,J^*$, and $\ell_{ij} = 0$ otherwise. Now, let $\bar{D} = \sum_{i=1}^{I} D_i/I$. The D_i are i.i.d. $N(\mu, \sigma_D^2)$. Noting that $\sum_{i=1}^{I} (D_i - \bar{D})^2/\sigma_D^2 \sim \chi^2_{n_1}$, it is easy to show that

$$\frac{\bar{\bar{Y}}^* - \bar{D}}{\sqrt{\Sigma(D_i - \bar{D})^2}}\sqrt{\frac{I(I-1)I^*}{(I + I^*)}} \sim t_{n_1}.$$

Thus we get an exact $(1-\alpha)$ level prediction interval for $\bar{\bar{Y}}^*$ given by

(3.2) $$I_{ED}: \bar{D} \pm t_{\alpha/2,n_1}\sqrt{\frac{I + I^*}{II^*}\,\frac{\Sigma(D_i - \bar{D})^2}{I - 1}}\,.$$

The idea of taking suitable linear combination of observations has been used by Burdick and Sielken [1] in the context of confidence intervals in variance components models.

Remark 3.1. It should be noted that the choice of the coefficients ℓ_{ij} in D_i is not unique unless $J^* = J$. When $J^* = J$, a unique interval I_{ED} is obtained based on minimal sufficient statistics. When $J^* < J$, one would expect I_{ED} to be less efficient as it does not depend on minimal sufficient statistics. In other words, we pay a price in obtaining an exact interval. This is brought out in Section 6 where I_{ED} and I_{AM} (to be defined in Section 4.2) are compared in terms of expected squared length.

4. APPROXIMATE PREDICTION INTERVALS

When R is known, we have an exact prediction interval for $\bar{\bar{Y}}^*$ which is easy to calculate. The need for approximate intervals arise when R is unknown. We consider three different methods, namely, (A) Plug-in Method, (B) Modified Large Sample Method, and (C) Satterthwaite Approximation Method.

4.1 Plug-in Method. Since $R = \gamma_1^2/\gamma_2^2 > 1$, we define $\hat{R} = \max(1, S_1^2/S_2^2)$ and obtain an approximately (1-α) level prediction interval for $\bar{\bar{Y}}^*$ given by

(4.1) $$I_{AP}: \bar{\bar{Y}} \pm t_{\alpha/2,n_1+n_2}\,\hat{S}_p\sqrt{M_2 + \hat{R}M_1},$$

where $\hat{S}_p$ is obtained by replacing R with $\hat{R}$ in S_p.

4.2 Modified Large Sample Method. This method is motivated by Graybill and Wang [3] who used such a procedure to obtain good approximate confidence intervals on nonnegative linear combinations of variances. The main idea is to modify a large sample interval so that it might be more exact for small or moderate sample sizes.

In our problem, as $I \to \infty$, we see that

$$\frac{\bar{\bar{Y}}^* - \bar{\bar{Y}}}{\sqrt{(M_1S_1^2 + M_2S_2^2)^+}} \sim N(0,1).$$

So a large sample ($I \to \infty$) prediction interval for $\bar{\bar{Y}}^*$ is given by

(4.2) $$\bar{\bar{Y}} \pm \sqrt{(z_{\alpha/2}^2 M_1S_1^2 + z_{\alpha/2}^2 M_2S_2^2)^+}$$

which has an associated coverage probability 1-α asymptotically. Now, the modified large sample method considers the interval (4.2) with the first $z_{\alpha/2}^2$ under the radical sign replaced by a constant A; this constant A is to be determined such that the associated coverage probability tends to $1-\alpha$ as $\sigma_A^2 \to \infty$. This yields the prediction interval

$$(4.3) \qquad I_{AM}: \bar{\bar{Y}} \pm \sqrt{(t^2_{\alpha/2,n_1} M_1S_1^2 + z^2_{\alpha/2} M_2S_2^2)^+}.$$

It should be noted that the interval I_{AM} is exact when $J^* = J$.

4.3 Satterthwaite Approximation Method. We first note that $Q = M_1S_1^2 + M_2S_2^2$ is an unbiased estimator of $\tau^2 = \text{Var}\,(\bar{\bar{Y}}^* - \bar{\bar{Y}})$. However, Q can be negative when $J^* > J$; in this case, we will use a modified estimator. We discuss the two cases separately.

Case (i): $J^* \leq J$. We want to find a nonnegative function $k(S_1^2, S_2^2)$ such that

$$(4.4) \qquad \Pr[|\bar{\bar{Y}}^* - \bar{\bar{Y}}| \leq k(S_1^2, S_2^2)\sqrt{Q}] = 1-\alpha.$$

Now, Q is a linear combination of independent but not necessarily identical chi-square variables. The Satterthwaite [6] approximation gives

$$(4.5) \qquad NQ/E(Q) \approx \chi^2_N$$

where

$$(4.6) \qquad N = 2[E(Q)]^2/\text{Var}\,(Q).$$

However, we do not know N as it depends on γ_1^2 and γ_2^2, unless $J^* = J$. So we use the estimates S_1^2 and S_2^2, respectively, leading to

$$(4.7) \qquad \hat{N} = [M_1S_1^2 + M_2S_2^2]^2 \Big/ \left[\frac{M_1^2S_1^4}{n_1} + \frac{M_2^2S_2^4}{n_2}\right].$$

Using the usual method of constructing a t-variable from a standard normal variate and a chi-square variate, we get an approximate $(1-\alpha)$ level prediction interval for $\bar{\bar{Y}}^*$ given by

$$(4.8) \qquad I_{AS}: \bar{\bar{Y}} \pm t_{\alpha/2,\hat{N}} \sqrt{Q}.$$

When $J^* = J$, we have $Q = M_1S_1^2$ and $\hat{N} = n_1$. In this case, $\hat{N}Q/E(Q)$ is exactly distributed as $\chi^2_{n_1}$ and consequently, I_{AS} is exact.

Case (ii): $J^* > J$. Note that Q can be written in the form

$$(4.9) \qquad Q = \frac{S_1^2}{IJ} + \frac{S_2^2}{I^*J^*} + \frac{(S_1^2 - S_2^2)}{I^*J}.$$

When $J^* > J$, we use the interval (4.7) with Q_+ in the place of Q, where

$$(4.10) \qquad Q_+ = \frac{S_1^2}{IJ} + \frac{S_2^2}{I^*J^*} + \frac{(S_1^2 - S_2^2)^+}{I^*J}.$$

Obviously, we can combine the two cases and write the prediction interval as

$$\text{(4.11)} \qquad I_{AS}: \bar{\bar{Y}} \pm t_{\alpha/2,\hat{N}}\sqrt{Q_+}$$

5. VALIDITY OF APPROXIMATE INTERVALS: SIMULATION STUDY

The prediction intervals I_{AP}, I_{AM}, and I_{AS} are constructed so that the coverage probability for each is approximately the specified level. The validity of any of these intervals depend on how close the actual coverage probability is to the specified level. This was investigated by a simulation study. In order to facilitate the estimation of the coverage probabilities, we provide below alternate forms for them. These can be easily verified by using the results in Section 2. Let $P(I_{AP})$, $P(I_{AM})$ and $P(I_{AS})$ denote the coverage probabilities of I_{AP}, I_{AM} and I_{AS}, respectively. Let $Z \sim N(0,1)$. Then the coverage probabilities can be written in the form

$$\text{(5.1)} \qquad P(I) = \Pr[|Z| \leq \psi(U_1,U_2;\ M_1M_2,\rho,\alpha)]$$

where the ψ-functions for the intervals (labeled accordingly) are as follows:

$$\psi_{AP} = \begin{cases} t_{\frac{\alpha}{2},n_1+n_2}\sqrt{\dfrac{M_1U_1\delta_1^2(\rho)}{n_1} + \dfrac{M_2U_2\delta_2^2(\rho)}{n_2}} & \text{if } \dfrac{n_2U_1\delta_1^2}{n_1U_2\delta_2^2} \geq 1 \\[2ex] t_{\frac{\alpha}{2},n_1+n_2}\sqrt{\dfrac{(M_1+M_2)[U_1\delta_1^2(\rho)+U_2\delta_2^2(\rho)]}{(n_1+n_2)}} & \text{otherwise;} \end{cases}$$

$$\psi_{AM} = \sqrt{\left(t^2_{\alpha/2,n_1}\frac{M_1U_1\delta_1^2(\rho)}{n_1} + z^2_{\alpha/2}\frac{M_2U_2\delta_2^2(\rho)}{n_2}\right)^+}\ ;$$

and

$$\psi_{AS} = \begin{cases} t_{\frac{\alpha}{2},\hat{N}}\sqrt{\dfrac{M_1U_1\delta_1^2(\rho)}{n_1} + \dfrac{M_2U_2\delta_2^2(\rho)}{n_2}} & \text{if } J^* \leq J, \\[2ex] t_{\frac{\alpha}{2},\hat{N}}\sqrt{\dfrac{U_1\delta_1^2(\rho)}{n_1IJ} + \dfrac{U_2\delta_2^2(\rho)}{n_2I^*J^*} + \dfrac{1}{I^*J}\left(\dfrac{U_1\delta_1^2(\rho)}{n_1} - \dfrac{U_2\delta_2^2(\rho)}{n_2}\right)^+} & \text{otherwise} \end{cases}$$

where

$$\hat{N} = \left[\frac{M_1U_1\delta_1^2(\rho)}{n_1} + \frac{M_2U_2\delta_2^2(\rho)}{n_2}\right]^2 \Big/ \left[\frac{M_1^2U_1^2\delta_1^4(\rho)}{n_1^3} + \frac{M_2^2U_2^2\delta_2^4(\rho)}{n_2^3}\right].$$

In order to estimate the coverage probabilities by simulation, 72 different sets of values of (I,J,I^*,J^*) were considered. These are the sets obtained when $I,J = 3,5,7$; and $I^*,J^* = 5,10,15$, omitting sets with $J = J^*$ (in which case, the intervals are exact). For each of these 72 sets of values, 1000 random sets of observations (Z,U_1,U_2) were generated by using IMSL subroutines. Based on these, the coverage probabilities of the three intervals were estimated for $\rho = 0.1\ (0.1)\ 0.9$; and $\alpha = .05, .10$. In the case of I_{AS}, $\hat{N}$ value was rounded down to an integer value. The range of the estimated probabilities over the chosen values of ρ are given in Table 1 only for $\alpha = 0.05$, since the pattern of the intervals is similar when $\alpha = .10$.

TABLE 1. Range of Estimated Coverage Probability
Specified Level = $1 - \alpha = .95$

I	J	I^*	J^*	I_{AP}	I_{AS}	I_{AM}
3	3	5	5	(85.2, 95.5)	(76.2, 97.9)	(92.9, 95.0)
		5	10	(85.4, 94.4)	(98.7, 99.5)	(92.1, 95.2)
		5	15	(85.3, 94.6)	(96.7, 99.5)	(91.0, 94.2)
		10	5	(83.4, 93.6)	(99.3, 99.7)	(92.8, 93.8)
		10	10	(85.3, 94.1)	(99.2, 99.7)	(93.8, 95.5)
		10	15	(82.9, 92.7)	(98.1, 99.8)	(92.4, 93.8)
		15	5	(84.7, 95.1)	(99.1, 99.9)	(93.6, 94.7)
		15	10	(85.5, 96.0)	(98.8, 99.7)	(94.4, 95.5)
		15	15	(83.9, 94.4)	(98.5, 99.7)	(94.0, 94.6)
3	5	5	10	(82.0, 91.3)	(98.2, 99.5)	(91.8, 93.0)
		5	15	(85.0, 93.7)	(97.7, 99.8)	(94.2, 95.6)
		10	10	(81.6, 91.7)	(99.0, 99.9)	(93.0, 93.9)
		10	15	(81.7, 92.3)	(97.9, 99.9)	(92.7, 94.3)
		15	10	(83.2, 92.9)	(98.3, 99.5)	(93.6, 94.0)
		15	15	(81.7, 92.7)	(98.6, 99.8)	(94.6, 95.7)
3	7	5	5	(81.4, 91.5)	(92.5, 94.1)	(94.2, 95.5)
		5	10	(82.6, 91.8)	(99.3, 99.9)	(94.0, 94.8)
		5	15	(84.2, 92.2)	(98.9, 99.7)	(93.9, 95.7)
		10	5	(82.5, 94.0)	(92.3, 94.7)	(94.3, 94.9)
		10	10	(81.9, 92.4)	(98.8, 99.8)	(93.9, 94.6)
		10	15	(83.3, 92.0)	(98.5, 99.9)	(94.4, 95.4)
		15	5	(82.5, 92.0)	(93.4, 94.5)	(94.4, 94.8)
		15	10	(82.9, 92.0)	(99.4,100.0)	(95.2, 95.7)
		15	15	(82.2, 90.9)	(98.9, 99.7)	(94.7, 95.3)
5	3	5	5	(90.2, 94.3)	(97.7,100.0)	(93.6, 95.5)
		5	10	(89.1, 94.1)	(97.4, 99.7)	(91.7, 95.5)
		5	15	(89.9, 94.5)	(97.3, 99.5)	(91.1, 95.3)
		10	5	(89.7, 94.5)	(97.4,100.0)	(94.9, 95.9)
		10	10	(89.4, 94.4)	(96.4, 99.7)	(93.2, 94.6)
		10	15	(89.4, 94.0)	(96.6, 99.7)	(93.3, 95.1)
		15	5	(90.9, 94.6)	(97.2, 99.3)	(95.3, 95.9)
		15	10	(90.4, 94.4)	(96.9, 99.8)	(93.9, 95.7)
		15	15	(90.1, 95.5)	(97.2, 99.7)	(94.8, 96.2)
5	5	5	10	(88.1, 91.4)	(95.6, 99.9)	(92.3, 93.7)
		5	15	(87.8, 92.5)	(97.1, 99.6)	(93.0, 94.8)
		10	10	(90.2, 94.5)	(96.7, 99.5)	(94.2, 95.6)
		10	15	(90.5, 94.1)	(97.0, 99.4)	(94.7, 95.5)
		15	10	(88.3, 92.9)	(95.9, 98.9)	(93.6, 94.2)
		15	15	(90.8, 94.7)	(96.8, 99.8)	(95.1, 95.7)

I	J	I^*	J^*	I_{AP}	I_{AS}	I_{AM}
5	7	5	5	(89.0, 93.2)	(95.0, 95.8)	(95.7, 96.2)
		5	10	(89.0, 92.5)	(97.7, 99.8)	(95.5, 96.3)
		5	15	(88.1, 90.7)	(95.7, 99.3)	(92.2, 93.6)
		10	5	(87.5, 92.0)	(94.0, 94.6)	(94.3, 95.0)
		10	10	(88.4, 92.2)	(96.3, 98.8)	(94.4, 94.8)
		10	15	(86.6, 91.5)	(96.7, 99.6)	(93.0, 94.9)
		15	5	(87.7, 92.4)	(94.4, 95.0)	(94.6, 95.3)
		15	10	(88.6, 92.7)	(96.5, 98.7)	(95.1, 95.4)
		15	15	(90.4, 93.6)	(97.7, 99.8)	(95.4, 96.2)
7	3	5	5	(91.7, 94.6)	(95.2, 99.6)	(93.0, 94.4)
		5	10	(89.7, 93.2)	(95.3, 99.9)	(91.9, 95.1)
		5	15	(89.2, 92.9)	(95.2, 99.7)	(89.7, 94.2)
		10	5	(92.0, 95.5)	(96.1, 99.4)	(94.2, 95.5)
		10	10	(90.2, 93.7)	(95.4, 99.4)	(93.0, 94.5)
		10	15	(91.3, 94.6)	(95.7, 99.7)	(92.5, 94.7)
		15	10	(90.4, 94.0)	(94.6, 99.4)	(92.8, 94.8)
		15	15	(90.6, 93.9)	(95.0, 99.7)	(92.1, 94.6)
7	5	5	10	(90.4, 93.2)	(95.5, 99.8)	(93.2, 94.6)
		5	15	(88.6, 92.1)	(95.6,100.0)	(91.4, 94.9)
		10	10	(93.1, 95.3)	(96.3, 99.4)	(95.1, 95.7)
		10	15	(90.8, 93.7)	(96.8, 99.7)	(94.3, 96.1)
		15	10	(90.4, 94.2)	(95.1, 98.9)	(93.7, 94.3)
		15	15	(90.3, 94.3)	(95.5, 99.0)	(93.6, 94.6)
7	7	5	5	(88.8, 92.1)	(93.7, 94.2)	(94.0, 94.3)
		5	10	(91.1, 94.0)	(96.0, 98.8)	(95.0, 95.7)
		5	15	(92.5, 93.8)	(96.4, 99.4)	(94.4, 96.0)
		10	5	(90.7, 93.2)	(94.2, 94.5)	(94.4, 94.8)
		10	10	(91.4, 92.9)	(96.2, 97.6)	(95.5, 95.7)
		10	15	(88.8, 91.3)	(96.2, 98.9)	(94.9, 95.3)
		15	5	(90.7, 94.0)	(95.4, 95.6)	(95.4, 95.6)
		15	10	(90.3, 93.4)	(96.0, 98.2)	(95.4, 95.6)
		15	15	(90.3, 93.3)	(96.2, 98.7)	(94.5, 95.5)

The results of the above study indicate beyond doubt that the modified large sample method is the most satisfactory in the sense that the actual coverage probability is quite close to the specified level over the entire range of the tables. Obviously, its competitors do not exhibit this behavior; in fact, sometimes they miss the mark by considerable margin.

6. SENSITIVITY COMPARISON OF I_{ED} AND I_{AM}.

From the validity point of view, we saw in Section 5 that I_{AM} was the most satisfactory approximate procedure. It still remains to compare I_{AM} with the exact procedure I_{ED} with regard to sensitivity. We will consider the expected squared length as our criterion. Letting L_{ED} and L_{AM} denote the half-lengths of these intervals, we will show that $E[L_{AM}^2] < E[L_{ED}^2]$ when $J^* < J$. Of course, the intervals are identical when $J^* = J$. Since $J^* < J$,

$$E[L_{AM}^2] = t_{\alpha/2,n_1}^2 M_1\gamma_1^2 + z_{\alpha/2}^2 M_2\gamma_2^2$$

$$= \sigma_\varepsilon^2\left[t_{\alpha/2,n_1}^2\left(\frac{1}{I} + \frac{1}{I^*}\right)\frac{1}{J} + z_{\alpha/2}^2\left(\frac{1}{J^*} - \frac{1}{J}\right)\frac{1}{I^*}\right]$$

$$+ \sigma_A^2 t_{\alpha/2,n_1}^2\left(\frac{1}{I} + \frac{1}{I^*}\right)$$

$$< \sigma_\varepsilon^2 t_{\alpha/2,n_1}^2\left(\frac{1}{IJ} + \frac{1}{I^*J^*}\right) + \sigma_A^2 t_{\alpha/2,n_1}^2\left(\frac{1}{I} + \frac{1}{I^*}\right),$$

since $t_{\alpha/2,n_1}^2 > z_{\alpha/2}^2$. Now, by replacing J on the right-hand side by J^*, we get

$$E[L_{AM}^2] < \sigma_\varepsilon^2 t_{\alpha/2,n_1}^2\left(\frac{1}{I} + \frac{1}{I^*}\right)\frac{1}{J^*} + \sigma_A^2 t_{\alpha/2,n_1}^2\left(\frac{1}{I} + \frac{1}{I^*}\right)$$

$$= E[L_{ED}^2].$$

Since I_{AM} is only approximately valid as opposed to I_{ED} which is exact, one might feel that the gain in sensitivity in using I_{AM} is only slight arising mainly due to the coverage probability being less than the specified level at times. A closer inspection will, however, show that this gain can be substantial. For example, $E(L_{AM}^2)/E(L_{ED}^2) \to 0$ as $\sigma_A^2 \to 0$ and $I, I^* \to \infty$.

7. AN ILLUSTRATIVE EXAMPLE

To illustrate our results, consider the following data on tread loss in (mils) after 20,000 miles for 4 brands of tires (Hicks [6], p. 52).

Brand			
A	B	C	D
17	14	12	13
14	14	12	11
13	13	11	10
13	8	9	9

Here I = 4 and J = 4. We wish to obtain 95% prediction interval for $\bar{\bar{Y}}^*$ based on $I^* = 4$, $J^* = 2$. For constructing I_{ED}, we take $\ell_1 = \ell_2 = 1/2$ and $\ell_3 = \ell_4 = 0$. From the data, we get

$$n_1 = 3, \quad n_2 = 12, \quad M_1 = 1/8, \quad M_2 = 1/16,$$

$$\bar{\bar{Y}} = 12.06, \quad S_1^2 = 10.23, \quad S_2^2 = 4.19,$$

$$\hat{R} = 2.44, \quad \hat{S}_p^2 = 4.19, \quad \hat{N} \approx 4,$$

The prediction intervals are:

I_{ED}: (9.55, 17.20), I_{AM}: (8.32, 15.80),

I_{AP}: (9.41, 14.71), I_{AS}: (8.61, 15.51).

ACKNOWLEDGEMENTS

The authors wish to thank the referee for useful comments. The research of the second author was partially supported by the Office of Naval Research Contract N00014-84-C-0167 at Purdue University, and he thanks Professor Shanti Gupta for this as well as other facilities and encouragements. The authors also thank CBS College Publishing for their permission to include the data for the illustrative example.

REFERENCES

[1] Burdick, R. K. and Sielken, R. L., Exact confidence intervals for linear combinations of variance components in nested classifications, J. Amer. Statist. Assoc. 73 (1978) 632-635.

[2] Graybill, F. A., Theory and Application of the Linear Model (Duxbury Press, North Scituate, Mass. 1976).

[3] Graybill, F. A. and Wang, C.-M., Confidence intervals on nonnegative linear combinations of variances, J. Amer. Statist. Assoc. 75 (1980) 869-873.

[4] Hahn, G. J., Simultaneous prediction intervals for a regression model, Technometrics 14 (1971) 203-214.

[5] Hicks, C. R., Fundamental Concepts in the Design of Experiments (Holt, Rinehart and Winston, New York, 1973).

[6] Satterthwaite, F. A., An approximate distribution of estimates of variance components, Biometrics Bull. 2 (1946) 110-114.

Original paper received: 10.03.85
Final paper received: 18.03.86

Paper recommended by D.C. Bowden

PROBABILITY AND STATISTICS
Essays in Honor of Franklin A. Graybill
J.N. Srivastava (Editor)

GOODNESS-OF-FIT AND g-SAMPLE OMNIBUS TESTS BASED ON COVERAGES

Paul W. Mielke, Jr.
Department of Statistics
Colorado State University
Fort Collins, CO 80523 USA

ABSTRACT

Goodness-of-fit and g-sample omnibus tests for analyses of continuous distributions are introduced. The goodness-of-fit tests are based on the positive powers of the absolute differences between coverages and their expected values. The totally new g-sample omnibus tests are based on the positive powers of the absolute differences between empirical coverages and their expected values. Unlike the Kolmogorov-Smirnov test which presents both definitional and technical difficulties for extensions involving $g \geq 3$, these new g-sample omnibus tests are conveniently defined for any $g \geq 2$.

1. INTRODUCTION

This paper introduces classes of goodness-of-fit tests and g-sample omnibus tests ($g \geq 2$) where the underlying distributions are continuous. The distribution of the test statistic associated with each of these tests is asymptotically normal. The goodness-of-fit tests presented in Section 2 are based on coverages and are alternatives to the Kolmogorov and Pearson goodness-of-fit tests. The well known Sherman test and Greenwood-Kimball test are special cases of these goodness of fit tests. In contrast with the subjective structure of the Pearson chi-square goodness-of-fit test, these tests are objective and include cases which may be better than the Kolmogorov goodness-of-fit test for routine applications. The completely new g-sample omnibus tests introduced in Section 3 are based on empirical coverages and are alternatives to the Pearson chi-square test for homogeneity and g-sample Kolmogorov-Smirnov tests. These new g-sample omnibus tests are applicable for any $g \geq 2$ and do not suffer from subjective structures associated with tests such as the Pearson chi-square test for homogeneity.

2. GOODNESS-OF-FIT TESTS

The goodness-of-fit tests considered and introduced here are based on n independent measurements $(x_1, \ldots, x_n)$ from a continuous distribution. The null hypothesis (H_0) for each goodness-of-fit test dictates that all n measurements are from a common population whose cumulative distribution function is specified by $F(x)$. The alternative hypothesis in question here states that $F(x)$ differs from the true cumulative distribution function.

The Kolmogorov goodness-of-fit test is the most commonly used test for this purpose. The Pearson chi-square goodness-of-fit test suffers from the fact that both the number of intervals and the placement of the intervals are subjective (i.e., different choices lead to different results) and is suspect for the present purpose. If $S(x)$ denotes the empirical cumulative distribution function, then the Kolmogorov goodness-of-fit test statistic is given by

$$M = n^{1/2} \sup_x |S(x) - F(x)|.$$

If n is large, then the approximate P-value for an observed value of M (say w) is given by

$$P(M \geq w) = 1 - G(w)$$

where

$$G(w) = \begin{cases} 1 - 2\Sigma_{j=1}^{\infty}(-1)^{j-1}\exp(-2j^2w^2) & \text{if } w > 0 \\ 0 & \text{if } w \leq 0 \end{cases}$$

is the asymptotic cumulative distribution function of M as $N \rightarrow \infty$ (Fisz, 1963, see pp. 390-403).

In contrast, the asymptotic distribution under H_0 of the following goodness-of-fit test statistics based on coverages is the normal distribution. The ith of n+1 coverages is defined by

$$C_i = F(x_{i,n}) - F(x_{i-1,n})$$

for i = 1,...,n+1 where $x_{i,n}$ is the ith order statistic, $F(X_{0,n}) = 0$ and $F(x_{n+1,n}) = 1$. As defined, $\Sigma_{i=1}^{n+1}C_i = 1$. Under H_0, the mean and variance of the subsequent statistics (designated A_v where v > 0) depend on (1) the C_i's being exchangeable random variables, (2) the probability density function of C_1 is

$$f_{C_1}(x) = n(1-x)^{n-1} \text{ for } 0 < x < 1,$$

and (3) the joint probability density function of C_1 and C_2 is

$$f_{C_1,C_2}(x,y) = n(n-1)(1-x-y)^{n-2} \text{ for } 0 < x < x+y < 1.$$

Since the expected value of C_i under H_0 is $\frac{1}{n+1}$ (i = 1,...,n+1), consider the goodness-of-fit test statistic based on coverages given by

$$A_v = (n+1)^{v-1}\Sigma_{i=1}^{n+1}|C_i - \frac{1}{n+1}|^v$$

where v > 0.

The Sherman goodness-of-fit test statistic occurs when v = 1, has the range $0 \leq A_1 \leq 2n/(n+1)$, and possesses the following properties (Sherman, 1950). The mean and variance of A_1 under H_0 are respectively given by

$$\mu_1 = 2(\frac{n}{n+1})^{n+1}$$

and

$$\sigma_1^2 = \frac{4\{2n(\frac{n}{n+1})^{n+1} + n(n-1)(\frac{n-1}{n+1})^{n+1} - (n+1)(n+2)(\frac{n}{n+1})^{2(n+1)}\}}{(n+1)(n+2)}.$$

If $n \to \infty$, then $\mu_1 \to 2/e$ and $n\sigma_1^2 \to 4(2e-5)/e^2$. If n is large, then the approximate P-value for an observed value of A_1 (say $A_{1,o}$) is

$$P\{Z \geq (A_{1,o} - \mu_1)/\sigma_1\}$$

where Z is a N(0,1) random variable. In addition to the preceding properties due to Sherman (1950), Rao (1976) obtained some exact percentage points of A_1 under H_0 for selected values of n.

The Greenwood-Kimball goodness-of-fit statistic occurs when v = 2 and has the range $0 \leq A_2 \leq n$ (Greenwood, 1946; Kimball, 1947). Moran (1947) established the following properties for A_2. The mean and variance of A_2 under H_0 are respectively given by

$$\mu_2 = n/(n+2)$$

and

$$\sigma_2^2 = 4(n+1)^2/\{(n+2)^2(n+3)(n+4)\}.$$

Here $\mu_2 \to 1$ and $n\sigma_2^2 \to 4$ as $n \to \infty$. If n is large, then the approximate P-value for an observed value of A_2 (say $A_{2,o}$) is

$$P\{Z \geq (A_{2,o} - \mu_2)/\sigma_2\}$$

where Z is a N(0,1) random variable. Along with the previously stated properties for A_2 due to Moran (1947), some exact percentage points of A_2 have been obtained for selected values of n (Burrows, 1979; Currie, 1981; Stephens, 1981).

If A_v is interpreted in the geometric context involving distance functions (i.e., $|C_i - \frac{1}{n+1}|^v$ is a distance function), then the congruence principle is satisfied only when v = 1 (Mielke, 1986). In this perspective, A_1 is the natural choice among this class for routine applications (A_2 being the only other choice among this class which has presently been considered). Also since the respective asymptotic ranges of statistics A_1 and A_2 are $[0,2)$ and $[0,\infty)$, the fact that the distribution of A_1 converges to normality much more rapidly than the distribution of A_2 should not be surprising. As a further indication of this convergence advantage, Table 1 lists the coefficient of variation ratio given by $(\sigma_1/\mu_1)/(\sigma_2/\mu_2)$ for selected values of n.

If $0 \leq \theta_1 \leq \ldots \leq \theta_{n+1} < 1$ denote the ordered arc length position values of the n+1 points relative to a specified starting point on the edge of a circle whose circumference has unit length, then the resulting n+1 coverages (arc lengths) are given by

$$C_1 = 1+\theta_1-\theta_{n+1} \text{ and } C_j = \theta_j-\theta_{j-1} \text{ for } j = 2,\ldots,n+1.$$

Table 1
Coefficient of variation comparisons between A_1 and A_2

n	$(\sigma_1/\mu_1)/(\sigma_2/\mu_2)$
1	0.64550
2	0.57054
5	0.45741
10	0.40093
20	0.36768
50	0.34582
100	0.33818
200	0.33430
500	0.33194
∞	0.33036

In this context, statistic A_v provides a test of the null hypothesis that the n+1 points are uniformly distributed on the circumference of a circle. Rao (1972, 1976) considered statistics A_1 and A_2 specifically for this purpose.

3. g-SAMPLE OMNIBUS TESTS

The g-sample omnibus tests introduced in this section are based on empirical coverages. These new tests correspond to the goodness-of-fit tests based on coverages in Section 2 in the same sense that a g-sample Kolmogorov-Smirnov test corresponds to the Kolmogorov goodness-of-fit test. Each test is based on $g \geq 2$ disjoint samples of independent measurements from continuous distributions where n_i denotes the size of the ith sample ($i = 1,\ldots,g$) and there are $N = \Sigma_{i=1}^{g} n_i$ measurements among the pooled collection of samples. The null hypothesis (H_0) for these g-sample omnibus tests specifies that all g samples are obtained from a common population. The alternative hypothesis here is that the g samples are not obtained from a common population.

Because the Pearson chi-square test for homogeneity suffers from the fact that both the number and placement of intervals are subjective when continuous measurements are considered, a g-sample Kolmogorov-Smirnov test would be preferred for this purpose. If $S_i(x)$ denotes the empirical cumulative distribution function for the ith sample and

$$H_i(x) = N_i^{-1}\Sigma_{j=1}^{i} n_j S_j(x)$$

denotes the pooled empirical cumulative distribution function for the first i of the g samples where $N_i = \Sigma_{j=1}^{i} n_j$ ($i = 1,\ldots,g$), then a g-sample Kolmogorov-Smirnov statistic is given by

$$M' = \max(D_1,\ldots,D_{g-1})$$

where

$$D_i = (n_{i+1}N_i/N_{i+1})^{1/2}\sup_x|S_{i+1}(x) - H_i(x)|$$

for $i = 1,\ldots,g-1$ (note that $N_g = N$). If all n_i's are large, then the approximate P-value for an observed value of M' (say w) is given by

$$P(M' \geq w) = 1 - \{G(w)\}^{g-1}$$

where G(w) is defined in the second paragraph of Section 2 (Fisz, 1963, see pp. 407–409). Incidentally, a disturbing feature of using M' is its dependence on the entry order of the g samples (i.e., if g = 3, then the value of M' may differ for the entry orders {1,2,3} and {2,3,1}). Some alternative g-sample variations of the Kolmogorov-Smirnov test suggested by Gihman (1957), Kiefer (1959), and Birnbaum and Hall (1960) which do not depend on the entry order are based on the statistics given by

$$M'' = (\Sigma_{i=1}^{g} n_i^{-1})^{-1/2}\sup_x \Sigma_{i=1}^{g} |S_i(x) - H_g(x)|,$$

$$M''' = \sup_x \Sigma_{i=1}^{g} n_i\{S_i(x) - H_g(x)\}^2,$$

and

$$M'''' = \sup_{x,i,j} |S_i(x) - S_j(x)|.$$

While the asymptotic distributions of M'', M''' and M'''' under H_0 are quite complicated, the following g-sample omnibus tests which are based on empirical coverages possess the advantages of (1) not depending on the entry orders of the g samples and (2) having statistics which are asymptotically distributed as a normal distribution under H_0.

Since the g-sample omnibus tests introduced here are based on empirical coverages, a description of the empirical coverages follows. Assign the value $\frac{1}{N+1}$ to each of the N+1 coverages associated with the N measurements comprising the pooled collection of the g samples (i.e., the N+1 coverages are assigned equal weights under H_0). In this context, the n_i+1 empirical coverages associated with the n_i measurements of the ith sample are denoted by

$$C_{j/i} = F_N(x_{j/i}) - F_N(x_{j-1/i})$$

for $j = 1,\ldots,n_i+1$ where $x_{j/i}$ is the jth order statistic of the ith sample, $F_N(X_{0/i}) = 0$, $F_N(x_{n_i+1/i}) = 1$ and $F_N(x_{j/i})$ = (number of measurements among the N pooled measurements being $\leq x_{j/i})/(N+1)$. Thus $\Sigma_{j=1}^{n_i+1} C_{j/i} = 1$ for $i = 1,\ldots,g$. For example, suppose g = 2, $n_1 = 2$, $n_2 = 4$ and $x_{1/1} < x_{1/2} < x_{2/1} < x_{2/2} < x_{3/2} < x_{4/2}$. Then the empirical coverages for this example are $C_{1/1} = 1/7$, $C_{2/1} = 2/7$, $C_{3/1} = 4/7$, $C_{1/2} = C_{2/2} = 2/7$ and $C_{3/2} = C_{4/2} = C_{5/2} = 1/7$. The

$N!/\Pi_{i=1}^{g} n_i!$ possible allocations of the N measurements to the g samples having fixed size structure $(n_1,\ldots,n_g)$ occur with equal chance under H_0. The mean and variance under H_0 of the subsequently defined test statistics (designated B_v where $v > 0$) are based on (1) the probability function of $C_{j/i}$ given by

$$P(k|i) = P(C_{j/i} = \frac{k}{N+1}|n_i) = \binom{N-k}{n_i-1}/\binom{N}{n_i}$$

where $1 \leq k \leq N-n_i+1$, (2) the joint probability function of $C_{j/i}$ and $C_{j'/i}$ when $j \neq j'$ given by

$$P(k,k'|i) = P(C_{j/i} = \frac{k}{N+1}, C_{j'/i} = \frac{k'}{N+1}|n_i) = \binom{N-k-k'}{n_i-2}/\binom{N}{n_i}$$

where $2 \leq k+k' \leq N-n_i+2$ and $n_i \geq 2$ [$P(k,k'|i)$ is omitted if $n_i = 1$], and (3) an approximate joint probability function of $C_{j/i}$ and $C_{j'/i'}$ when $i \neq i'$ given by

$$P(k,k'|i\neq i') = P(C_{j/i} = \frac{k}{N+1}, C_{j'/i'} = \frac{k'}{N+1}|n_i,n_{i'})$$

$$= \binom{N-k}{n_i-1}\binom{N-k'}{n_{i'}-1}/\left[\binom{N}{n_i}\binom{N}{n_{i'}}\right]$$

where $1 \leq k \leq N-n_i+1$ and $1 \leq k' \leq N-n_{i'}+1$. The g-sample omnibus tests corresponding to the goodness-of-fit test statistic A_v of Section 2 are based on the statistic given by

$$B_v = (N+1)^{v-1}\Sigma_{i=1}^{g}\Sigma_{j=1}^{n_i+1}|C_{j/i} - \frac{1}{n_i+1}|^v$$

where $v > 0$. The mean and variance of B_v under H_0 are respectively given by

$$\mu_{B_v} = (N+1)^{v-1}\Sigma_{i=1}^{g}(n_i+1)t(1,i)$$

and

$$\sigma_{B_v}^2 = \mu_{B_v^2} - \mu_{B_v}^2$$

where

$$\mu_{B_v^2} = (N+1)^{2(v-1)}\{\Sigma_{i=1}^{g}(n_i+1)t(2,i) + \Sigma_{i=1}^{g}n_i(n_i+1)t(3,i) +$$

$$2\Sigma_{i<i'}(n_i+1)(n_{i'}+1)t(4,i\neq i')\},$$

$$t(1,i) = \Sigma_{k=1}^{N-n_i+1} \left|\frac{k}{N+1} - \frac{1}{n_i+1}\right|^v P(k|i),$$

$$t(2,i) = \Sigma_{k=1}^{N-n_i+1} \left|\frac{k}{N+1} - \frac{1}{n_i+1}\right|^{2v} P(k|i),$$

$$t(3,i) = \Sigma_{k=1}^{N-n_i+1} \Sigma_{k'=1}^{N-n_i+2-k} \left|\frac{k}{N+1} - \frac{1}{n_i+1}\right|^v \left|\frac{k'}{N+1} - \frac{1}{n_i+1}\right|^v P(k,k'|i)$$

when $n_i \geq 2$ $[t(3,i) = t(2,i)$ if $n_i = 1]$, and

$$t(4,i\neq i') = \Sigma_{k=1}^{N-n_i+1} \Sigma_{k'=1}^{N-n_{i'}+1} \left|\frac{k}{N+1} - \frac{1}{n_i+1}\right|^v \left|\frac{k'}{N+1} - \frac{1}{n_{i'}+1}\right|^v P(k,k'|i\neq i').$$

If $n_1,\ldots,n_g$ are not too small, then the approximate P-value for an observed value of B_v (say $B_{v,o}$) is

$$P\{Z \geq (B_{v,o} - \mu_{B_v})/\sigma_{B_v}\}$$

where Z is a N(0,1) random variable. The approximations given by

$$P(k|i) \sim \exp\{(N-k+0.5)\ln(N-k+1) - (n_i-0.5)\ln(n_i)$$

$$- (N-k-n_i+1.5)\ln(N-k-n_i+2) - (N+0.5)\ln(N+1)$$

$$+ (n_i+0.5)\ln(n_i+1) + (N-n_i+0.5)\ln(N-n_i+1)\},$$

$$P(k,k'|i) \sim \exp\{(N-k-k'+0.5)\ln(N-k-k'+1) - (n_i-1.5)\ln(n_i-1)$$

$$- (N-k-k'-n_i+2.5)\ln(N-k-k'-n_i+3) - (N+0.5)\ln(N+1)$$

$$+ (n_i+0.5)\ln(n_i+1) + (N-n_i+0.5)\ln(N-n_i+1)\}$$

when $n_i \geq 2$ $[P(k,k'|i)$ is omitted if $n_i = 1]$, and

$$P(k,k'|i\neq i') \sim \exp\{(N-k+0.5)\ln(N-k+1) + (n-k'+0.5)\ln(N-k'+1)$$

$$- (n_i-0.5)\ln(n_i) - (n_{i'}-0.5)\ln(n_{i'}) - (N-k-n_i+1.5)\ln(N-k-n_i+2)$$

$$- (N-k'-n_{i'}+1.5)\ln(N-k'-n_{i'}+2) - 2(N+0.5)\ln(N+1) + (n_i+0.5)\ln(n_i+1)$$

$$+ (n_{i'}+0.5)\ln(n_{i'}+1) + (N-n_i+0.5)\ln(N-n_i+1)$$

$$+ (N-n_{i'}+0.5)\ln(N-n_{i'}+1)\}\}$$

yield an efficient algorithm for computing P-values.

With respect to the geometric context involving distance functions in Section 2, B_1 is the recommended choice among this class of g-sample omnibus tests since only v = 1 satisfies the congruence principle. Corresponding to the distributional relationship between statistics A_1 and A_2 in Section 2, the distribution of B_1 approaches normality much faster than the distribution of B_2.

REFERENCES

[1] Birnbaum, Z.W. and R.A. Hall (1960). Small sample distributions for multi-sample statistics of the Smirnov type. Ann. Math. Statist. 31, 710-720.

[2] Burrows, P.M. (1979). Selected percentage points of Greenwood's statistic. J. Roy. Statist. Soc. Ser. A 142, 256-258.

[3] Currie, I.D. (1981). Further percentage points of Greenwood's statistic. J. Roy. Statist. Soc. Ser. A 144, 360-363.

[4] Fisz, M. (1963). Probability Theory and Mathematical Statistics. 3rd ed. John Wiley and Sons, New York.

[5] Gihman, I.I. (1957). On a certain non-parametric test of homgeniety of k-samples. Teoriya Veroyatnostey 2, 380-384.

[6] Greenwood, M. (1946). The statistical study of infectious diseases. J. Roy. Statist. Soc. Ser. A 109, 85-103.

[7] Kiefer, J. (1959). K-sample analogues of the Kolmogorov-Smirnov and Cramer-von Mises tests. Ann. Math. Statist. 30, 420-447.

[8] Kimball, B.F. (1947). Some basic theorems for developing tests of fit for the case of non-parametric probability distribution function, I. Ann. Math. Statist. 18, 540-548.

[9] Mielke, P.W. (1986). Non-metric statistical analyses: Some metric alternatives. J. Statist. Plann. Inference 13, 377-387.

[10] Moran, P.A.P. (1947). The random division of an interval. J. Roy. Statist. Soc. Ser. B 9, 92-98. Corrigendum (1981): J. Roy. Statist. Soc. Ser. A 144, 388.

[11] Rao, J.S. (1972). Bahadur efficiencies of some tests for uniformity on the circle. Ann. Math. Statist. 43, 468-479.

[12] Rao, J.S. (1976). Some tests based on arc lengths for the circle. Sankhya Ser. B 38, 329-338.

[13] Sherman, B. (1950). A random variable related to the spacing of sample values. Ann. Math. Statist. 21, 339-361.

[14] Stephens, M.A. (1981). Further percentage points of Greenwood's statistic. J. Roy. Statist. Soc. Ser. A 144, 364-366.

Original paper received: 18.02.85
Final paper received: 15.05.86

Paper recommended by M.M. Siddiqui

PROBABILITY AND STATISTICS
Essays in Honor of Franklin A. Graybill
J.N. Srivastava (Editor)

STATISTICAL SAMPLING IN AUDITING: A REVIEW

John Neter
James Godfrey

University of Georgia

In this paper, the use of statistical sampling methods in auditing for examining account balances is reviewed. We first trace the historical development of the use of statistical sampling procedures in auditing. Research findings that demonstrate the frequent inapplicability of classical large-sample inference procedures in auditing for sample sizes commonly used by auditors are then reviewed. Research into the special characteristics of accounting populations that contribute to the problems with classical inference procedures is discussed next. The principal source of difficulty is the low error rate in many accounting populations. As a result, there will often be no errors in the sample, for typical sample sizes used in auditing. Auxiliary variable estimators are then of limited use in making inferences about the population total error amount.

New inference procedures that have been developed to overcome the limitations of classical procedures for auditing applications are described and discussed. These procedures include dollar unit sampling, in which indivdual recorded dollar units are the sampling units, and Bayesian methods of deriving inferences from dollar unit samples.

Key Words

Accounting population
Dollar unit sampling
Stringer bound
Multinomial bound
Bayesian bound

INTRODUCTION

The use of statistical sampling methods in auditing for examining account balances has grown rapidly in recent years. However, classical estimation and testing procedures, based on large-sample normal distribution theory, frequently are not appropriate for the sample sizes used by auditors.

In this paper, we trace the development of the use of statistical sampling procedures in auditing, review the research that demonstrates the frequent inapplicability of classical large-sample inference procedures for sample sizes used in auditing, and consider alternate inference procedures that have been developed and utilized.

We begin by considering in Section 1 the nature of the audit process. We continue in Section 2 by reviewing briefly the historical origins of statistical sampling in auditing. In Section 3 we discuss the statistical inference procedures that were initially used and their limitations. In Section 4 we describe new information on characteristics of accounting populations relevant to the use of statistical procedures. In Sections 5 and 6 we review some new inference procedures that have been developed to overcome the limitations of the classical procedures. Finally, in Section 7 we present some closing comments.

1. NATURE OF AUDIT PROCESS

A primary function of the professional auditor is to provide an independent appraisal of the quality of financial statements published by publicly-held business firms so that current and potential owners have assurance that the financial statements prepared by the managers accurately reflect the financial status of the firm.

The accounting system that underlies the financial statements is the primary focus of the auditor. An integral component of the accounting system is a carefully documented set of procedures that prescribe how the accounting system should operate. These procedures contain "internal controls" that are designed to minimize the chance of unintentional and intentional errors in the operation of the system. For example, checks written above a certain amount may require approval by more than one person and certain inventory requisitions may require special authorization.

Although an accounting system may be well designed, errors still can occur. The auditor's investigation usually begins with a general review of the client's accounting system, including its internal controls. The results of that review dictate the quantity and intensity of the remaining audit work. If the general review reveals that the client's accounting system and internal controls are poor and not reliable, much subsequent audit work will be required. On the other hand, if the general review reveals that the accounting system and internal controls are well-designed, then much less subsequent audit work will be required. Still, it is necessary for the auditor to check to see if the internal controls are actually being implemented as prescribed because the general review is not specific enough to provide this type of compliance information.

Auditors call the checking of the operation of the internal controls "compliance testing." Sampling is widely used for compliance testing since the various internal controls may be activated many times in an accounting period. The nature of the sampling is attribute-oriented--i.e., whether or not a control was implemented properly. Thus, the sampling objective is to determine how large the process error rate for a control may be; a Poisson approximation to the binomial distribution is often used for sample evaluation. Based on the sample evaluation, the auditor decides whether or not the error rate for the control is too large and develops the plan for subsequent audit work accordingly.

The use of statistical sampling and inference procedures for compliance testing has not involved any special problems, and consequently not much research is currently being done in this area [see Roberts (1978) for a summary of compliance testing in auditing].

A second area in auditing where statistical sampling procedures are utilized is in the "subsequent audit work" referred to earlier. This audit work, called "substantive testing," consists of the detailed checking of specific account balances to ascertain their correctness. Typical examples of substantive testing are the audits of accounts receivable and inventory populations. Each of these populations often consists of thousands of individual items that may have a total value of millions of dollars. For such large populations, it is not economical for the auditor to audit every population item. Consequently, samples are frequently selected and audited to aid in making the final audit conclusion.

The form of the inference in "substantive testing" may be an estimate of the total audit amount for the population or, equivalently, of the total error amount in the population. At other times, the inference may involve a test as to whether or not the total error amount in the population equals or exceeds a critical amount called the "materiality amount."

In this review of statistical sampling in auditing, we shall focus on substantive testing applications because challenging problems have been encountered in this area in using statistical sampling and inference procedures.

2. HISTORICAL ORIGINS OF STATISTICAL SAMPLING IN AUDITING

In this section, we present a brief review of statistical sampling applications in auditing through the 1960s. A discussion of more recent developments is contained in Sections 5 and 6.

The earliest suggested applications of sampling in auditing viewed the desired inferences as tests of hypotheses. For example, in the Journal of Accountancy in the early 1940s [Editorial (1942)], editorial support was given for an approach in which "very small" percentages of a population could be used to test account balances and internal control procedures. Prytherch (1942) considered the importance of testing whether an account balance is materially misstated; he also emphasized the importance of checking the large dollar items in a population on a 100 percent basis. Neter (1949, 1952, 1954) gave greater statistical structure to the proposed applications in auditing; he emphasized the need to consider the two types of risks of making incorrect decisions and utilized the accounts receivable audit as a primary example.

Vance (1949, 1951, 1952), working on compliance testing applications, emphasized testing of error rates, for which acceptable and unacceptable error rates had to be specified. Kunz (1956) also considered hypothesis testing of error rates.

In the mid-1950s, a change in direction appeared to take place. From the mid-1950s through the 1960s, the emphasis turned to estimation inferences. Trueblood and Cyert (1954) discussed an application in which sampling was used to "age" accounts receivable. The objective of the aging of accounts receivable was to determine how old the accounts are so that an estimate could be obtained of the amount that will be uncollectible. Neter (1956) at this time also took an estimation approach in the accounts receivable area, considering the determination of the sample size required to achieve desired precision and confidence levels when sampling individual accounts. Trueblood (1957) also discussed confidence level and precision with an estimation objective and Rudell (1957) and Obrock (1958) took an estimation approach for total audit inventory value.

In the January 1960 issue of the Accounting Review, the clash between the two orientations was highlighted. Vance (1960), in a review of sampling applications in auditing and accounting, concluded that the main applications in auditing should have estimation as the objective. Stephan (1960), in contrast, suggested that estimation may be the wrong objective. He concluded that for accounts receivable audits, an hypothesis test is the proper objective.

In another development in the 1960s, Cyert, Hinckley, and Monteverde (1960) introduced the prospect of greater sampling efficiency through stratified sampling and use of the combined ratio, separate ratio, or mean-per-unit estimators. Their sampling objective was still estimation. Nigra (1963) also offered more sophistication in estimation through use of the ratio and difference estimators. He discussed the determination of sample size based upon precision and confidence level requirements. The auxiliary information estimators proposed at this time had much appeal in auditing applications since there was the prospect of much greater sampling efficiency. However, as we shall discuss in Section 3, serious problems were discovered with the use of auxiliary information estimators for sample sizes commonly used in auditing.

In 1962 the American Institute of Certified Public Accountants (AICPA, Committee on Statistical Sampling, 1962) published a report in which the discussion centered on estimation and the concepts of precision and confidence level. No suggestions for possible uses of testing inferences were included.

A few years later, Stettler (1966) stated that estimation was not adequate, because the auditor really wants to conclude whether or not the difference between the recorded account balance and the unknown, audit amount is a material (untolerable) amount or more. Still, an AICPA publication at about the same time (AICPA, Professional Development Division, 1967) entitled, "Introduction to Statistical Concepts and Estimation of Dollar Values," emphasized estimation as the primary objective in auditing. Indeed, the emphasis on estimation pervaded the literature into the early 1970s. The AICPA Statistical Sampling Committee in the early 1970s [AICPA, Committee on Statistical Sampling (1972)] published a report describing the uses of "Ratio and Difference Estimation," so that in effect the AICPA endorsed the use of auxiliary information estimators for estimation.

By the mid-1970s a series of research studies had appeared that caused much concern about the validity of many of the classical sampling methods that had been used in auditing for sample sizes employed by auditors. A discussion of these studies and other, more recent, developments are contained in the following sections.

3. CLASSICAL SAMPLING METHODS AND INFERENCE PROCEDURES

In this and succeeding sections, we review the sampling methods that have been used for substantive testing in auditing, describe problems encountered with these methods, and discuss relevant research findings. We begin by presenting in Table 1 the notation to be used throughout the section. "Book value" refers to the value of a population item recorded by an auditor's client. It is the value that the client purports to be the correct value. Book values are usually known to the auditor. The "audit value" of an item refers to the unknown true value. For example, in the audit of a client's accounts receivable balance, the auditor selects a sample of customer accounts and confirms whether the book value is

TABLE 1
Notation

Population:

Y_i = known book value of the ith item

X_i = unknown audit (true) value of the ith item

$Y = \Sigma Y_i$, total book value

$X = \Sigma X_i$, total audit value

$D_i = Y_i - X_i$, error amount of ith item

$D = \Sigma D_i$, total error amount

$i = 1, \ldots, N$ (number of population items)

Sample:

The notation for sample items is similar to that utilized for population items except that lower case is used.

correct for each account. For a correct account, the audit value is equal to the book value. For accounts with incorrect book values, the correct (audit) value is determined by additional audit procedures. To avoid additional complexities, it is frequently assumed that the audit values determined by the auditor contain no measurement error.

We begin with classical sampling methods and inference procedures because these were initially the ones used in audit applications. Our discussion is divided into two parts, estimation and testing, to correspond to the historical development and to aid in understanding how statistical methods were interwoven into auditing applications.

Estimation Approach

With estimation the objective, the auditor estimates the total audit (true) amount, X, for the population. Comparing the point estimate, $\hat{X}$, to the known total book amount, Y, provides a point estimate of the total error amount in the population, $\hat{D} = Y - \hat{X}$.

A classical confidence interval for X or D, given the specified confidence level, is calculated based on large-sample theory. Usually, the confidence interval desired for the total error amount D is a one-sided upper confidence interval based on the upper confidence bound B:

$$B = \hat{D} + z(1-\alpha)s(\hat{D}) \tag{1}$$

where $z(1-\alpha)$ is the $100(1-\alpha)$ percentile of the standard normal distribution and $s(\hat{D})$ is the estimated standard error of the point estimator $\hat{D}$.

The classical sampling designs most frequently used in auditing applications have been simple random sampling and stratified random sampling of individual accounts or line items. The estimators most frequently used have been mean-per-unit, ratio, difference, and regression. The formulas for these estimators and their estimated variances under each of the two sample designs are contained in the appendix.

Limitations. The use of sampling in auditing in the 1950s and 1960s resulted in research to refine the sampling methods being applied and to develop new methods. The mean-per-unit estimator was simple to understand and was widely used. It was soon recognized, however, that this estimator with simple random sampling of line items is very inefficient and that it would be easy to use one of the more efficient auxiliary information estimators since all of the population line item book values, Y_i, are usually known. Use of a more efficient estimator is appealing to auditors because auditing is a highly competitive field and audit efficiency can be important in a pricing strategy.

The findings of three studies [Kaplan (1973), Neter and Loebbecke (1975), and Beck (1980)] caused much doubt about the validity of using the classical estimators, particularly the auxiliary information estimators, in auditing applications. These studies used simulations to examine the behavior of various classical estimators for sample sizes commonly used in auditing. Kaplan used an assortment of simulated populations and sample sizes of 100 and 200 to obtain coverages for the confidence limits (i.e., the proportion of confidence limits that are correct). The nominal confidence level of 95 percent used in his study was not achieved, partly, he believed, because of a lack of normality of the estimator and partly because of a lack of independence between the estimator and the estimated standard error.

Neter and Loebbecke used four actual accounting populations and from these constructed twenty study populations by varying the error rate from .005 to .30.

They used sample sizes of 100 and 200 with both unstratified and stratified sampling of line items and checked the performance of confidence interval estimates for a variety of estimators and confidence levels. They found that the ratio and difference estimators, with both unstratified and stratified sampling, had coverages considerably lower than the nominal level. The mean-per-unit estimator with stratified sampling performed near the nominal confidence level, but, as expected, had substantially greater variability than the auxiliary information estimators.

Beck used a methodology similar to that of Neter and Loebbecke to study the performance of the regression estimator. Beck found, as Neter and Loebbecke did for the ratio and difference estimators, that the coverages for the regression estimator were lower than the nominal level. His results were based on unstratified and stratified sampling of line items with sample sizes of 200 and 600.

Testing Approach

Elliott and Rogers (1972) criticized the existing emphasis on estimation in the application of statistical sampling in auditing. They chose to describe the objectives of auditing in an hypothesis testing context. Their logic was that the auditing objective goes beyond estimating the true (audit) amount of an account, requiring the auditor to either accept the client's balance as not being materially misstated or to reject the balance as being materially misstated. Thus, the auditor is exposed to two types of decision errors.

Elliott and Rogers proposed the following hypothesis structure:

H_0: The financial statement amount is correct

H_1: The financial statement amount is materially in error

Letting α and β be the probabilities of the two types of inferential errors, respectively, and M the material amount, Elliott and Rogers suggested that the required sample size be calculated using large-sample normal distribution theory such that $P(H_1|D=0) = \alpha$ and $P(H_1|D=M) = 1 - \beta$. They also proposed that the decision rule be expressed in terms of the $1 - \alpha$ confidence interval for the total error amount D, obtained after the sample has been selected and audited, in the form:

If $\hat{D} \pm z(1-\alpha/2)s(\hat{D})$ includes 0, conclude H_0
Otherwise, conclude H_1

Limitations. Elliott and Rogers' suggested implementation of the testing approach, based on classical inference procedures and relying on large-sample normal distribution theory, has limitations in auditing similar to those found in estimation applications. Duke (1980) and Beck (1980) conducted simulation studies investigating the power characteristics of several classical testing procedures. Duke investigated unstratified, stratified, and pps sampling of line items and considered the mean-per-unit and difference estimators, among others, with sample sizes of 100, 300, and 500. As reported earlier, Beck studied the regression estimator with unstratified and stratified sampling of line items, based on sample sizes of 200 and 600. The results of both studies revealed very poor control over the specified risks and unsatisfactory power characteristics for the various test procedures. The lack of control over the specified risks is a function of the lack of normality of the sampling distribution of the test statistic, for the sample sizes considered.

4. CHARACTERISTICS OF ACCOUNTING POPULATIONS

As the research results demonstrating the inapplicability of large-sample theory in auditing applications for sample sizes commonly used in auditing became available, some special features of accounting populations came to be recognized. Accounting populations frequently contain relatively few errors. Hence, the book (auxiliary) value of a line item is often equal to the audit (primary) value--in particular, in accounts receivable populations where the incidence of equal book and audit values is often 95 percent or more. This differs from applications for which the auxiliary information estimators were originally developed. For these applications, the value of the primary variable is rarely equal to the value of the auxiliary variable--for example, for estimating the output of an orchard based on the number of trees in the orchard.

When the error rate in the population is low, many sample items will have a book value equal to the audit value. Not infrequently, all sample items will have their book value equal to the audit value. For example, if 99 percent of the population items have a book value equal to the audit value, the probability of obtaining a sample containing no errors with a simple random sample of 100 is .366. When this occurs, the standard error estimate for the auxiliary information estimators is zero, and the large-sample confidence limits used for interval estimation or hypothesis testing are inapplicable. Even if there are a few errors in the sample, the standard error estimate can be very volatile, depending upon the number and size of errors.

Information on error rates in accounting populations was more readily available than data on the magnitudes of errors. The latter was difficult to obtain, largely because of the need for independent auditors to keep clients' data files confidential. Some data on errors did become available in the late 1970s. Ramage, Krieger, and Spero (1979); Johnson, Leitch, and Neter (1981); and Neter, Johnson, and Leitch (1985) analyzed these data, which came from over 80 audits containing large numbers of errors that had been accumulated by one of the national auditing firms. Approximately two-thirds of the audits were accounts receivable audits and one-third inventory audits. The data base included the sample error rate, the book and audit values for all sample items with errors, and various characteristics of the book amounts in each of the accounting populations.

Johnson, Leitch, and Neter found great variation in the error rates for both accounts receivable and inventory audits, with accounts receivable having error rates considerably lower than those for inventory populations. There was evidence for both types of audit populations that the error rate may increase with the size of a line item. Also, most receivables errors were overstatements (the book value exceeds the audit value), while errors in inventory tended to be more balanced between overstatements and understatements. The distributions of overstatement and understatement error amounts tended to be unimodal.

Neter, Johnson, and Leitch later analyzed these data also on the basis of the characteristics of dollar units. The impetus for this analysis came from new sampling applications in auditing in which the population sampling units are individual dollar units rather than line items. This new sampling approach will be discussed in the next section.

More recently, a data set provided by another national auditing firm was studied by Ham, Losell, and Smieliauskas (1985). This data set consists of samples from five annual audits for each of 20 companies. The data set contains the book and audit values for samples drawn from five different types of accounts, including inventory and accounts receivable. The study considered error rate and error amount characteristics on both a line item basis and a dollar unit basis.

Ham, Losell, and Smieliauskas' findings were, in general, consistent with the findings cited earlier. One notable exception was that Ham, Losell, and Smieliauskas found a larger line item error rate, on average, for accounts receivable populations than for inventory populations.

Although it cannot be claimed that these two sample data sets on errors in accounting populations are representative of all accounting populations, to date they are the only major data sets available publicly.

5. DOLLAR-UNIT-SAMPLING

Dollar-unit-sampling (DUS) is a sample selection method in which the individual book value dollars in a population are the sampling units. The sampling methods previously discussed use line items as the sampling units, such as individual customer accounts in an accounts receivable population or individual parts in an inventory population. Random sampling of dollar units is equivalent to sampling line items with probability proportional to book amount, when both are done with replacement, and this has much appeal to auditors. Larger line items have a greater exposure to overstatement errors and these items are more likely to be included in the sample with pps selection.

In order to motivate the introduction of DUS in auditing and the bounds for the total error amount in the population that have been developed for use with this sampling procedure, we utilize an approach presented by Fienberg, Neter, and Leitch (1977). Recall that the total error amount in a population is denoted by $D = Y - X$ (total book amount minus total audit amount). We assume that all errors are overstatement errors so that $0 \le D \le Y$. Letting Y_L denote the largest line item book value in the population and p the proportion of line items in error, a bound on the total overstatement error D is:

$$C = NY_L p \tag{2}$$

Assuming that the population size N is large relative to the simple random sample size n, a $1 - \alpha$ upper confidence bound for C is:

$$NY_L p_u(1-\alpha;n,m) \tag{3}$$

where $p_u(1-\alpha;n,m)$ is the $1 - \alpha$ upper confidence bound for the population proportion p when m $(m=0,1,\ldots,n)$ errors are found in the random sample of n, based on the binomial distribution. Since C is an upper bound for D, it follows that the confidence coefficient is at least $1 - \alpha$ such that:

$$D \le NY_L p_u(1-\alpha;n,m) \tag{4}$$

When there is much variability in the population line item book values, this confidence bound will tend to be very large. Suppose the population contains $N = 10{,}000$ accounts with total book amount $Y = \$1{,}000{,}000$, $Y_L = \$1{,}000$, $p = .01$, and $D = \$1{,}000$. The upper bound C then is $C = \$100{,}000$. If a sample of size $n = 100$ is taken and $m = 1$ error is found, the 95 percent upper confidence bound for D is \$466,000. Obviously, the upper confidence bound is very high when compared to the total book amount of \$1 million and the total error amount of \$1,000.

This bound can be tightened by using stratified rather than simple random sampling of line items. Letting N_h denote the stratum size of the hth $(h = 1, \ldots, L)$ stratum, Y_{hL} the largest book value in the hth stratum, and p_h the overstatement error rate in the hth stratum, an upper bound for the total population overstatement error is:

$$C = \Sigma N_h Y_{hL} p_h \tag{5}$$

Fienberg, Neter, and Leitch showed that a $1 - \alpha$ upper confidence interval for D, when the strata error rates are not large, the strata sample sizes are not small, and the strata sample sizes are allocated according to:

$$n_h = \frac{N_h Y_{hL}}{\Sigma N_h Y_{hL}} n \tag{6}$$

is as follows:

$$D \le \frac{\Sigma N_h Y_{hL}}{n} \lambda_u(1-\alpha;m) \tag{7}$$

where $\lambda_u(1-\alpha;m)$ is the $1 - \alpha$ upper confidence bound for the Poisson parameter λ when m errors are found in the sample. To illustrate the calculation of this confidence limit, consider the same numerical example as before with the following stratification and sample allocation:

Book Value	N_h	Y_{hL}	$N_h Y_{hL}$	n_h
\$ 1 - \$100	5,900	\$ 100	\$ 590,000	37
101 - 200	3,900	200	780,000	50
201 - 1,000	200	1,000	200,000	13
	10,000		\$1,570,000	100

If one overstatement error is found in the sample, the 95 percent confidence interval for D is:

$$D \le \left(\frac{1,570,000}{100}\right)(4.744) = \$74,481$$

Thus, stratification has reduced the upper confidence bound considerably from the simple random sampling situation in which the bound was \$466,000.

If the strata are narrowed so that all book values in stratum h are close to Y_{hL} and the strata sample sizes are allocated proportional to $N_h Y_{hL}$, the population items are sampled approximately with probability proportional to book amount. One of the largest national auditing firms, Haskins & Sells (1970), described such an approach in which line items are sampled with probability proportional to book value.

Anderson and Teitlebaum (1973) suggested the use of dollar unit sampling in auditing. A sample dollar leads to the audit of the line item to which it belongs. Any error found in the book value of the line item is prorated to each book value dollar. Thus if the jth population dollar ($j = 1, \ldots, Y$) belongs to the ith ($i = 1, \ldots, N$) line item, its prorated error amount is D_i/Y_i.

The upper confidence interval for the total overstatement error amount D with dollar unit sampling can be obtained by the same approach as for simple random sampling of line items. In this case, $Y_L = \$1$ and the total population size is Y:

$$D \le Y p_u(1-\alpha;n,m) \tag{8}$$

For the previous numerical example, suppose that one overstatement error is found in a sample of n = 100 dollar units. The 95 percent upper confidence bound then is:

$$D \le 1,000,000(.0466) = \$46,600$$

This confidence bound is significantly tighter than the one for stratified sampling. Indeed, dollar unit sampling can be viewed as a limiting case for stratification when all line items in a stratum have the same book value.

The confidence bounds just discussed are rigorous. They only utilize sample information on the error rate and assume conservatively that all overstatement errors are 100 percent overstatements. Several evaluation approaches for DUS emerged in the mid-1970s which recognize that not all errors may be 100 percent overstatement errors and seek to tighten the bound further. We now discuss three of the early and best known of these approaches.

Stringer Bound

This bound is attributed to K. Stringer who did not publish any full explanation of the approach [see Anderson and Teitlebaum (1973) for a detailed discussion and Goodfellow, Loebbecke, and Neter (1974 A and B) for an explanation of the rationale]. The bound utilizes the prorated errors observed in the sample of dollar units, t_k, where $t_k = d_k/y_k$, $(k = 1, \ldots, m)$, and m is the number of errors observed in the n sampled dollar units. For the case of overstatement errors and no negative book and audit amounts, $0 < t_k \leq 1.0$. The prorated errors t_k are commonly referred to as taints.

The Stringer bound reduces in a marginal fashion the upper confidence bound given in (8) that is based on all taints being 100 percent. The decrements are calculated for the ordered observed taints $t_1 \geq t_2 \geq \ldots \geq t_m$. For example, the Stringer bound adjusted for the smallest observed taint t_m is:

$$Yp_u(1-\alpha;n,m) - Y[p_u(1-\alpha;n,m) - p_u(1-\alpha;n,m-1)](1-t_m)$$

where the factor $1 - t_m$ reflects how far the observed taint t_m is from the maximum taint of 1.0. Similar adjustments are made for the other observed taints. The Stringer bound B_S can therefore be expressed as follows:

$$B_S = Yp_u(1-\alpha;n,m) - Y\sum_{k=1}^{m}[p_u(1-\alpha;n,k) - p_u(1-\alpha;n,k-1)](1-t_k) \qquad (9)$$

This bound is usually expressed in the following equivalent form:

$$B_S = Yp_u(1-\alpha;n,0) + Y\sum_{k=1}^{m}[p_u(1-\alpha;n,k) - p_u(1-\alpha;n,k-1)]t_k \qquad (10)$$

Returning to the previous numerical example, if one error with $t_1 = .4$ is observed in a random sample of 100 dollar units, the 95 percent upper Stringer bound is:

$$B_S = 1{,}000{,}000(.0295) + 1{,}000{,}000(.0466-.0295)(.4) = \$36{,}340$$

Recall that in the previous example of DUS evaluation where one error was observed but was treated to be a 100 percent error, the 95 percent upper confidence bound was $46,600.

Cell Bound

The cell bound for evaluation of dollar unit samples was proposed as a less conservative approach than the Stringer bound by Leslie, Teitlebaum, and Anderson (1979). It involves a change in the sample selection procedure and a different method of sample evaluation. The sample selection procedure consists of first dividing the population of book value dollars into n cells with an equal number of dollars in each cell. One dollar unit is then randomly selected from each cell, the cell selections being independent, which produces the desired sample size.

In calculating the cell bound, an upper error limit component, UEL_k, is determined for the kth ordered sample taint ($t_1 \geq t_2 \geq \ldots \geq t_m$) as follows:

$$UEL_k = \max[UEL_k(1), UEL_k(2)]$$

where:

$$UEL_k(1) = UEL_{k-1} + t_k$$

$$UEL_k(2) = p_u(1-\alpha;n,k)$$

and:

$$UEL_0 = p_u(1-\alpha;n,0)$$

The upper cell bound B_c is then determined as follows:

$$B_c = \frac{Y}{n}(UEL_m) \qquad (11)$$

Multinomial Bound

The multinomial bound was developed by Fienberg, Neter, and Leitch (1977) as an alternative to the Stringer bound for evaluating DUS samples. The objective was to find a bound that would achieve the nominal confidence coefficient but be tighter than the Stringer bound. If the taints are expressed to the nearest cent, there are 101 possible outcomes (including zero) for a sample observation. Hence, Fienberg, Neter, and Leitch used the multinomial distribution to develop a joint confidence region for the p_i, the population proportions of taints of i cents.

The bound is then taken to be the maximum value of the total error amount within the joint confidence region. The bound B_m is obtained as follows:

$$\text{Maximize} \quad B_m = \frac{Y}{100}\sum_{i=1}^{100} ip_i \qquad (12)$$

$$\text{subject to} \quad \sum_S \frac{n!}{Z_0!Z_1! \ldots Z_{100}!} \prod_{i=0}^{100} p_i^{Z_i} \geq \alpha$$

where $1 - \alpha$ = the desired confidence coefficient

S = the set of vectors $Z = (Z_0, Z_1, \ldots, Z_{100})$ which are as extreme as, or less extreme than, the observed counts $W = (W_0, W_1, \ldots, W_{100})$, as defined in Neter, Leitch, and Fienberg (1978).

Evaluation of the Stringer, Cell, and Multinomial Bounds

The Stringer and cell bounds involve heuristic adjustments to the rigorous confidence bound (8) based on assumed 100 percent taints, and no confidence properties for these bounds have yet been proved. The S set chosen by Neter, Leitch, and Fienberg for computational feasibility provides only a partial ordering of all possible sample outcomes; hence, there may be occasions when this bound will have a confidence level somewhat below the specified one.

A number of simulation studies have therefore been conducted about the behavior of the Stringer, cell, and multinomial bounds for a wide variety of population settings; see Reneau (1978); Duke, Neter, and Leitch (1982); Leitch, et.al. (1982); and Plante, Neter, and Leitch (1985). A general conclusion from these studies is that the Stringer bound always achieves a coverage level larger than the nominal level. For example, when the nominal level is 95 percent, the actual coverage level achieved by the Stringer bound is usually 99 percent or higher. Thus, when using the Stringer bound in practice, there is only a very small risk that the true population total error amount is larger than the bound for nominal confidence levels of 95 percent or higher.

The simulation studies also found that the cell bound achieves coverage levels larger than the nominal confidence level, though lower than the coverages for the Stringer bound. The multinominal bound was found to have coverage levels that are close to the nominal confidence level, but are lower than the coverage levels for the Stringer and cell bounds.

The same studies found that the Stringer bound tends to be much larger than the multinomial bound and the cell bound to be somewhat larger. In the Plante, Neter, and Leitch study, for instance, the average Stringer and cell bounds were about 30 and 13 percent larger than the average multinomial bound, respectively, over four different populations.

The lack of tightness of the Stringer bound has caused concern by auditors. The typical way of using this bound is to compare it to a materiality amount M, i.e., the smallest error amount in the population that is considered intolerable. When the bound equals or exceeds the materiality amount, then the population is rejected; otherwise, the population is accepted. If the true population error amount, D, equals M or more, the auditor would like to reject the population, and the Stringer bound then provides excellent assurance of this conclusion. However, if the true error amount is less than M, the auditor would like to accept the population, but the Stringer bound may cause rejection too frequently.

Several recent studies investigated this concern by auditors. Duke, Neter, and Leitch (1982) considered sample sizes of 100, 300, and 500 and examined the power characteristics of the Stringer bound. Plante, Neter, and Leitch (1985) considered a sample size of 100 and examined the power characteristics of the Stringer, cell, and multinomial bounds. In the four study populations considered by Duke, Neter, and Leitch, the probabilities of rejection for the Stringer bound ranged from .75 to 1.00 when the true total error amount is one-half of the materiality amount. In the four populations considered by Plante, Neter, and Leitch, the probabilities of rejection for the Stringer bound ranged from .59 to .86 when the true total error amount is near one-half of the materiality amount. The corresponding ranges for the rejection probabilities for the cell and multinomial bounds were .29 to .67 and .15 to .49, respectively. Clearly, the cell and multinomial bounds provide substantially better protection to auditors against rejecting a satisfactory population, while at the same time these bounds control the risks of accepting an unsatisfactory population at or near specified levels.

6. BAYESIAN METHODS

A Bayesian structure for the auditor's inference problem has a natural appeal. The auditor acquires a reasonable amount of experience by auditing a client over a period of time. This experience can provide the basis for specifying relevant prior distributions. Use of prior distributions is particularly appealing for low error rate populations where the sample will frequently not provide any information about error amounts. A Bayesian approach can utilize the experience of the auditor through a prior distribution on error amounts that can impact the evaluation of the sample results even when the sample contains no or only a few errors.

Despite the appeal of the Bayesian approach, relatively little progress was made until recently in the development of Bayesian methods for application to audit sampling. Scott (1973) and Kinney (1975 A and B) proposed decision theoretic models of the entire audit process which have implications for the determination of the sampling plan. However, both of these models are mainly conceptual and offer few guidelines for implementation in practice.

Felix and Grimlund Model

Felix and Grimlund (1977) proposed a Bayesian model for determining the posterior distribution of the population total error amount. The model decomposes the population total error amount into two independent variables. One variable is the error rate for population items, for which a beta prior distribution and Bernoulli sampling are assumed. The other pertains to the error amounts for items in error, for which a normal-gamma joint prior distribution is assumed for the mean (μ) and variance (σ^2) of the process generating the error amounts. Sample error amounts are assumed to be generated from a normal process. The sampling units are individual population line items. The assumed prior distributions and sampling process lead to a beta posterior distribution for the error rate and a normal-gamma posterior distribution for the error amounts. Finally, these two posterior distributions combine into a beta-normal posterior distribution for the population total error amount.

The Felix and Grimlund model is primarily conceptual, and provides little guidance for implementation. In addition, it is formulated in terms of line item sampling while the emphasis in audit practice has now turned to DUS.

Cox and Snell Model

Cox and Snell (1979) proposed a Bayesian model based on DUS that decomposes the total error amount into two independent variables, error rate and mean error amount. The error rate is for dollar units, and the mean error amount is the mean taint for dollar units in error. The Cox and Snell model is designed for overstatement errors and is intended primarily for low error rate populations.

The Cox and Snell model assumes a gamma prior distribution for the dollar unit error rate π and a Poisson process for the number of errors in the sample, m, which gives a gamma posterior distribution for π:

$$f(\pi|m) = \frac{(a/\pi_0 + n)^{a+m}\pi^{a+m-1}\exp[-(a/\pi_0 + n)\pi]}{\Gamma(a+m)} \tag{13}$$

where π_0 is the mean of the prior density function and a is a second parameter which is a function of the variance of the prior distribution.

The Cox and Snell model also assumes that the reciprocal of the mean taint μ (i.e., μ^{-1}) follows a gamma prior distribution, and the sample taints, t_k, are assumed to be m random observations from an exponential distribution. This gives a gamma posterior distribution for μ^{-1}, or equivalently, an inverse gamma distribution for μ:

$$f(\mu|t_1,\ldots,t_m) = \frac{[(b-1)\mu_0+m\bar{t}]^{b+m}\mu^{-(b+m+1)}\exp\{-[(b-1)\mu_0+m\bar{t}]/\mu\}}{\Gamma(b+m)} \tag{14}$$

where μ_0 is the expected value of μ in the prior distribution and b is a second parameter which is a function of the variance of the prior distribution.

The product, $\tau = \pi\mu$, represents the overall expected taint in a population dollar unit. Given the independence of π and μ, it follows that the posterior distribution of τ is a multiple of the F distribution:

$$f(\tau|m\bar{t}) = \left[\frac{(b-1)\mu_0 + m\bar{t}}{a/\pi_0 + n}\right] \left[\frac{a + m}{b + m}\right] F[2(a + m),2(b + m)] \quad (15)$$

where F(c,d) denotes the density of the F distribution with c degrees of freedom in the numerator and d degrees of freedom in the denominator.

Given the assessed values for the prior parameters, the observed sample values m and $\bar{t}$, and the (1 - α)100 percentile of the F distribution with 2(a + m) and 2(b + m) degrees of freedom, the Cox and Snell 1 - α upper bound for τ can be easily obtained:

$$UB(1-\alpha) = \left[\frac{(b-1)\mu_0 + m\bar{t}}{a/\pi_0 + n}\right] \left[\frac{a + m}{b + m}\right] F[1-\alpha;2(a + m),2(b + m)] \quad (16)$$

where F(1-α;c,d) denotes the (1 - α)100 percentile of the F distribution with c and d degrees of freedom. UB(1-α) constitutes the 1 - α upper bound on the expected total overstatement error per dollar unit in the population, and [UB(1-α)]Y is the 1 - α upper bound on the total overstatement error amount in the population.

Evaluation of Bayesian Bounds

Since the Stringer bound is widely used in practice, most studies of Bayesian bounds have used it as a benchmark for comparison. Godfrey and Neter (1984), Neter and Godfrey (1985), and Phillips (1985) studied the behavior of the Cox and Snell bound. Godfrey and Neter examined the effects of modifying the Cox and Snell model to make it more representative of the actual auditing situation. They truncated at 1.0 the prior distributions for the error rate and mean taint. They also truncated the exponential distribution at 1.0 for the sample taints, and made still other alterations. They found that these modifications had relatively little effect on the magnitude of the bound compared to modifications in the prior parameter values for sample sizes used in auditing.

Since auditors can be called upon to defend their auditing procedures in court, the substantial dependence of the Cox and Snell bound on the prior parameter values for sample sizes commonly used in auditing can affect the acceptance of this bound by auditors. A defense based on subjective judgments may be difficult, particularly if in hindsight it can be shown that the Cox and Snell bound for the given prior parameter values and sample size had little chance of detecting a material total error amount from the population under study. This led Neter and Godfrey to examine the coverage properties of a variety of Cox and Snell bounds in repeated sampling from a number of different populations. This frequency approach is related to the Bayesian frequency calculation studies described by Rubin (1984) and to the work by Stein (1982). Rubin describes a variety of approaches for examining the robustness of prior distributions when the true model might be different.

Godfrey and Neter used ten different prior parameter settings for the Cox and Snell bound, ranging from very low error expectations to moderately high ones. They employed 21 realized populations, with total error amounts ranging from very small to large. They compared the coverage and tightness of the Cox and Snell bounds in 500 replications to the Stringer and cell bounds over the 21 populations. Godfrey and Neter found that as the population total error amount increased, each Cox and Snell bound tended to have coverages below the nominal 1 - α level. However, one of the Cox and Snell bounds with higher error expectation approximately achieved the nominal level for all but one of the 21 populations. As expected, the Stringer and cell bounds had coverages greatly exceeding the nominal confidence level for all 21 populations. This study also found that the Cox and Snell bound tended to be considerably smaller than the Stringer and cell bounds. In particular, this was true for the one Cox and Snell bound that approximately achieved the nominal $1 - \alpha$ level for all but one of the 21 populations.

In view of the results just described, Neter and Godfrey launched a search for robust prior parameter values for the Cox and Snell bound, i.e., parameter values for which the Cox and Snell bound approximately achieves the nominal $1 - \alpha$ level over a wide variety of populations. They extended the number of populations from 21 to 30 to include populations with very large total error amounts. They found a region of prior parameter value settings for the Cox and Snell bound that yielded robust prior parameter values. Three particular prior parameter settings that are robust were studied. These produced bound values that tended to be substantially smaller than the Stringer and cell bounds, particularly for low error amount populations.

Phillips (1985) extended the Neter and Godfrey study of robust Cox and Snell bounds by considering a more extensive set of study populations based upon actual accounting populations. He created 96 study populations by altering the error rate, concentration of errors, mean taint size, and the taint distribution. Phillips found that two of the three robust bounds originally identified by Neter and Godfrey were robust over all 96 study populations, with substantially smaller average values than the Stringer bound.

Menzefricke and Smieliauskas (1984) also studied the performance of the Stringer bound and the Cox and Snell bound. Their analysis focused on the power performance of the bounds. The study confirmed that the Stringer bound tends to provide poor protection against rejecting populations whose total error amount is less than materiality. The analysis of the Cox and Snell bound was restricted to a single setting of prior parameter values; hence, the significance of the results will need to be assessed after further studies have been completed.

Other Bayesian Models

We now briefly describe four other bounds that have been proposed for DUS. Tsui, Matsumura, and Tsui (1985) assumed a Dirichlet prior distribution for the 101 different possible taints (defined in cents) with the multinomial sampling model. Since the Dirichlet distribution is a conjugate prior for the multinomial distribution, this leads to a Dirichlet posterior distribution. This posterior distribution is then utilized to calculate an upper bound for the total error amount.

McCray (1984) assumed a discrete prior sampling model for the total error amount. He also assumed a multinomial sampling model for the monetary unit taints. Since the prior distribution on the total error amount does not permit an independent prior on the multinomial parameters, McCray selected those multinomial parameter values for a given total error amount that maximize the likelihood of the observed taints.

Garstka (1977) proposed a compound Poisson model in which a discrete prior distribution for the mean error taint is used to obtain an upper bound for the total error amount.

While Garstka incorporated the Bayesian approach for the mean error taint only, Vanacek (1978) utilized a beta prior distribution for the population error rate only. He then obtained Bayesian posterior bounds for the population error rate for different numbers of errors in the sample and developed a Stringer-type bound based on the Bayesian bounds for the error rate.

7. CLOSING COMMENTS

Recent research has been directed primarily towards overstatement errors and low error rate populations. Many accounting populations, however, contain a significant number of understatement errors and have relatively large overall error rates. The Stringer, cell, and multinomial bounds can accommodate understatement errors if a maximum understatement taint can be specified. Some heuristics have been suggested for combining the separate estimates of overstatement and understatement errors but the performance of these heuristics is only now being investigated.

Dworin and Grimlund (1984) have proposed a moment bound for dollar unit sampling that accommodates overstatement and understatement errors simultaneously. This bound is based on a three-moment representation of the sampling distribution for the mean error. It combines classical sampling concepts, prior information about errors, and heuristics to obtain a bound for the net total error amount without relying on any large-sample normal distribution assumptions.

Dworin and Grimlund conducted an extensive simulation study of the moment bound and compared it to the Stringer bound. It was found that the coverage of the moment bound is usually approximately at or above the nominal confidence level and that it is substantially tighter than the Stringer bound.

Other research in developing Bayesian methods for both overstatement and understatement errors is now being undertaken.

The problems in applying statistical methods to sampling in auditing are challenging. No methods have been developed yet that provide superior power protection for moderate sample sizes. Many of the methods developed are heuristic. Simulation studies that support these methods need to be supplemented by theoretical underpinnings.

REFERENCES

[1] AICPA, Committee on Statistical Sampling (1962), "Statistical Sampling and the Independent Auditor: A Special Report," Journal of Accountancy, 113, 60-62.

[2] _____, Professional Development Division (1967), Introduction to Statistical Concepts and Estimation of Dollar Values. New York: AICPA.

[3] _____, Committee on Statistical Sampling (1972), Ratio and Difference Estimation. New York: AICPA.

[4] Anderson, R. and Teitlebaum, A. D. (1973), "Dollar-Unit Sampling," Canadian Chartered Accountant, 102, No. 4, 30-39.

[5] Beck, P. J. (1980), "A Critical Analysis of the Regression Estimator in Audit Sampling," Journal of Accounting Research, 18, 16-37.

[6] Cox, D. R. and Snell, E. J. (1979), "On Sampling and the Estimation of Rare Errors," Biometrika, 66, 125-132.

[7] Cyert, R. M., Hinckley, G. M., and Monteverde, R. J. (1960), "Statistical Sampling in the Audit of the Air Force Motor Vehicle Inventory," The Accounting Review, 35, 667-73.

[8] Duke, G. L. (1980), An Empirical Investigation of the Power of Statistical

Sampling Procedures used in Auditing Under Different Models of Change of Error Patterns, Ph.D. Dissertation, University of Georgia.

[9] ________, Neter, J., and Leitch, R. A. (1982), "Power Characteristics of Test Statistics in the Auditing Environment: An Empirical Study," Journal of Accounting Research, 20, 42-67.

[10] Dworin, L. and Grimlund, R. A. (1984), "Dollar Unit Sampling for Accounts Receivable and Inventory," The Accounting Review, 59, 218-241.

[11] Editorial (1942), "Testing 'very small' Percentages," Journal of Accountancy, 73, 2-3.

[12] Elliott, R. K. and Rogers, J. R. (1972), "Relating Statistical Sampling to Audit Objectives," Journal of Accountancy, 134, 46-55.

[13] Felix, W. L. and Grimlund, R. A. (1977), "A Sampling Model for Audit Tests of Composite Accounts," Journal of Accounting Research, 15, 23-41.

[14] Fienberg, S., Neter, J., and Leitch, R. A. (1977), "Estimating the Total Overstatement Error in Accounting Populations," Journal of American Statistical Association, 72, 295-302.

[15] Garstka, S. J. (1977), "Models for Computing Upper Error Limits in Dollar-Unit Sampling," Journal of Accounting Research, 15, 179-92.

[16] Godfrey, J. and Neter, J. (1984), "Bayesian Bounds for Monetary Unit Sampling in Accounting and Auditing," Journal of Accounting Research, 22, 497-525.

[17] Goodfellow, J., Loebbecke, J., and Neter, J. (1974A), "Some Perspectives on CAV Sampling Plans, Part I," CA Magazine, 105, No. 4, 23-30.

[18] ________ (1974B), "Some Perspectives on CAV Sampling Plans, Part II," CA Magazine, 105, No. 5, 47-53.

[19] Ham, J., Losell, D., and Smieliauskas, W. (1985), "An Empirical Study of Error Characteristics in Accounting Populations," The Accounting Review, 60, 387-406.

[20] Haskins & Sells (1970), Audit Sampling: A Programmed Instruction Course, New York: Haskins & Sells.

[21] Johnson, J. R., Leitch, R. A., and Neter, J. (1981), "Characteristics of Errors in Accounts Receivable and Inventory Audits," The Accounting Review, 56, 270-293.

[22] Kaplan, R. S. (1973), "Statistical Sampling in Auditing with Auxiliary Information Estimators," Journal of Accounting Research, 11, 238-258.

[23] Kinney, W. R. (1975A), "A Decision Theory Approach to the Sampling Problem in Auditing," Journal of Accounting Research, 13, 117-132.

[24] ________ (1975B), "Decision Theory Aspects of Internal Control System Design/Compliance and Substantive Tests," Journal of Accounting Research, 13, Supplement, 14-37.

[25] Kunz, E. J. (1956), "Application of Statistical Sampling to Inventory Audits," Internal Auditor, 13, 38-49.

[26] Leitch, R. A., Neter, J., Plante, R. A., and Sinha, P. (1982), "Modified

Multinomial Bounds for Larger Numbers of Errors in Audits," The Accounting Review, 57, 384-400.

[27] Leslie, D. A., Teitlebaum, A. D., and Anderson, R. J. (1979), Dollar-Unit Sampling: A Practical Guide for Auditors, Toronto: Copp, Clark, Pitman.

[28] McCray, J. H. (1984), "A Quasi-Bayesian Audit Risk Model for Dollar Unit Sampling," The Accounting Review, 59, 35-51.

[29] Menzefricke, U. and Smieliauskas, W. (1984), "A Simulation Study of the Performance of Parametric Dollar Unit Sampling Statistical Procedures," Journal of Accounting Research, 22, 588-604.

[30] Neter, J. (1949), "An Investigation of the Usefulness of Statistical Sampling Methods in Auditing," Journal of Accountancy, 87, 390-398.

[31] ________ (1952), "Sampling Tables: An Important Statistical Tool for Auditors," The Accounting Review, 27, 475-483.

[32] ________ (1954), "Problems in Experimenting with the Application of Statistical Techniques in Auditing," The Accounting Review, 29, 591-600.

[33] ________ (1956), "Applicability of Statistical Sampling Techniques to the Confirmation of Accounts Receivable," The Accounting Review, 31, 82-94.

[34] ________ and Godfrey, J. (1985), "Robust Bayesian Bounds for Monetary-Unit Sampling in Auditing," The Journal of the Royal Statistical Society, Series C, Applied Statistics, 34, 157-168.

[35] ________, Johnson, J., and Leitch, R. A. (1985), "Characteristics of Dollar-Unit Taints and Error Rates in Accounts Receivable and Inventory," The Accounting Review, 60, 488-499.

[36] ________, Leitch, R. A., Fienberg, S. E. (1978), "Dollar Unit Sampling: Multinomial Bounds for Total Overstatement and Understatement Errors," The Accounting Review, 53, 77-93.

[37] ________ and Loebbecke, J. K. (1975), Behavior of Major Statistical Estimators in Sampling Accounting Populations - An Empirical Study, New York: AICPA.

[38] Nigra, A. L. (1963), "Statistical Sampling with Variables," Internal Auditor, 20, 25-37.

[39] Obrock, R. F. (1958), "A Case Study of Statistical Sampling," Journal of Accountancy, 105, 53-59.

[40] Phillips, J. J. (1985), Bayesian Bounds for Monetary-Unit Sampling Using Actual Accounting Populations, Ph.D. Dissertation, University of Georgia.

[41] Plante, R., Neter, J., and Leitch, R. A. (1985), "Comparative Performance of Multinomial, Cell, and Stringer Bounds," Auditing, 5, 40-56.

[42] Prytherch, R. H. (1942), "How Much Test Checking is Enough?," Journal of Accountancy, 74, 525-530.

[43] Ramage, J. G., Krieger, A. M., and Spero, L. L. (1979), "An Empirical Study of Error Characteristics in Audit Populations," Journal of Accounting Research, 17, Supplement, 72-102.

[44] Reneau, J. H. (1978), "CAV Bounds in Dollar Unit Sampling: Some Simulation

Results," The Accounting Review, 53, 669-680.

[45] Roberts, D. M. (1978), Statistical Auditing, New York: AICPA.

[46] Rubin, D. B. (1984), "Bayesianly Justifiable and Relevant Frequency Calculations for the Applied Statistician," The Annals of Statistics, 12, 1151-1172.

[47] Rudell, A. L. (1957), "Applied Sampling Doubles Inventory Accuracy, Halves Cost," N.A.A. Bulletin, 39, Section 1, 5-11.

[48] Scott, W. R. (1973), "A Bayesian Approach to Asset Valuation and Audit Size," Journal of Accounting Research, 11, 304-330.

[49] Stein, C. (1982), "The Coverage Probability of Confidence Sets Based on a Prior Distribution," Stanford University Technical Report #180.

[50] Stephan, F. F. (1960), "Faulty Advice About Statistical Sampling - Some Comments on 'A Simplified Statistical Technique for Use in Verifying Accounts Receivable'," The Accounting Review, 35, 29-32.

[51] Stettler, H. F. (1966), "Some Observations on Statistical Sampling in Auditing," Journal of Accountancy, 121, 55-60.

[52] Trueblood, R. M. (1957), "Auditing and Statistical Sampling," Journal of Accountancy, 103, 48-52.

[53] ____________ and Cyert, R. M. (1954), "Statistical Sampling Applied to Aging of Accounts Receivable," Journal of Accountancy, 97, 293-298.

[54] Tsui, K., Matsumura, E. M., and Tsui, K. (1985), "Multinomial-Dirichlet Bounds for Dollar-Unit Sampling in Auditing," The Accounting Review, 60, 76-96.

[55] Vanacek, M. T. (1978), Bayesian Dollar Unit Sampling in Auditing, Ph.D. Dissertation, The University of Texas at Austin.

[56] Vance, L. L. (1949), "Auditing Uses of Probabilities in Selecting and Interpreting Test Checks," Journal of Accountancy, 88, 214-217.

[57] __________ (1951), "How Much Test-Checking is Enough?," The Accounting Review, 26, 22-30.

[58] __________ (1952), "An Experience with Small Random Samples in Auditing," Accounting Review, 27, 472-474.

[59] __________ (1960), "Review of Developments in Statistical Sampling for Accountants," The Accounting Review, 35, 19-28.

Original paper received: 23.05.85
Final paper received: 26.03.86

Paper recommended by D.C. Bowden

APPENDIX

Formulas for Classical Estimators

Simple Random Sampling

Mean-per-unit: $\hat{X} = N\bar{x}$

$$s^2(\hat{X}) = N^2 \frac{N - n}{Nn(n-1)} \Sigma(x_j - \bar{x})^2$$

Ratio: $\hat{X} = \frac{\bar{x}}{\bar{y}} Y$

$$s^2(\hat{X}) = N^2 \frac{N - n}{Nn(n-1)} \Sigma(x_j - \frac{\bar{x}}{\bar{y}} y_j)^2$$

Difference: $\hat{X} = Y + N(\bar{x} - \bar{y})$

$$s^2(\hat{X}) = N^2 \frac{N - n}{Nn(n-1)} \Sigma[(x_j - y_j) - (\bar{x} - \bar{y})]^2$$

Regression: $\hat{X} = N\bar{x} + \left(\frac{\Sigma x_j y_j - n\bar{x}\bar{y}}{\Sigma y_j^2 - n\bar{y}^2}\right)(Y - N\bar{y})$

$$s^2(\hat{X}) = N^2 \left(\frac{N - n}{Nn(n-2)}\right) \left[\Sigma x_j^2 - n\bar{x}^2 - \frac{(\Sigma x_j y_j - n\bar{x}\bar{y})^2}{\Sigma y_j^2 - n\bar{y}^2}\right]$$

Stratified Sampling

Mean-per-unit: $\hat{X} = \Sigma N_h \bar{x}_h$

$$s^2(\hat{X}) = \Sigma N_h (N_h - n_h) \frac{s_h^2(X)}{n_h}$$

where
- $\bar{x}_h$ = sample mean audit value per item in stratum h
- $s_h^2(X)$ = sample estimate of variance of audit values in stratum h
- N_h = number of population items in stratum h
- n_h = number of sample items in stratum h
- $h = 1, \ldots, L$

Ratio (combined): $\hat{X} = \left(\frac{\Sigma N_h \bar{x}_h}{\Sigma N_h \bar{y}_h}\right) Y$

$$s^2(\hat{X}) = \Sigma N_h (N_h - n_h) \frac{s_h^2(r)}{n_h}$$

where $\bar{y}_h$ = sample mean book value per item in stratum h

$s_h^2(r)$ = sample estimate of variance of ratio deviations in stratum h

Difference: $\hat{X} = Y + \Sigma N_h(\bar{x}_h - \bar{y}_h)$

$$s^2(\hat{X}) = \Sigma N_h(N_h - n_h)\frac{s_h^2(d)}{n_h}$$

where $s_h^2(d)$ = sample estimate of variance of the difference deviations in stratum h

Regression (combined)

$$\hat{X} = \Sigma N_h\bar{x}_h + b(Y - \Sigma N_h\bar{y}_h)$$

$$s^2(\hat{X}) = \Sigma N_h(N_h - n_h)\frac{s_h^2(H)}{n_h}$$

where $$b = \frac{\Sigma N_h(N_h - n_h)\dfrac{s_h(XY)}{n_h}}{\Sigma N_h(N_h - n_h)\dfrac{s_h^2(Y)}{n_h}}$$

$s_h^2(H)$ = sample estimate of variance of the regression deviations in stratum h

$s_h^2(Y)$ = sample estimate of variance of book values in stratum h

$s_h(XY)$ = sample estimate of covariance of book and audit values in stratum h

PROBABILITY AND STATISTICS
Essays in Honor of Franklin A. Graybill
J.N. Srivastava (Editor)

SUMS OF INDEPENDENT RANDOM VARIABLES WITH THE EXTREME TERMS EXCLUDED

William E. Pruitt
School of Mathematics
University of Minnesota
Minneapolis, MN 55455

The subject of this study is the asymptotic behavior of the sum of independent, identically distributed random variables when the r summands which are largest in absolute value are discarded. The primary interest will be when the number of terms discarded goes to infinity with n but not as fast as ϵn. Both central limit theorem and law of the iterated logarithm behavior prevail quite generally.

1. Introduction. The problem we wish to discuss here is the asymptotic behavior of the sum of independent and identically distributed random variables when the r summands which are largest in absolute value are discarded. We will call these trimmed sums. The object of this procedure is that one hopes to improve the convergence properties of the sums (when the summands have infinite variance). A closely related problem which is easier to study mathematically and may also be easier to deal with in practice considers throwing away all summands which exceed a level ξ_n in absolute value where typically ξ_n will increase with n. We will call these truncated sums (although it is really the summands that are being truncated). But the problem is mathematically more interesting for the trimmed sums so we will only consider it although the trimmed sums will be approximated by truncated sums in

---

AMS 1980 Subject classification, 60F05; 60F15.
Key words and phrases, Random walk, trimmed sums, truncation, stochastic compactness, central limit theorem, law of the iterated logarithm.
This work was partially supported by NSF Grant MCS 83-01080

the proofs.

We will start by describing some of the known results. The general principle seems to be that trimming (with r fixed) does not improve weak convergence results but may improve strong convergence in borderline cases. For example, Arov and Bobrov (1960) showed that if the summands are in the domain of attraction of a non-normal stable law, then the trimmed sum will still converge weakly with the same normalization but the limit law is no longer stable. Maller (1982) showed the trimmed sum, properly normalized, could not converge weakly to the normal unless the original sum already did; he assumed symmetry of the distribution of the summands but this was later removed by Mori (1984). Mori (1976) found necessary and sufficient conditions for the strong law of large numbers to apply to these trimmed sums; here, for every r there are distributions where the strong law holds if r terms are trimmed but does not if only $r-1$ are trimmed. Feller (1968) obtained an (improved) generalized law of the iterated logarithm for an example in the domain of attraction of the normal by trimming one term. Some further results of this type are in Maller (1985).

The next question that arises is about what happens if one lets $r = r_n$ grow with n instead of keeping it fixed. If one goes to the opposite extreme and considers $r_n \sim qn$, $0 < q < 1$, then this is very close to truncating at a fixed level ξ such that the probability that the absolute summand exceeds ξ is q. Since the truncated summands have a fixed finite variance one would expect the central limit theorem (CLT) and the law of the iterated logarithm (LIL) to hold for the trimmed sums with normalizers $n^{1/2}$ and $(n \log \log n)^{1/2}$ respectively no matter what the distribution is for the summands. This is essentially correct. The CLT was proved in this context by Stigler (1973) for the related problem of discarding the pn largest and qn smallest values where $p + q < 1$, but here it turns out that flat stretches in the distribution function can lead to non-normal limits in the CLT.

Our basic interest in the present paper is to examine the middle ground where $r_n \to \infty$ but $n^{-1}r_n \to 0$. The basic question is how fast does r_n have to grow to obtain the CLT(LIL). It turns out that these results are true quite generally as soon as $r_n \to \infty$, no matter how slowly, although symmetry of the distribution of the summands plays a role in the case of the CLT. In the absence of this symmetry, the question of whether one has weak convergence is quite delicate. Of course, the norming sequences will depend on the distribution of the summands. The CLT was recently obtained by Csörgő, Horváth, and Mason (1986) in this setting when the summands are in the domain of attraction of a stable law and one discards the r_n largest and r_n smallest summands. But, surprisingly, it turns out that the CLT may fail if one discards the r_n summands which are largest in absolute value even when in the domain of attraction of a (non-normal) symmetric stable law. Kuelbs and Ledoux (1987,a,b) have considered both the CLT and LIL and even allow the summands to take values in a Banach space. However, they use a hybrid between trimming and truncating which appears to be closer to the truncated sums.

The basic context in which we will work in this paper is where the summands have a distribution which will make the sums stochastically compact. This class was introduced by Feller (1967) and is more general than the domain of attraction setting; it will be defined in the next section. It was suggested to me by Philip Griffin that the CLT might hold for any symmetric distribution in this class as soon as $r_n \to \infty$. This is true and will be proved in the next section. But, as pointed out above, even relatively slight perturbations from symmetry may destroy the convergence. It also appears that the LIL holds in this context (even without symmetry) whenever $r_n \to \infty$. This has been proved by Griffin (1985 a,b) when r_n grows more slowly than $(\log\log n/\log\log\log n)^{1/2}$ and the summands are not in the domain of partial attraction of the normal or when r_n grows at least as fast as $\log\log n$. An outline of these proofs is in section 3.

A couple of further observations about these results should be made here. For the CLT, when the summands are not symmetric, the trimmed sums are still stochastically compact but may converge along subsequences to non-normal limits. Since asymmetry does not affect the LIL for the trimmed sums this provides fairly natural examples where there is LIL behavior without the CLT. In the LIL result for the smaller range of r_n the large values of the trimmed sums are produced by the single largest summand that remains. This is reminiscent of the work of Darling (1952) and Teugels (1981) who consider the ratio of the trimmed sum to the largest remaining summand when r is fixed. But when r_n exceeds log log n the large values of the trimmed sum are due to the contribution of many moderately large summands as in the classical LIL.

To give some feeling for the rate of growth we will examine the norming sequences for the case of a symmetric stable distribution of index $\alpha < 2$ for the summands. If r is fixed the norming sequence for weak convergence is $n^{1/\alpha}$. When $r_n \to \infty$ it becomes $(nr_n^{-1})^{1/\alpha} r_n^{1/2}$. Thus when r_n is slowly varying it is still fairly close to $n^{1/\alpha}$ but as r_n gets closer to n it decreases to $n^{1/2}$. The norming sequence for the LIL when r_n exceeds log log n is the same as for weak convergence multiplied by the usual additional factor of $(\log \log n)^{1/2}$. But for the smaller values of r_n the behavior is different. As an example if $r_n = [\ell_k n]$ with $k \geq 3$, the logarithm function iterated k times, then the norming sequence is

$$\left[\frac{n}{\ell_k n} \right]^{1/\alpha} \exp \left[\frac{\ell_2 n + \ell_3 n + \ldots + \ell_{k-1} n}{\alpha(r_n+1)} \right].$$

In the course of this work, I have benefitted from several very useful conversations with Philip Griffin.

(In the two years that have elapsed since this manuscript was written considerable progress has been made in this area. Necessary and sufficient conditions for asymptotic normality of the trimmed sum when the trimming is done in terms of absolute value and the summands has a symmetric distribution are given in Griffin and Pruitt (1987). Also the entire class of possible subsequential limit laws is given along with domain of attraction conditions. Similar results for trimming the r_n largest and s_n smallest appear in Griffin and Pruitt (1988); here symmetry is not assumed. The differences in the two types of trimming become even more apparent in these papers. Some of these results are also in Csörgő, Haeusler, Mason (1986) in a different form. Griffin (1985b) solved problem 3.1 below in his final version; thus the LIL behavior of trimmed sums is well understood when S_n is stochastically compact. Some results on LIL behavior for the other type of trimming appear in Haeusler and Mason (1985).)

2. <u>Weak Theorems</u> Let $\{X_k\}$ be a sequence of independent, identically distributed, nondegenerate random variables taking values in $\mathbb{R}^1$ and $S_n = X_1 + \ldots + X_n$. In order to define the trimmed sums we let

$$M_n(j) = \text{card}\{i\in[1,n]: |X_i|>|X_j|\}+\text{card}\{i\in[1,j]: |X_i|=|X_j|\},\ 1 \le j \le n,$$

and

$$X_n^{(r)} = \Sigma_{j=1}^n X_j\, 1\{M_n(j) = r\}.$$

(We will use $1\{A\}$ to denote the indicator function of the event A.) Thus $|X_n^{(r)}|$ is the rth largest of $|X_1|, \ldots, |X_n|$ where we have ordered any ties in their natural order. The trimmed sum is

$$^{(r)}S_n = S_n - \Sigma_{k=1}^r X_n^{(k)} .$$

Now we will give Feller's condition for stochastic compactness of S_n. Let X be a random variable with the same distribution as X_1 and define for $x > 0$

$$G(x) = P\{|X|>x\},\quad K(x) = x^{-2}\, EX^2 1\{|X| \le x\}.$$

Feller (1967) proved that the analytic condition

$$\limsup_{x\to\infty} \frac{G(x)}{K(x)} < \infty \tag{2.1}$$

is equivalent to the stochastic compactness of S_n, i.e. the existence of norming and centering sequences $\{\beta_n\}$, $\{\gamma_n\}$ such that $\{\beta_n^{-1} S_n - \gamma_n\}$ is tight and all its subsequential limit laws are nondegenerate. Thus the class of distributions satisfying (2.1) includes all those which are in some domain of attraction where the entire sequence converges. In fact, in the domain of attraction case one has

$$\lim_{x\to\infty} \frac{G(x)}{K(x)} = \frac{2-\alpha}{\alpha}$$

where α is the index of the limiting stable law. Thus it is apparent that the class of distributions under consideration here is considerably larger than the class of those distributions in the domain of attraction of some stable law.

Now we are ready to prove a version of the CLT. We are going to assume that the distribution of X is continuous. This not only avoids ties when ordering the X_k but also simplifies the formulas for the norming and centering sequences as well as the proof of the theorem. Our basic assumptions in addition to (2.1) will be that

$$G \text{ is continuous},\quad r_n \to \infty,\quad n^{-1}r_n \to 0. \tag{2.2}$$

It will also simplify matters slightly to assume that

(2.3) $$EX = 0 \quad \text{if} \quad EX^2 < \infty .$$

We will first prove a few easy lemmas.

Lemma 2.1. Assume (2.2) and define $a_n(j)$ by

(2.4) $$G(a_n(j)) = n^{-1}(r_n - j\delta r_n^{1/2})$$

where δ is a fixed positive constant. (If the solution to (2.4) is not unique, any solution will suffice.) Then

$$\lim_{n\to\infty} P\{|X_n^{(r_n)}| > a_n(j)\} = 1 - \Phi(j\delta)$$

where Φ is the standard normal distribution function.

Proof. Observe that

(2.5) $$\{|X_n^{(r_n)}| > a_n(j)\} = \{V_n(j) \geq r_n\}$$

where

(2.6) $$V_n(j) = \Sigma_{k=1}^n 1\{|X_k| > a_n(j)\}.$$

Also

$$EV_n(j) = r_n - j\delta r_n^{1/2}, \quad Var(V_n(j)) = nG(a_n(j))(1-G(a_n(j))) \sim r_n$$

and Lyapounov's condition is easy to check so that

$$\frac{V_n(j) - EV_n(j)}{r_n^{1/2}} \Rightarrow N(0,1) .$$

Using this in (2.5) completes the proof.

Lemma 2.2. Assume (2.2) and define $a_n(j)$ as in (2.4). Let

$$N_n(j) = \Sigma_{k=1}^n 1\{a_n(j) < |X_k| \leq a_n(j+1)\} .$$

Then $P\{N_n(j) > 2\delta r_n^{1/2}\} \to 0$.

Proof. Since

$$EN_n(j) = \delta r_n^{1/2} , \quad Var(N_n(j)) \sim \delta r_n^{1/2} ,$$

this follows from Chebyshev's inequality.

Lemma 2.3. Assume (2.1), (2.2) and let

$$\beta_n(j) = a_n(j)(nK(a_n(j)))^{1/2} = (nEX^2 1\{|X| \leq a_n(j)\})^{1/2}.$$

We will write $\beta_n = \beta_n(0)$, $a_n = a_n(0)$. Then for j fixed

$$\beta_n(j) \sim \beta_n \quad \text{as} \quad n \to \infty .$$

Proof. For simplicity we will consider $j > 0$. Note that $a_n(j)$ increases with j. Since $x^2K(x) \uparrow$ we have $\beta_n \leq \beta_n(j)$ and

$$\begin{aligned} a_n^2(j)K(a_n(j)) &\leq a_n^2K(a_n) + a_n^2(j)\, P\{a_n < |X| \leq a_n(j)\} \\ &= a_n^2K(a_n) + a_n^2(j)\, n^{-1} j\delta r_n^{1/2} \end{aligned}$$

and so

$$1 \le \frac{\beta_n^2}{\beta_n^2(j)} + \frac{j\delta r_n^{1/2}}{nK(a_n(j))} .$$

Now by (2.1) we have for some $c > 0$ and all large n

(2.7) $$nK(a_n(j)) \ge cnG(a_n(j)) \sim cr_n$$

so the last term approaches zero. For future reference note that (2.7) implies: for some $c > 0$ and all large n

(2.8) $$\beta_n \sim \beta_n(j) \ge cr_n^{1/2} a_n(j) .$$

<u>Lemma 2.4</u>. Assume (2.1) - (2.3) and let

(2.9) $$T_n(j) = \Sigma_{k=1}^n X_k 1\{|X_k| \le a_n(j)\}.$$

Then with β_n as defined in Lemma 2.3

$$\beta_n^{-1}(T_n(j) - ET_n(j)) \Rightarrow N(0,1).$$

<u>Proof</u>. First (see e.g. page 11 of Pruitt (1981))

$$\mathrm{Var}(X1\{|X| \le a_n(j)\}) \sim a_n^2(j) K(a_n(j)) ;$$

this is where (2.3) is used. Thus $\mathrm{Var}(T_n(j)) \sim \beta_n^2(j) \sim \beta_n^2$ by Lemma 2.3. To check Lyapounov's condition,

$$nE|X|^3 1\{|X| \le a_n(j)\} \le n\, a_n(j)\, a_n^2(j)\, K(a_n(j)) = \beta_n^3(j)(nK(a_n(j)))^{-1/2}$$

and this is $o(\beta_n^3)$ by Lemma 2.3 and (2.7).

<u>Lemma 2.5</u>. Assume (2.1) and (2.2). Let $T_n(j)$ be as in (2.9) with $T_n = T_n(0)$. For fixed j

$$\beta_n^{-1}(T_n(j) - ET_n(j) - T_n + ET_n) \to 0 \quad \text{in probability.}$$

<u>Proof</u>. Assume $j > 0$. By Chebyshev,

$$\begin{aligned} P\{|T_n(j)-ET_n(j)-T_n+ET_n| > \epsilon\beta_n\} &\le (\epsilon\beta_n)^{-2} \mathrm{Var}(T_n(j) - T_n) \\ &\le (\epsilon\beta_n)^{-2} n\, EX^2 1\{a_n < |X| \le a_n(j)\} \\ &\le (\epsilon\beta_n)^{-2} n a_n^2(j)\, P\{a_n < |X| \le a_n(j)\} \\ &= (\epsilon\beta_n)^{-2}\, a_n^2(j)\, j\delta r_n^{1/2} \end{aligned}$$

and this approaches zero by (2.8).

Now we are ready to prove the central limit theorem.

<u>Theorem 2.1</u>. Assume (2.1) - (2.3). Let

$$\beta_n = a_n(nK(a_n))^{1/2} \quad , \quad \gamma_n = \beta_n^{-1}\, n\, EX1\{|X| \le a_n\}$$

where a_n is defined by $G(a_n) = n^{-1} r_n$. Then the sequence $\{\beta_n^{-1}\, {}^{(r_n)}S_n - \gamma_n\}$ is stochastically compact. If, in addition, X is symmetric the sequence converges weakly to a standard normal random variable.

Proof. Fix $\epsilon > 0$ and let

$$E_n(j) = \{a_n(j) < |X_n^{(r_n)}| \le a_n(j+1)\} .$$

By Lemma 2.1 we can find M so that for large n

$$P(\cup_{|j|>M} E_n(j)) < \epsilon/2$$

and then by Lemma 2.2

$$P(\cup_{|j|\le M} \{N_n(j) > 2\delta r_n^{1/2}\}) < \epsilon/2 .$$

Thus by excluding a set Λ of probability ϵ we may assume that one of the events $E_n(j)$ with $|j| \le M$ occurs and that $N_n(j) \le 2\delta r_n^{1/2}$ for $|j| \le M$. On the event $E_n(j)\backslash\Lambda$ we have

$$(2.10) \quad |{}^{(r_n)}S_n - T_n(j)| \le a_n(j+1)\, N_n(j) \le 2\delta\, a_n(j+1)\, r_n^{1/2} \le 2\delta c^{-1}\beta_n$$

by (2.8). Now, on the event $E_n(j)\backslash\Lambda$, write

$${}^{(r_n)}S_n - T_n = ({}^{(r_n)}S_n - T_n(j)) + (T_n(j) - ET_n(j) - T_n + ET_n) + (ET_n(j) - ET_n).$$

After dividing by β_n, the first term is small by (2.10) and the second approaches zero in probability by Lemma 2.5. In the symmetric case, the last term vanishes so that $\beta_n^{-1}({}^{(r_n)}S_n - T_n)$ goes to zero in probability which in conjunction with Lemma 2.4 gives the asymptotic normality. In the absence of symmetry, the term $ET_n(j) - ET_n$ is the troublesome one. But (for $j > 0$)

$$|ET_n(j) - ET_n| \le nE|X|1\{a_n < |X| \le a_n(j)\} \le na_n(j)\, P\{a_n < |X| \le a_n(j)\}$$
$$= a_n(j)\, j\delta\, r_n^{1/2} = O(\beta_n)$$

by (2.8) so the sequence is tight. To see that any subsequential limit is nondegenerate, we only look at the behavior on $E_n = E_n(0)$ which asymptotically has probability $\Phi(\delta) - 1/2$ by Lemma 2.1. It is not hard to check that E_n and T_n are asymptotically independent so we have

$$\beta_n^{-1}({}^{(r_n)}S_n - ET_n) = \beta_n^{-1}({}^{(r_n)}S_n - T_n) + \beta_n^{-1}(T_n - ET_n)$$

with the first term being bounded as in (2.10), after discarding an exceptional set of probability approaching zero, and the second is asymptotically normal and so unbounded. Thus the support of any subsequential limit law must be the entire line.

We should point out here the feature that makes the difference in the domain of attraction setting between throwing away the r_n

terms largest in absolute value and discarding the r_n largest and r_n smallest. In the (non-normal) domain of attraction case the tail of the distribution is nice enough so that $a_n \sim a_n(j)$. (This need not happen under (2.1).) Then (for $j > 0$)

$$ET_n(j)-ET_n = nEX1\{a_n<X\le a_n(j)\} + nEX1\{-a_n(j)\le X<-a_n\} = A_n+B_n ,$$

say, and since $a_n \sim a_n(j)$

(2.11) $A_n \sim na_nP\{a_n < X \le a_n(j)\}$, $B_n \sim -na_nP\{-a_n(j) \le X < -a_n\}$.

Thus, recalling (2.4), $A_n-B_n \sim j\delta a_nr_n^{1/2}$. In order to have a normal limit we would need $A_n+B_n \sim ja_nr_n^{1/2}c$ for some constant c and this may fail, for example, if all the mass is on the positive axis for an interval, then on the negative axis for an interval, etc. This can be arranged even in the domain of attraction setting if the lengths of these intervals are chosen appropriately. For the other problem where one is trimming the r_n largest and r_n smallest summands, one picks $a_n(j), b_n(j)$ to satisfy

$$P\{X < a_n(j)\} = n^{-1}(r_n-j\delta r_n^{1/2}) = P\{X > b_n(j)\}$$

and in place of $T_n(j)$ uses

$$\Sigma_{i=1}^n X_i1\{a_n(j) \le X_i \le b_n(k)\} .$$

Now when one must estimate the difference of the expectations there will be a term involving k from the right end and one involving j from the left end and, one has, e.g. $nP\{b_n < X \le b_n(k)\} = k\delta r_n^{1/2}$, i.e. the probability in each tail is determined. This leads to a normal limit.

What can be said if the stochastic compactness assumption (2.1) is dropped? By Lemma 2.1, $|X_n^{(r_n)}|$ will be spread throughout $[a_n(-M), a_n(M)]$. The $T_n(j)$ will be asymptotically normal with norming sequence $\beta_n(j)$ provided only that $nK(a_n(j)) \to \infty$. Since $nG(a_n(j)) \sim r_n \to \infty$, this may happen even if $K(a_n(j))$ is small compared to $G(a_n(j))$. However, it seems likely that to even obtain stochastic compactness for ${}^{(r_n)}S_n$ the $\beta_n(j)$ must be of the same order of magnitude as j changes. The proof of Lemma 2.3 suggests that this requires that $nK(a_n(j)) \ge r_n^{1/2}$ and one presumably needs

$nK(a_n(j))r_n^{-1/2} \to \infty$ for asymptotic normality of ${}^{(r_n)}S_n$. (This is correct; see Theorem 3.10 and Lemma 3.11 of Griffin and Pruitt (1987).) But making this weaker assumption makes the comparison of ${}^{(r_n)}S_n$ and $T_n(j)$ more difficult since β_n will not be as large; this seems to require spacing the $a_n(j)$ more closely.

Next we want to examine what happens if the stochastic compactness is assumed but the symmetry is dropped. Of course if the distribution is close enough to being symmetric then it may happen that for all j, $\beta_n^{-1}(ET_n(j)-ET_n) \to 0$. In this case, the given proof of asymptotic normality is still valid. Otherwise, the proof shows that on the event $E_n(j)$

$$\beta_n^{-1}\,{}^{(r_n)}S_n - \gamma_n = \beta_n^{-1}(T_n-ET_n) + \beta_n^{-1}(ET_n(j) - ET_n) + \eta_n$$

where η_n is small. Since $E_n(j)$ and T_n are asymptotically independent, if we knew that

$$\beta_n^{-1}(ET_n(j) - ET_n) \to y_j \tag{2.12}$$

then we would essentially know that

$$\beta_n^{-1}\,{}^{(r_n)}S_n - \gamma_n \Rightarrow H$$

where H is approximately the sum of a $N(0,1)$ random variable and an independent random variable Y with distribution

$$P\{Y = y_j\} = \Phi((j+1)\delta) - \Phi(j\delta).$$

In the domain of attraction setting, if one knows in addition that the ratio of the probabilities in (2.11) approaches p, then $y_j \sim cj$ so Y is itself approximately normal! But since these probabilities are quite small relative to the tail probability this is not implied by the domain of attraction conditions. It is easy to construct examples satisfying (2.1) where the limits in (2.12) exist along a subsequence $\{n_k\}$ and give a non-normal limit; for example, the absolute value of a normal random variable is a possibility for Y along a subsequence. But in the examples, one does not have convergence in (2.12) along the entire sequence. This leads us to

Conjecture 2.1. Only normal limits are possible for ${}^{(r_n)}S_n$, suitably normalized, under (2.1) and (2.2) when the sequence actually converges weakly. Here one may be able to replace (2.1) by

$nK(a_n(j))r_n^{-1/2} \to \infty$ for all j.

We had planned to conjecture a necessary and sufficient condition here for the CLT for the trimmed sums. But the problem has turned out to be so much more delicate than expected that this seems impossible at the present time.

We will conclude this section with a version of the weak law of large numbers to give some indication of the order of magnitude of the trimmed sums when the CLT fails.

Theorem 2.2. Assume (2.2) and suppose $\{\beta_n\}$ is any given sequence. Define $a_n(j)$ as in (2.4) and abuse the notation slightly by letting

$$T_n(a) = \Sigma_{k=1}^n X_k 1\{|X_k| \le a\} .$$

If

$$\beta_n^{-2} \operatorname{Var}(T_n(a)) \to 0 \text{ and } \beta_n^{-1}(ET_n(a) - ET_n(a_n)) \to 0 \tag{2.13}$$

for all $a \in [a_n(-j), a_n(j)]$ and all j

then

$$\beta_n^{-1}({}^{(r_n)}S_n - ET_n(a_n)) \to 0 \quad \text{in probability.} \tag{2.14}$$

The proof requires a finer partition than that given by the $a_n(j)$ to allow a good enough approximation of the trimmed sum by a truncated sum. In fact, one needs a partition of intervals having mass $n^{-1}r_n^{1/4}$ instead of $n^{-1}r_n^{1/2}$.

We can actually get by with a little less than (2.13). Let

$$A_n(\epsilon) = \{a : \beta_n^{-1}|ET_n(a) - ET_n(a_n)| > \epsilon\}.$$

Then we have

Conjecture 2.2. The conditions

$$\beta_n^{-2} \operatorname{Var}(T_n(a)) \to 0 \quad \text{for all } a \in [a_n(-j), a_n(j)] \text{ and all } j$$

and

$$P\{|X_n^{(r_n+1)}| \in A_n(\epsilon)\} \to 0 \quad \text{for all } \epsilon > 0$$

is necessary and sufficient for (2.14).

3. Strong Theorems. In this section we will describe Griffin's results on the LIL for trimmed sums. The first result will cover the case when the large values of ${}^{(r_n)}S_n$ are given by the single term $X_n^{(r_n+1)}$. This does not happen in the classical LIL where the large values of S_n are of order $(n \log\log n)^{1/2}$ but

$X_n^{(1)} = o(n^{1/2})$. Feller (1968) showed that the large values of S_n could result from the single term $X_n^{(1)}$ for an example in the domain of attraction of the normal. However, in his example the large values of ${}^{(1)}S_n$ did not result from the single term $X_n^{(2)}$. One can construct examples in the domain of attraction of the normal where the large values of ${}^{(r_n)}S_n$ are given by the single term $X_n^{(r_n+1)}$, even with $r_n \to \infty$, but in order to simplify matters we will assume that X is not even in the domain of partial attraction of the normal. Analytically, this means that

$$\liminf_{x\to\infty} \frac{G(x)}{K(x)} > 0 .$$

When we combine this with our basic assumption of stochastic compactness of S_n we will have G and K comparable for large x.

Theorem 3.1. (Griffin) Assume that G is continuous, $G(x) \approx K(x)$ for large x and

$$(3.1) \qquad r_n \to \infty , \quad r_n \le (\log \log n/ \log \log \log n)^{1/2} .$$

Then there exist β_n, γ_n such that with probability one

$$(3.2) \quad 0 < \limsup_{n\to\infty} \beta_n^{-1} |X_n^{(r_n+1)}| = \limsup_{n\to\infty} |\beta_n^{-1}\, {}^{(r_n)}S_n - \gamma_n| < \infty.$$

In fact,

$$(3.3) \qquad \beta_n^{-1}\, {}^{(r_n+1)}S_n - \gamma_n \to 0 \quad \text{a.s.}$$

Remarks. The upper bound on r_n in (3.1) may not be the best possible for this result. (Griffin (1985b) shows that $r_n = o((\log \log n)^{1/2})$ suffices here.) Maller (1985) has essentially shown that for fixed r (3.2) can only happen when X is in the domain of partial attraction of the normal. Also see Mori (1977) for related results for fixed r.

Outline of the Proof. We start with

$$P\{|X_n^{(r_n+1)}| > \lambda\beta_n\} = \Sigma_{i=r_n+1}^{n} \binom{n}{i} (G(\lambda\beta_n))^i (1-G(\lambda\beta_n))^{n-i} .$$

If β_n is such that

$$(3.4) \qquad nG(\lambda\beta_n) \to 0$$

then the usual estimation by comparing with a geometric series and using Stirling's formula shows that

$$(3.5)\qquad P\{|X_n^{(r_n+1)}| > \lambda\beta_n\} \approx \frac{1}{r_n^{1/2}}\left[\frac{neG(\lambda\beta_n)}{r_n+1}\right]^{r_n+1} .$$

(Note that the normal approximation used in Lemma 2.1 is not valid here.) The assumption that $G \approx K$ implies that we can find $\lambda_1 < 1 < \lambda_2$ such that for large x

$$G(\lambda_1 x) \geq eG(x) , \quad G(\lambda_2 x) \leq e^{-1}G(x).$$

This means that by adjusting the λ we can obtain a factor of e^{r_n} or e^{-r_n} in (3.5). Of course if r_n were fixed this would be irrelevant. But since $r_n \to \infty$ we can arrange that this factor will make the difference between convergence and divergence of the sequence in (3.5) when summed along a geometrically spaced subsequence. Of course, the more slowly r_n grows the more delicate this becomes. The idea is to choose β_n to satisfy

$$(3.6)\qquad \left[\frac{neG(\beta_n)}{r_n+1}\right]^{r_n+1} = \frac{1}{\ell_1 n\ \ell_2 n \ldots \ell_{j_n} n}$$

where $\ell_j n$ is the logarithm function iterated j times and j_n is determined by the growth rate of r_n. Note that (3.4) is satisfied due to the upper bound on r_n. Fairly standard Borel-Cantelli arguments now show the first part of (3.2). For (3.3) we let ξ_n be the solution of (3.6) with r_n+1 replaced by r_n+2. The above argument shows that for appropriate λ and large n

$$|X_n^{(r_n+2)}| \leq \lambda\xi_n \qquad \text{a.s.}$$

Then we let

$$T_n = \Sigma_{k=1}^n X_k\, 1\{|X_k| \leq \lambda\, \xi_n\} , \quad \gamma_n = \beta_n^{-1}\, ET_n .$$

It is clear that

$$|{}^{(r_n+1)}S_n - T_n| \leq (r_n+1)\,\lambda\xi_n$$

and then one uses exponential bounds such as Lemma 3.1 in Pruitt (1981) and Borel-Cantelli arguments to show that $\beta_n^{-1}(T_n-ET_n) \to 0$ a.s. Finally the assumed upper bound on r_n is used to show that $r_n\xi_n = o(\beta_n)$; this fact is also used in applying the exponential bounds.

The other result of Griffin is concerned with the case when r_n

grows at least as fast as log log n. In this situation the large values of the trimmed sum are produced by the cumulative effect of many summands as in the classical LIL.

Theorem 3.2. (Griffin) Assume that G is continuous, S_n is stochastically compact and

$$\liminf_{n\to\infty}(\log\log n)^{-1}r_n > 0, \quad n^{-1}r_n \to 0.$$

Define a_n, β_n, γ_n by

$$G(a_n) = n^{-1}r_n, \ \beta_n = a_n(nK(a_n)\log\log n)^{1/2}, \ \gamma_n = \beta_n^{-1}nEX1\{|X|\le a_n\}.$$

Then

$$-\infty < \liminf(\beta_n^{-1}\,{}^{(r_n)}S_n-\gamma_n) < 0 < \limsup(\beta_n^{-1}\,{}^{(r_n)}S_n-\gamma_n) < \infty \quad \text{a.s.}$$

The idea of the proof is to approximate ${}^{(r_n)}S_n$ by $T_n(a_n)$ where

$$T_n(a) = \Sigma_{k=1}^n X_k 1\{|X_k| \le a\}.$$

The principle difficulty is that the truncation point a_n can change fairly quickly. To get around this one needs a maximal inequality which is essentially of the form

$$P\{\sup_{a_{n_k}\le a\le a_{n_{k+1}}} \max_{n_k\le n\le n_{k+1}} |T_n(a)-ET_n(a)| > M\beta_{n_k}\} \le k^{-2}$$

where n_k grows exponentially. This requires a fair amount of work and uses exponential bounds. Furthermore one requires a similar maximal inequality with $T_n(a)$, β_n replaced by

$$V_n(a) = \Sigma_{k=1}^n 1\{|X_k| > a\}, \quad (r_n \log\log n)^{1/2},$$

respectively. This gives the required estimate for ${}^{(r_n)}S_n-T_n(a_n)$ and allows the completion of the proof of finiteness. The other half of the argument is more straightforward.

Although I have no conjectures to close this section, there are clearly some interesting problems that remain.

Problem 3.1. What happens when S_n is stochastically compact and $(\log\log n/\log\log\log n)^{1/2} \le r_n = o(\log\log n)$? This problem seems to require better estimates for the large tail of the distribution of $T_n(a)$ in a range where the known exponential bounds are too crude.

Problem 3.2. What happens if one drops the stochastic

compactness? Since the LIL is completely general when $r_n \sim qn$, the class of distributions for which it holds should get larger as r_n increases more rapidly.

Problem 3.3. What happens when r_n is small and X is in the domain of partial attraction of the normal?

References

[1] Arov, D.Z. and Bobrov, A.A. (1960). The extreme terms of a sample and their role in the sum of independent variables. Theor. Prob. Appl. 5, 377-396.

[2] Csörgő, S., Horváth, L., and Mason, D.M. (1986). What portion of the sample makes a partial sum asymptotically stable or normal? Probab. Th. Rel. Fields 72, 1-16.

[3] Darling, D.A. (1952). The influence of the maximum term in the addition of independent random variables. Trans. Amer. Math. Soc. 73, 95-107.

[4] Feller, W. (1967). On regular variation and local limit theorems. Proc. Fifth Berkeley Symp. Math. Statist. Probab. Vol II, Part 1, 373-388. University of California Press, Berkeley.

[5] Feller, W. (1968). An extension of the law of the iterated logarithm to variables without variance. J. Math. Mech. 18, 343-354.

[6] Griffin, P.S. (1985a). The influence of extremes on the law of the iterated logarithm. Preprint.

[7] Griffin, P.S. (1985b). Non-classical law of the iterated logarithm behaviour for trimmed sums. Preprint.

[8] Hall, P. (1978). On the extreme terms of a sample from the domain of attraction of a stable law. J. London Math. Soc. 18, 181-191.

[9] Kuelbs, J. and Ledoux, M. (1987a). Extreme values and a Gaussian central limit theorem. Probab. Th. Rel. Fields 74, 341-355.

[10] Kuelbs, J. and Ledoux, M. (1987b). Extreme values and the law of the iterated logarithm. Probab. Th. Rel. Fields 74, 319-340.

[11] Maller, R.A. (1982). Asymptotic normality of lightly trimmed means - a converse. Math. Proc. Camb. Phil. Soc. 92, 535-545.

[12] Maller, R.A. (1984). Relative stability of trimmed sums. Z. Wahr. Verw. Geb. 66, 61-80.

[13] Maller, R.A. (1985). Some almost sure properties of trimmed sums. Preprint.

[14] Mori, T. (1976). The strong law of large numbers when extreme terms are excluded from sums. Z. Wahr. Verw. Geb. 36, 189-194.

[15] Mori, T. (1977). Stability for sums of i.i.d random variables when extreme terms are excluded. Z. Wahr. Verw. Geb. 40, 159-167.

[16] Mori, T. (1984). On the limit distributions of lightly trimmed sums. Math. Proc. Camb. Phil. Soc. 96, 507-516.

[17] Pruitt, W.E. (1981). General one-sided laws of the iterated logarithm. Ann. Probab. 9, 1-48.

[18] Stigler, S.M. (1973). The asymptotic distribution of the trimmed mean. Ann. Statist. 1, 472-477.

[19] Teugels, J.L. (1981). Limit theorems on order statistics. Ann. Probab. 9, 868-880.

(Added in proof)

[20] Griffin, P.S. and Pruitt, W.E. (1987). The central limit problem for trimmed sums, Math. Proc. Camb. Phil. Soc. (to appear).

[21] Griffin, P.S., and Pruitt, W.E. (1988). Asymptotic normality and subsequential limits of trimmed sums. Ann. Probab. (to appear).

[22] Csörgő, S., Haeusler, E., and Mason, D.M. (1986). The asymptotic distribution of trimmed sums. Preprint.

[23] Haeusler, E., and Mason, D.M. (1985). Laws of the iterated logarithm for sums of the middle portion of the sample. Preprint.

Original paper received: 25.02.85
Final paper received: 10.06.87

Paper recommended by M.M. Siddiqui

PROBABILITY AND STATISTICS
Essays in Honor of Franklin A. Graybill
J.N. Srivastava (Editor)

UNIFORM RATES OF CONVERGENCE TO EXTREME VALUE DISTRIBUTIONS

S. I. Resnick*, Colorado State University

Abstract

When the distributions of normalized maxima of iid random variables converge to a limit distribution, it is important to assess the speed of convergence. Using representations for distributions in domains of attraction of extreme value distributions we construct bounds on the maximum discrepancy between the distribution of the maximum of a sample of size n and the limit distribution. Numerical examples are given to illustrate slow rates of convergence.

Keywords: *Extreme values, maxima, rates of convergence, regular variation, limit distributions.*

1980 AMS subject classification: 60F99, 60E05.

1. Introduction

Suppose $\{X_n, n \geq 1\}$ are independent, identically distributed (iid) random variables with common distribution function F(x). Set $M_n = \bigvee_{i=1}^{n} X_i$ and suppose there exist $a_n > 0$ and $b_n \in \mathbb{R}$ such that

$$P[a_n^{-1}(M_n - b_n) \leq x] = F^n(a_n x + b_n) \to G(x)$$

weakly, where G is non degenerate. Then we say F (or $\{M_n\}$) is in the domain of attraction of G (written $F \in D(G)$) and it is well known that G must be one of the following (Gnedenko, 1943; de Haan, 1970):

$$\Lambda(x) = \exp\{-e^{-x}\} \quad , \quad x \in \mathbb{R}$$

$$\Phi_\alpha(x) = \begin{cases} 0 & x < 0 \\ \exp\{-x^{-\alpha}\} & x \geq 0 \end{cases}$$

$$\Psi_\alpha(x) = \begin{cases} e^{-|x|^\alpha} & x < 0 \\ 1 & x \geq 0 \end{cases}$$

We survey some results which give bounds for $\sup_x |F^n(a_n x + b_n) - G(x)|$. These bounds are necessary for judging how useful G is as an approximation to the distribution function of M_n.

Such problems have been considered in the literature dating back to Fisher and Tippett (1928). We have found the following papers most instructive: Anderson (1971), Davis (1982), Hall (1979), Hall and Wellner (1979), Cohen (1982a,b) and Smith (1982). In particular, when $F \in D(\Phi_a)$, Smith relates uniform rates of convergence to the concept of slow variation with remainder (c.f. Goldie, 1980, Goldie and Smith, 1984).

In contrast to the approach based on slow variation with remainder, our results center around represention results for distributions $F \in D(G)$ (de Haan, 1970, Balkema and de Haan, 1972).

---

*Partial support is gratefully acknowledged from Erasmus University, Rotterdam, a Lady Davis Fellowship at the Technion, Haifa, Israel and from NSF Grant MCS 8202335. Thanks are due to the Econometric Institute, Erasmus University and to the Faculty of Industrial and Management Engineering, Technion for hospitality during the 1981-1982 academic year.

For most of this paper, in the interest of usability, we are less than completely general and assume F(x) has a convenient number of derivatives and that sufficient conditions for $F\epsilon D(G)$, known as Von Mises conditions, are satisfied. Under these conditions more convenient representations of $F\epsilon D(G)$ exist. We feel this is the appropriate level of generality but in any event our methods can be generalized by making use of general representations for F. Cf. Cohen 1982b.

Several authors (Smith, 1982; Cohen, 1982a,b) have concentrated on obtaining the order of convergence and have obtained excellent results. We have tried, where possible, to get an actual bound since for rates of convergence questions behavior for finite values of n is of interest as well as asymptotic methods.

Portions of this paper were worked out in conjunction with L. de Haan and A. A. Balkema while visiting Erasmus University, Rotterdam in Fall, 1981. The methods to be discussed have also proved useful in problems of tail estimation; see Davis and Resnick (1984).

2. Uniform Rates of Convergence to Φ_α.

Write $F = \exp\{-e^{-\phi}\}$ and suppose F has a continuous derivative F'. The Von Mises condition guaranteeing $F\epsilon D(\Phi_\alpha(x))$ is

$$h(x) := x\phi'(x) - \alpha = \frac{xF'(x)}{F(x)(-\log F(x))} - \alpha \to 0$$

and there exists a non-increasing continuous function g such that

$$|h(x)| \le g(x) \downarrow 0 \tag{2.1}$$

as $x \to \infty$. Typically we take $g(x) = \sup_{y\ge x}|h(y)|$. Set $\exp\{-\phi(a_n)\} = n^{-1}$ so that

$$\phi(a_n x) - \phi(a_n) = \int_1^x t^{-1}(\alpha + h(a_n t))dt \to \alpha \log x,\ n \to \infty,$$

for $x > 0$ showing $F \epsilon D(\Phi_\alpha)$. The rate of convergence will be given in terms of g.

We begin with the following simple lemma:

Lemma 2.1: For $\alpha_1, \alpha_2 > 0$,

$$\sup_{x>0}|\Phi_{\alpha_1}(x) - \Phi_{\alpha_2}(x)| \le (.2701)|\alpha_1 - \alpha_2|/(\alpha_1\wedge\alpha_2)$$

Proof: Observe for $x > 0$, $\frac{\partial}{\partial\beta}\Phi_\beta(x) = \Phi_\beta(x)x^{-\beta}\log x$ and so assuming $\alpha_1 < \alpha_2$

$$\sup_{x>0}|\Phi_{\alpha_1}(x) - \Phi_{\alpha_2}(x)| \le \sup_{x>0}\ \sup_{\beta\epsilon[\alpha_1,\alpha_2]}|\Phi_\beta(x)x^{-\beta}\log x||\alpha_2 - \alpha_1|$$

$$= \sup_{y>0}\ \sup_{\beta\epsilon[\alpha_1,\alpha_2]}|e^{-y}y\log y|\beta^{-1}|\alpha_2 - \alpha_1|$$

$$\le \left(\sup_{y>0}|e^{-y}y\log y|\right)\alpha_1^{-1}|\alpha_2 - \alpha_1|.$$

The supremum is found numerically to four decimal places. □

It is now easy to obtain the rate of convergence on the interval $[1, \infty)$.

Proposition 2.2: If (2.1) holds and $\exp\{-\phi(a_n)\} = n^{-1}$ then

$$\sup_{x\geq 1}|F^n(a_n x) - \Phi_\alpha(x)| \leq (.2701)(\alpha - g(a_n))^{-1}g(a_n) = O(g(a_n)).$$

Proof: For $x \geq 1$

$$\phi(a_n x) - \phi(a_n) = \int_1^x (\alpha + h(a_n t))t^{-1}dt \leq \int_1^x (\alpha + g(a_n t))t^{-1}dt$$
$$\leq (\alpha + g(a_n))\log x.$$

Obtaining a lower bound in a similar way we get

$$(\alpha - g(a_n))\log x \leq \phi(a_n x) - \phi(a_n) \leq (\alpha + g(a_n))\log x$$

and taking negative exponentials twice gives for $x \geq 1$

$$\Phi_{(\alpha - g(a_n))}(x) \leq F^n(a_n x) \leq \Phi_{(\alpha + g(a_n))}(x) \tag{2.2}$$

and an application of Lemma 2.1 gives the desired result. □

On the region $(-\infty, 1)$, more care must be taken. We first present a method which works quite generally and then we show that if more is assumed about g a better bound can be obtained.

If $x < 1$ we find in the same way as (2.2) was obtained that

$$\Phi_{(a + g(a_n x))}(x) \leq F^n(a_n x) \leq \Phi_{(\alpha - g(a_n x))}(x) \tag{2.3}$$

and hence $|F^n(a_n x) - \Phi_\alpha(x)| \leq (.2701)(\alpha - g(a_n x))^{-1}g(a_n x)$ from Lemma 2.1. Now suppose $\{x_n\}$ is a sequence (to be specified) satisfying $x_n \to 0$ and $a_n x_n \to \infty$. Then we conclude

$$\sup_{x_n\leq x<\infty} |F^n(a_n x) - \Phi_\alpha(x)| \leq (.2701)(\alpha - g(a_n x))^{-1}g(a_n x_n).$$

For $x \leq x_n$ observe from (2.3) that

$$F^n(a_n x) \leq F^n(a_n x_n) \leq \Phi_{(\alpha - g(a_n x_n))}(x_n)$$

and for $x_n \leq 1$

$$\Phi_\alpha(x_n) \leq \Phi_{(\alpha - g(a_n x_n))}(x_n)$$

and so the uniform bound becomes

$$\sup_x |F^n(a_n x) - \Phi_\alpha(x)| \leq (.2701)\ (\alpha - g(a_n x_n))^{-1}g(a_n x_n)\vee \Phi_{(\alpha - g(a_n x_n))}(x_n). \tag{2.4}$$

The way to choose x_n so that the right side of (2.4) is minimized is to pick x_n to satisfy

$$(.2701)(\alpha - g(a_n x_n))^{-1}g(a_n x_n) = \Phi_{(\alpha - g(a_n x_n))}(x_n)$$

or equivalently

$$a_n x_n(-\log((.2701)(\alpha-g(a_n x_n))^{-1}g(a_n x_n)))^{(\alpha-g(a_n x_n))^{-1}} = a_n.$$

To get an expression for x_n it is convenient to switch to a continous variable. Define

$$a(t) = \left(\frac{1}{-\log F}\right)^{\leftarrow}(t) = \inf\{u: 1/(-\log F(u)) \geq t\}$$

and define a non-decreasing function $\rho(t)$ by

$$\rho(a(t)) = a(t)x(t)$$

where $x(t)$ is an unknown function decreasing to zero, while $a(t)x(t)$ increases to ∞. Then we have

$$\rho(a(t))\{-\log(.2701)(\alpha - g(\rho(a(t))))^{-1} g(\rho(a(t)))\}^{(\alpha - g(\rho(a(t))))^{-1}} = a(t).$$

Change variables replacing $a(t)$ by t. If we let $\rho^{\leftarrow}(t)$ be the inverse of ρ we obtain

$$\rho^{\leftarrow}(t) = t(-\log(.2701)(\alpha - g(t))^{-1} g(t))^{(\alpha - g(t))^{-1}}. \tag{2.5}$$

It may be difficult to invert this expression but an asymptotic inversion can usually be performed. Note that it is clear that

$$\lim_{t\to\infty} \rho^{\leftarrow}(t)/t = \infty$$

and hence

$$\lim_{t\to\infty} \rho(t)/t = 0$$

so from the definition of ρ we get $x(t)\downarrow 0$. Since $\rho(t)\uparrow\infty$ we get $a(t)x(t)\uparrow\infty$ and so $x(t)$ has the desired properties.

We now summarize our findings.

Theorem 2.3: Suppose (2.1) holds so that $F \in D(\Phi_\alpha)$. Then

$$\sup_{x\in\mathbb{R}} |F^n(a_n x) - \Phi_\alpha(x)| \leq (.2701)(\alpha - g(\rho(a_n)))^{-1} g(\rho(a_n))$$

where ρ is given in terms of its inverse by (2.5).

Remark: One sometimes gets the impression from results in the literature that $O(n^{-1})$ is the best convergence rate possible. This is not the case. Since the function g can be any function converging monotonically to zero, a wide variety of convergence rates are to be expected. For example, suppose $g(t) = e^{-t}$ and we define a distribution F by

$$F(x) = \begin{cases} 0 & x < 1 \\ \exp\{-\exp\{-\int_1^x (1 + e^{-u})\, u^{-1} du\}\}, & x \geq 1 \end{cases}$$

Then $\rho^{\leftarrow}(t) \sim t(-\log g(t))^{(1-g(t))^{-1}} \sim t^2$ so that $\rho(t) \sim t^{1/2}$. Since $\log t = \int_1^{a(t)} (1+e^{-u}) u^{-1} du$ we have $\log a^{\leftarrow}(t) = \int_1^t (1+e^{-u}) u^{-1} du = \log t + c + o(1)$ and thus $a^{\leftarrow}(t) = te^c(1 + o(1))$ and $a(t) = te^{-c}(1+o(1))$. So an order of convergence $g(\rho(a_n))$ is of the form $\exp\{-k\, n^{-1/2}\}$, $k > 0$.

The convergence rate on $[1,\infty)$ is $g(a_n)$ but the above technique gives the overall rate $g(\rho(a_n))$. When g satisfies growth conditions of regular variation type it is possible to improve the bound from $O(g(\rho(a_n)))$ to $O(g(a_n))$ as is done in Smith (1982). Indeed when g is regularly varying with index $-\beta$ we have

$$\lim_{n\to\infty} \frac{g(\rho(a_n))}{g(a_n))} = \lim_{n\to\infty} \frac{g(a_n(\rho(a_n)/a_n))}{g(a_n)} = \lim_{n\to\infty} (\rho(a_n)/a_n)^{-\beta} = 0$$

(recall $\rho(t)/t \to 0$ as $t \to \infty$) so that $O(g(a_n))$ is a significant improvement over $O(g(\rho(a_n)))$.

We show $O(g(a_n))$ is a valid convergence rate under the following assumption: Pick δ so large that $g(\delta) < \alpha$. Then for n such that $a_n^{-1}\delta < 1$ we assume

$$\frac{g(a_n x)}{g(a_n)} \leq x^{-\beta}, \quad a_n^{-1}\delta \leq x \leq 1, \tag{2.6}$$

where $\beta > 0$.

Remark: There are two circumstances where (2.6) is easy to verify. The first is where y is of the form $g(x) = x^{-\beta}$ which occurs in several common cases. The other situation is where g is regularly varying, differentiable and satisfies $tg'(t)/g(t) \to -\hat{\beta}$, as $t\to\infty$. In this case g is of the form

$$g(t) = c \exp\{-\int_1^t t^{-1}\beta(t)\,dt\}$$

where $\beta(t) = \hat{\beta} + \hat{h}(t)$ and there exists a non-increasing function $\hat{g} \geq 0$ such that

$$|\hat{h}(t)| \leq \hat{g}(t) \downarrow 0$$

as $t\to\infty$. Then for $a_n^{-1}\delta \leq x \leq 1$

$$\frac{g(a_n x)}{g(a_n)} = \exp\{\int_x^1 u^{-1}\,\beta(a_n u)du$$

$$\leq \exp\{\int_x^1 u^{-1}\,(\hat{\beta} + \hat{g}(a_n u))du\}$$

$$\leq \exp\{-\hat{\beta}\log x + \int_x^1 u^{-1}\hat{g}\,(a_n u)du\}$$

$$\leq \exp\{-(\hat{\beta} + \hat{g}(\delta))\log x\} = x^{-(\hat{\beta} + \hat{g}(\delta))}$$

and so (2.6) is satisfied with $\beta = \hat{\beta} + \hat{g}(\delta)$.

We need a variant of Lemma 2.1.

Lemma 2.4: Suppose (2.6) holds and $\delta > 0$ is chosen so that $g(\delta) < \alpha$. We have for n such that $a_n^{-1}\delta < 1$

$$\sup_{a_n^{-1}\delta\leq x\leq 1} (|\Phi_{(\alpha-g(a_n x))}(x)-\Phi_\alpha(x)| \vee |\Phi_{(\alpha+g(a_n x))}(x)-\Phi_\alpha(x)| \tag{2.7}$$

$$\leq c(\alpha,\beta,\delta)\, g(a_n)$$

where

$$c(\alpha,\beta,\delta) = \beta^{-1}\theta \sup_{s\geq 1} \{s^{1+\theta}(\log s)\, e^{-s}\}$$

and

$$\theta = \beta/(\alpha-g(\delta)).$$

Proof: Since $\Phi_\gamma(x) = \Lambda(\gamma \log x)$ for $x > 0$ and $\Lambda'(y)$ is increasing for $y < 0$ we have

$$\sup_{a_n^{-1}\delta\le x\le 1} |\Phi_\alpha(x) - \Phi_{(\alpha-g(a_nx))}(x)|$$

$$\le \sup_{a_n^{-1}\delta\le x\le 1} \int_{\alpha-g(a_nx)}^{\alpha} \Lambda'(\gamma \log x)|\log x|\, d\gamma$$

$$\le \sup_{a_n^{-1}\delta\le x\le 1} \Lambda'((\alpha-g(a_nx))\log x)\ |\log x|\ g\ (a_nx)$$

$$\le \sup_{a_n^{-1}\delta\le x\le 1} \Lambda'((\alpha-g(\delta))\log x)\ |\log x|\ g\ (a_nx)$$

$$\le \sup_{o<x\le 1} \Lambda'((\alpha-g(\delta)\log x)\ |\log x|x^{-\beta})g(a_n)$$

$$= \sup_{0<y<1} \{y^{-1}e^{-y^{-1}}|\log y|\ y^{-\theta}\ \theta\beta^{-1}\}\ g(a_n)$$

$$= c(\alpha,\beta,\delta)\ g(a_n).$$

The bound for $|\Phi_{\alpha+g(a_nx)}(x) - \Phi_\alpha(x)|$ is dominated by the one just presented so we are done.

Remark: The constant $c(\alpha,\beta,\delta)$ must be computed numerically once g, δ, β are specified. See the example below and the following table:

θ	.25	.5	1	2	3	4
$\sup_{s\ge 1}\{s^{1+\theta}(\log s)e^{-s}\}$	.2372	.2976	.4928	1.6392	6.8703	35.058

Theorem 2.5: Suppose (2.6) holds and $\delta > 0$ is chosen so large that $g(\delta) < \alpha$. Then for n such that $a_n^{-1}\delta < 1$

$$\sup_x |F^n(a_nx)-\Phi_\alpha(x)| \le (.2701)(\alpha-g(a_n))^{-1}g(a_n) \vee c(\alpha,\beta,\delta)g(a_n) \vee F^n(\delta) \vee \Phi_\alpha(a_n^{-1}\delta).$$

For n sufficiently large, the dominant term in the bound is $O(g(a_n))$.

Proof: Proposition 2.2 is still applicable. From equation (2.3) and Lemma 2.4 we have

$$\sup_{a_n^{-1}\delta\le x\le 1} |F^n(a_nx)-\Phi_\alpha(x)| \le c(\alpha,\beta,\delta)g(a_n).$$

Finally

$$\sup_{x\le a_n^{-1}\delta} |F^n(a_nx)-\Phi_\alpha(x)| \le F^n(\delta) \vee \Phi_\alpha(a_n^{-1}\delta). \qquad \square$$

For many distributions it is convenient to work with $1-F$ rather than $-\log F$. In cases where the convergence rate is slower than $1/n$, the following result is useful. Set $\bar{F} = 1-F$.

Proposition 2.5: Suppose (2.1) holds and set

$$B(x) = \sum_{k=1}^{\infty} \frac{\bar{F}^k(x)}{k(k+1)} \quad \text{so that}$$

$$\bar{F}(x)/2 \leq B(x) \leq \frac{1}{2}\bar{F}(x)(1 + \bar{F}(x))$$

and $B(x) \sim \frac{1}{2}\bar{F}(x)$ as $x \to \infty$. Then $-F \log F = \bar{F}(1-B)$ and hence

(2.8) $$\frac{x F'(x)}{F(x)(-\log F(x)} - \alpha = \frac{x F'(x)}{\bar{F}(x)} - \alpha + \frac{xF'(x) B(x)}{\bar{F}(x)(1-B(x))}$$

so that

(2.9) $$\frac{x F'(x)}{F(x)(-\log F(x)} - \alpha = \frac{x F'(x)}{\bar{F}(x)} - \alpha + c(x)\bar{F}(x)$$

where $c(x) \to \alpha/2$.

Unless $(xF'(x)/\bar{F}(x)) - \alpha$ goes to zero slower than $\bar{F}(x)$, the use of this formula will lead to a convergence rate of $O(\frac{1}{n})$. This will be the case, for instance, with the Cauchy distribution.

Example (cf. Smith, 1982, example 1): Suppose for $x \geq 1$

$$\bar{F}(x) = c\, x^{-\alpha} + d\, x^{-\beta-\alpha}$$

where $c > 0$, $d > 0$, $0 < \beta < \alpha$, $c + d = 1$. We find

$$\frac{x F'(x)}{\bar{F}(x)} = \alpha \left(\frac{1 + d(\alpha+\beta)\, c^{-1}\alpha^{-1}x^{-\beta}}{1 + d\, c^{-1}x^{-\beta}}\right)$$

and so

$$\frac{x F'(x)}{\bar{F}(x)} - \alpha = \frac{d\beta c^{-1}x^{-\beta}}{1+dc^{-1}x^{-\beta}}$$

which implies

$$\left|\frac{x F'(x)}{\bar{F}(x)} - \alpha\right| \leq c_1\, x^{-\beta}$$

where $c_1 = d\beta c^{-1}$. Set $c_2 = \alpha + d(\alpha + \beta)\, c^{-1}$ and for $x > 1$ we have

$$\left|\frac{x F'(x)}{\bar{F}(x)} \frac{B(x)}{1-B(x)}\right| \leq c_2\frac{B(x)}{1-B(x)} \leq \frac{c_2 \frac{1}{2}\bar{F}(x)(1+\bar{F}(x))}{1-\frac{1}{2}\bar{F}(x)(1+\bar{F}(x))} \leq c_2\bar{F}(x)/(1-\bar{F}(x) \leq 2\, c_2\, \bar{F}(x)$$

provided $\bar{F}(x) < \frac{1}{2}$ which will be the case if $x \geq x_o$ and $x_o = (4c)^{1/\alpha} \vee (4d)^{1/(\alpha+\beta)} \vee 1$ is a suitable and convenient choice.

According to (2.8) we have for $x \geq x_o$

$$\left|\frac{x F'(x)}{\bar{F}(x)(-\log F(x))} - \alpha\right| \leq x^{-\beta}[c_1 + 2c_2(cx^{-(\alpha-\beta)} + dx^{-\alpha})] \leq k\, x^{-\beta} =: g(x)$$

where $k = c_1 + 2c_2\,(c+d) = c_1 + 2c_2$

We do not find a_n but instead compute $\alpha_n \leq a_n$ which will be more convenient but still give a valid bound $O(g(\alpha_n))$. Recall a_n is the solution of $-\log F(x) = n^{-1}$.

Let a_n' be the solution of $\bar{F}(x) = n^{-1}$. Since $-\log F \geq \bar{F}$ we have $a_n \geq a_n'$. Also $\bar{F}(x) \geq c\, x^{-\alpha}$ so if we set $\alpha_n = (cn)^{1/\alpha}$ we have $\alpha_n \leq a_n$ and also $\alpha_n \sim a_n' \sim a_n$ as $n \to \infty$.

If we pick $\delta \geq x_o$ we then have for all n such that $\alpha_n > \delta$ (i.e. $n > c^{-1}\delta^{\alpha}$)

$$\sup_x |F^n(a_n x) - \Phi_\alpha(x)|$$

$$\leq (.2701)(\alpha - g(\alpha_n))^{-1} g(\alpha_n) \vee c(\alpha,\beta,\delta)g(\alpha_n) \vee F^n(\delta) \vee \Phi_\alpha(\alpha_n^{-1}\delta)$$

where $g(x) = k\, x^{-\beta}$, $\alpha_n = (cn)^{1/\alpha}$. The order of convergence is $O(n^{-\beta/\alpha})$.

To get a better feel for the method, suppose $\alpha=1/2$, $\beta=1/4$, $c=3/4$, $d=1/4$. Then we find $x_o=9$, $c_1=.0833$, $c_2=.75$. $k=1.5833$. We pick δ to give a reasonable value for θ and hence for $c(\alpha,\beta,\delta)$. If $\theta=1$ then $\delta=1608.9012$ and $c(\alpha,\beta,\delta) = 1.9712$. The condition $a_n \geq \delta$ requires $n \geq 54$ and on this range the dominant term in the bound is $c(\alpha,\beta,\delta)g\,(\alpha_n)$ showing the dependence of the bound on $c(\alpha,\beta,\delta)$. Some values for the bounds are given in the table below to four decimal places.

n	$(.2701)(\alpha-g(\alpha_n))^{-1}g(\alpha_n)$	$c(\alpha,\beta,\delta)g(\alpha_n)$	$F^n(\delta)$	$\Phi_\alpha(\frac{\delta}{\alpha_n})$
54	.2675	.4904	.0393	.0000
75	.1974	.4161	.0112	.0000
100	.1557	.3604	.0024	.0000
150	.1150	.2943	.0000	.0000
300	.0723	.2081	.0000	.0000
500	.0528	.1612	.0000	.0000
1000	.0353	.1140	.0000	.0000
5000	.0148	.0510	.0000	.0000
10000	.0102	.0360	.0000	.0000

3. Uniform Rates of Convergence to $\Lambda(x)$.

Again write $F = \exp\{-e^{-\phi}\}$ and suppose F is twice differentiable. The Von Mises condition guaranteeing $F \in D(\Lambda)$ is

$$h(x) = (1/\phi'(x))' = -\log F(x) - \left\{\frac{-F(x)F''(x)\log F(x)}{(F'(x))^2}\right\} + 1 \to 0$$

as $x \to x_o := \sup\{y: F(y) < 1\}$. There exists a non-increasing function g with

$$|h(x)| \leq g(x) \downarrow 0 \tag{3.1}$$

as $x \to x_o$. Set $f(x) = 1/\phi'(x)$ and define b_n by $F(b_n) = \exp\{-n^{-1}\}$ so $\phi(b_n)=\log n$ and define a_n by $a_n = f(b_n) = n^{-1}F(b_n)/F'(b_n)$. Observe that

$$\phi(a_n x + b_n) - \phi(b_n) = \int_0^x \frac{f(b_n)}{f(a_n \upsilon + b_n)}\, d\upsilon.$$

Since $f' = h \to 0$ we have

$$\frac{f(b_n + f(b_n)\upsilon)}{f(b_n)} \to 1$$

locally uniformly in υ as $n\to\infty$ and thus we see

$$\phi(a_n x + b_n) - \phi(b_n) \to x$$

as $n\to\infty$, showing $F \in D(\Lambda)$. The function g will again yield the rate of convergence.

Proposition 3.1: For a positive real number g define the distribution functions

$$F(g,x) = \begin{cases} 0 & \text{if } x < -g^{-1} \\ \exp\{-(1+gx)^{-g^{-1}}\} & \text{if } x > -g^{-1} \end{cases}$$

$$F(-g,x) = \begin{cases} \exp\{-(1-gx)^{g^{-1}}\} & \text{if } x < g^{-1} \\ 1 & \text{if } x > g^{-1}. \end{cases}$$

Then for $0 < g < 1$

$$\sup_{x \in R} |F(\pm g, x) - \Lambda(x)| \leq e^{-1}g \approx .3679g.$$

Remark: It is possible the constant e^{-1} can be improved using techniques of Hall and Wellner (1979) but efforts to do this have so far been unsuccessful.

Proof: The method follows Ailam (1968) and Hall and Wellner (1979). We consider only the bound on $F(-g, x) - \Lambda(x)$, the other case being similar. We have

$$\sup_{x \in \mathbb{R}} |F(-g,x) - \Lambda(x)|$$

$$= \sup_{x<0} |F-\Lambda| \vee \sup_{0 \leq x \leq g^{-1}} |F-\Lambda| \vee \sup_{x > g^{-1}} |F-\Lambda|$$

$$= A \vee B \vee C.$$

Now

$$C = \sup_{x \geq g^{-1}} |1 - e^{-e^{-x}}| = 1 - \exp\{-e^{-g^{-1}}\} \leq e^{-g^{-1}}$$

and by the mean value theorem

$$B \leq \sup_{0 \leq x \leq g^{-1}} |(1-gx)^{g^{-1}} - e^{-x}| \leq e^{-1}g$$

by a result of Ailam (1968). Note $e^{-g^{-1}} \leq e^{-1}g$ for $0 < g < 1$. For A we have

$$A = \sup_{x<0} |\exp\{-(1-gx)^{g^{-1}}\} - \exp\{-e^{-x}\}|$$

$$= \sup_{y<0} (\exp\{-(1+y)^{g^{-1}}\} - \exp\{-e^{y/g}\}) = \sup_{y>0} q(y).$$

Check that the supremum of $q(y)$ can be found by solving $q'(y) = 0$ for the non-zero root. Since $q'(y) = 0$ gives

$$\exp\{-(1+y)^{g^{-1}}\}(1+y)^{g^{-1}-1} = \exp\{-e^{y/g}\}\, e^{y/g}$$

we have

$$A \leq \sup_{y>0} (\exp\{-e^{y/g}\}\, e^{y/g}(1+y)^{1-g^{-1}} - \exp\{-e^{y/g}\})$$

$$= \sup_{y>0} (e^{y/g} \exp\{-e^{y/g}\}((1+y)^{1-g^{-1}} - e^{-y/g}))$$

$$\leq e^{-1} \sup_{y>0}((1+y)^{1-g^{-1}} - e^{-y/g}) = e^{-1} \sup_{y>0} \bar{q}(y)$$

since $\sup_{y<0} ye^{-y} = e^{-1}$. Again check $\sup \bar{q}(y)$ is achieved at the non-zero root of $\bar{q}'(y) = 0$. The equation $\bar{q}'(y) = 0$ yields

$$e^{-y/g} = (1 - g)(1 + y)^{-g^{-1}}$$

so

$$\sup_{y>0} \bar{q}(y) \leq \sup_{y>0}((1 + y)^{1-g^{-1}} - (1 - g)(1 + y)^{-g^{-1}})$$

$$= \sup_{y>0}((1 + y)^{-g^{-1}}(y + g)) = g$$

since the supremum is achieved at $y = 0$. The proof is complete. □

On the region $[0, \infty)$ we have the following result.

Proposition 3.2: If (3.1) holds and a_n, b_n are as specified after (3.1) then

$$\sup_{x \geq 0} |F^n(a_n x + b_n) - \Lambda(x)| \leq e^{-1} g(b_n).$$

Proof: Recalling that $f = 1/\phi'$ we have for $v > 0$

$$\left|\frac{f(b_n + a_n v) - f(b_n)}{f(b_n)}\right| \leq \int_{b_n}^{b_n + a_n v} \frac{|f'(u)| du}{f(b_n)} \leq \frac{g(b_n) a_n v}{f(b_n)} = g(b_n) v.$$

Therefore for $v > 0$

$$1 - g(b_n) v \leq \frac{f(b_n + a_n v)}{f(b_n)} \leq 1 + g(b_n) v$$

and taking reciprocals we have assuming $g(b_n) v < 1$ that

$$\frac{1}{1 + g(b_n) v} \leq \frac{f(b_n)}{f(b_n + a_n v)} \leq \frac{1}{1 - g(b_n) v} .$$

For x such that $x > 0$ and $g(b_n) x < 1$ we get by integrating

$$-\log(-\log F(g(b_n), x)) \leq \phi(a_n x + b_n) - \phi(b_n) \leq -\log(-\log F(-g(b_n), x)).$$

Taking negative exponentials twice the following is true for $x > 0$:

$$F(g(b_n), x) \leq F^n(a_n x + b_n) \leq F(-g(b_n), x). \tag{3.2}$$

The desired result follows by means of Proposition 3.1. □

We now obtain a bound on the region $(-\infty, 0)$ which will be generally applicable. For $x < 0$, the analogue of (3.2) is

$$F(g(a_n x + b_n), x) \leq F^n(a_n x + b_n) \leq F(-g(a_n x + b_n), x) \tag{3.3}$$

and so

$$|F^n(a_n x + b_n) - \Lambda(x)| \leq e^{-1} g(a_n x + b_n) \tag{3.4}$$

by an appeal to Proposition 3.1. Let $\{x_n\}$ satisfy

$$x_n \downarrow -\infty \text{ and } a_n x_n + b_n \to \infty. \tag{3.5}$$

Combining (3.4) and Proposition 3.2 gives

$$\sup_{x \geq x_n} |F^n(a_n x + b_n) - \Lambda(x)| \leq e^{-1} g(a_n x_n + b_n)$$

and using (3.3) we obtain

$$\sup_x |F^n(a_n x + b_n) - \Lambda(x)| \leq e^{-1} g(a_n x_n + b_n) \vee F(-g(a_n x_n + b_n), x_n) \vee \Lambda(x_n).$$

It is easy to check that

$$F(-g(a_n x_n + b_n), x_n) \geq \Lambda(x_n)$$

so that the uniform bound becomes

$$e^{-1} g(a_n x_n + b_n) \vee F(-g(a_n x_n + b_n), x_n).$$

At this point we see the bound is minimized if we pick $\{x_n\}$ to satisfy

$$e^{-1} g(a_n x_n + b_n) = F(-g(a_n x_n + b_n), x_n). \tag{3.6}$$

It is necessary to check this choice of $\{x_n\}$ satisfies (3.5). Suppose to get a contradiction, that $x_n \not\to \infty$ so that for a subsequence $\{n'\}$ and a number $K, x_{n'} \geq K$. Then $a_{n'} x_{n'} + b_{n'} \to \infty$ and the left side of (3.6) converges to zero as $n' \to \infty$. However the right side is of the order of $\Lambda(x_{n'}) \to 0$ which gives the desired contradiction. Next suppose $a_n x_n + b_n \not\to \infty$ so that for a subsequence $\{n'\}$ and $M < \infty$ we have $a_{n'} x_{n'} + b_{n'} \leq M$. Then $g(a_{n'} x_{n'} + b_{n'}) \geq g(M) > 0$ and $g(a_{n'} x_{n'} + b_{n'}) x_{n'} \to -\infty$. So the left side of (3.6) does not converge to zero but the right side does which again gives a contradiction.

We summarize our findings.

<u>Theorem 3.3</u>: Suppose (3.1) holds so that $F \in D(\Lambda)$ and suppose a_n, b_n are chosen as specified after (3.1). Then with $\{x_n\}$ chosen as in (3.6) we have

$$\sup_x |F^n(a_n x + b_n) - \Lambda(x)| \geq e^{-1} g(a_n x_n + b_n).$$

This bound may be compared with the more attractive bound $e^{-1} g(b_n)$ valid for $x \in [0, \infty)$. When g satisfies conditions of regular variation type we may extend the bound $g(b_n)$ to cover all $x \in R$. If $|f'|$ is regularly varying then $|f'(x)| \sim g(x) = \sup_{y \geq x} f'(y)|$ as $x \to x_0$ and so by Karamata's Theorem

$$\frac{f(x)}{xg(x)} \to \text{constant}$$

as $x \to x_0$. So with the regular variation case in mind we assume there exists $k \in (0, \infty)$ such that for $n \geq n_0$

$$\frac{f(b_n)}{b_n g(b_n)} \leq k \tag{3.7}$$

and for $c < k^{-1}$ and $n \geq n_0$, $\beta > 0$, $\gamma > 0$

$$g(b_n(1 - ck)) \leq \gamma(1 - ck)^{-\beta} g(b_n). \tag{3.8}$$

<u>Theorem 3.4</u>: If (3.7) and (3.8) hold then for $n \geq n_0$

$$\sup_x |F^n(a_n x+b_n)-\Lambda(x)| \leq F^n(a_n(\frac{c}{g(b_n)})+b_n) \vee \Lambda(-\frac{c}{g(b_n)}) \vee (\gamma(1-kc)^{-\beta} \vee 1)e^{-1}g(b_n).$$

Remarks: From (3.4) and (3.8) we see that

$$F^n(a_n(-c/g(b_n)) + b_n) \leq \Lambda(-c/g(b_n)) + \gamma\, e^{-1}g(b_n)(1 - ck)^{-\beta}$$

for $n \geq n_0$. Hence the order of convergence in Theorem 3.4 is $O(g(b_n))$. If g is regularly varying, we have that $g(b_n)$ is a slowly varying function of n (since b_n is slowly varying) and hence the bound converges to zero at a very slow rate.

Proof of Theorem 3.4: From (3.4)

$$\sup_{\frac{-c}{g(b_n)} \leq x \leq 0} |F^n(a_n x + b_n) - \Lambda(x)|$$

$$\leq \sup_{\frac{-c}{g(b_n)} \leq x \leq 0} e^{-1}g(a_n x + b_n)$$

$$\leq e^{-1}g(a_n(-\frac{c}{g(b_n)}) + b_n)$$

$$\leq e^{-1}g(b_n(1 - \frac{cf(b_n)}{b_n g(b_n)}))$$

and using (3.7) and the fact that g is non-increasing we have the above bounded by

$$\leq e^{-1}g(b_n(1 - ck))$$

and from (3.8) this is

$$\leq \gamma\, e^{-1}g(b_n)(1 - ck)^{-\beta}.$$

Combining this with Proposition 3.2 gives the result. □

As in section 2, we often prefer to work with $\bar{F}: = 1 - F$ rather than $-\log F$. Recall $B(x) = \sum_{k=1}^{\infty} \bar{F}^k(x)/(k(k+1))$ and $B(x) \sim \bar{F}(x)/2$.

Proposition 3.5: Set $\rho(x) = \bar{F}(x)/F'(x)$. Then since $f = 1/\phi' = \rho(1-B)$

$$(3.9) \qquad (\frac{1}{\phi'(x)})' = \rho'(x)(1 - B(x)) + \sum_{k=1}^{\infty} \bar{F}^k(x/(k+1)$$

$$= \rho'(x)c_1(x) + \tfrac{1}{2}\bar{F}(x)(1 + c_2(x))$$

where $c_1(x) \to 1$, $c_2(x) \to 0$.

Note

$$\sum_1^{\infty} \bar{F}^k/(k+1) = \tfrac{1}{2}\bar{F} + \bar{F} \sum_{k=2}^{\infty} \bar{F}^{k-1}/(k+1)$$

$$(3.10) \qquad \leq \tfrac{1}{2}\bar{F} + \tfrac{1}{3}\bar{F}(\bar{F}/F) \leq \bar{F}$$

<u>Example</u>: Weibull. Suppose for $x > 0$, $\beta > 0$, $\beta \neq 1$

$$\bar{F}(x) = \exp\{-x^{\beta}\}.$$

Then $F'(x) = \bar{F}(x)\beta x^{\beta-1}$ and

$$\rho(x) = \frac{\bar{F}(x)}{F'(x)} = \frac{\bar{F}(x)}{\bar{F}(x)\beta x^{\beta-1}} = \frac{1}{\beta}\, x^{-(\beta-1)}$$

$$\rho'(x) = -\beta^{-1}(\beta-1)x^{-\beta}.$$

Using (3.9) and (3.10) gives

$$\left|\left(\frac{1}{\phi'(x)}\right)'\right| \leq |\beta-1|\beta^{-1}x^{-\beta} + e^{-x^{\beta}} =: g(x)$$

for x such that $\bar{F}(x) \leq 3/5$; ie $x \geq (\log 5/3)^{1/\beta}$. We have $f(x) = \beta^{-1}x^{-(\beta-1)}(1-B(x))$ and so

$$\frac{f(x)}{xg(x)} \leq \frac{\beta^{-1}x^{-(\beta-1)}}{x|\beta-1|\beta^{-1}x^{-\beta}} = \frac{1}{|\beta-1|} =: k.$$

For $x \geq x_0 \geq 1$ (where x_0 is the solution of $x_0^{\beta} e^{-x_0^{\beta}} = \delta$) we have $\exp\{-x^{\beta}\} \leq \delta\, x^{-\beta}$ and therefore

$$g(x) \leq (|\beta-1|^{\beta-1} + \delta)x^{-\beta}.$$

So for $x \geq (x_0 \vee (\log 5/3)^{1/\beta})/(1-c)$

$$\frac{g(x(1-c))}{g(x)} \leq \left[\frac{\delta + \beta^{-1}|\beta-1|}{\beta^{-1}|\beta-1|}\right](1-c)^{-\beta}$$

and so we get $\gamma = (\delta + \beta^{-1}|\beta-1|)/(\beta^{-1}|\beta-1|)$.

For concreteness, suppose $\beta = 2$. Then $k = 1$, $b_n = (-\log(1-e^{-n^{-1}}))^{1/2}$, $a_n = f(b_n) \geq \frac{1}{2}\, b_n^{-1}(1-\frac{1}{2}e^{-b_n^2}(1+e^{-b_n^2})) =: \alpha_n$. A moderate value of c must be chosen else the very slow decrease of $g(b_n)$ will prevent $\Lambda(-c/g(b_n))$ from being small for reasonable sample sizes. We choose $c = .1$, $x_0 = 1.75$ so that $\delta = .1432$, $\gamma = 1.2864$ and the bound in Theorem 3.4 is valid for $n \geq 44$. Some typical values are given below

n	$e^{-1}\gamma(1-c)^{-2}g(b_n)$	$\Lambda(-c/g(b_n))$	$F^n(\alpha_n(-c/g(b_n))+b_n)$
44	.0901	.1477	.1548
75	.0753	.1139	.1220
100	.0692	.0976	.1060
250	.0552	.0561	.0644
500	.0482	.0346	.0418
1000	.0429	.0201	.0258
10000	.0318	.0019	.0032

□

In the previous Theorem 3.4 we considered the situation where $f'(x) \to 0$ roughly like a negative power of x. We consider now what happens when f' decays to zero roughly like an exponential function. More precisely we suppose $0 \le f'(x) \downarrow 0$ and f' is in de Haan's class Γ (de Haan, 1970) which means f' is integrable in a neighborhood of ∞ and

$$\lim_{x\to\infty} f'(x) \int_x^\infty f(u)du/f^2(x) = 1. \tag{3.11}$$

Equivalently (de Haan, 1970) (3.11) asserts that $f' \in \Gamma$ with auxiliary function f/f' so that locally uniformly in $x \in R$ we have

$$\lim_{t\to\infty} f'(t + x(f(t)/f'(t)))/f'(t) = e^{-x}. \tag{3.12}$$

With this case in mind, we state the final result.

Theorem 3.5: Suppose $F \in D(\Lambda)$ and for $\epsilon > 0$, $c > 0$ and $n \ge n_0$ we have

$$g(b_n - ca_n/g(b_n)) \le e^{c+\epsilon} g(b_n). \tag{3.13}$$

Then for $n \ge n_0$

$$\sup_x |F^n(a_n x + b_n) - \Lambda(x)|$$

$$\le F^n(a_n(-c/g(b_n)) + b_n) \vee \Lambda(-c/g(b_n)) \vee e^{c+\epsilon-1} g(b_n).$$

The proof is virtually identical to the proof of Theorem 3.4. Again the order of convergence is $O(g(b_n))$. If $0 \le f'(x) \downarrow 0$ and (3.11) holds then

$$g(b_n) \sim f^2(b_n)/\int_{b_n}^\infty f(u)du.$$

Note if we change variables $u = b(s) := \phi^{\leftarrow}(\log s)$ then

$$\int_{b(n)}^\infty f(u)du = \int_n^\infty f(b(s))b'(s)ds$$

and since

$$b'(s) = 1/\{\phi'(\phi^{\leftarrow}(\log s))s\} = 1/\{\phi'(b(s))s\}$$

$$= f(b(s))/s =: a(s)/s$$

we have

$$g(b_n) \sim a^2(n)/\int_n^\infty a^2(s)s^{-1}ds.$$ □

According to de Haan (1970), $\int_x^\infty a^2(s)s^{-1}ds = \pi(x)$, being the integral of a -1 varying function is Π-varying with auxiliary function $a^2(\cdot)$. So $g(b_n) \to 0$ like the reciprocal of a Π-varying function divided by its auxiliary function. Both π and $a^2(\cdot)$ are slowly varying so again the convergence rate is rather slow.

References

[1] Ailam, G. On probability properties of random sets and the asymptotic behavior of empirical distribution functions. J. Appl. Prob., 5, 196-202.

[2] Anderson, C. W. (1971). Ph.D. Thesis, London University.

[3] Balkema, A. A. and de Haan, L. (1972). On R. Von Mises' condition for the domain of attraction of $\exp\{-e^{-x}\}$.

[4] Cohen, J. P. (1982a). The penultimate form of approximation to normal extremes. Adv. Appl. Prob., 14, 324-339.

[5] Cohen, J. P. (1982b). Convergence rates for the ultimate and penultimate approximations in extreme value theory. Adv. Appl. Prob., 14, 324-339.

[6] Davis, R. (1982). The rate of convergence in distribution of the maxima. Statistica Neerlandica, 36, 31-35.

[7] Davis, R. and Resnick, S. (1984). Tail estimates motivated by extreme value theory. Ann. Statist., 12, 1467-1487.

[8] Fisher, R. A. and Tippett, L. H. C. (1928). Limiting forms of the frequency distributions of largest or smallest member of a sample. Proc. Camb. Phil. Soc., 24, 180-190.

[9] Gnedenko, B. V. (1943). Sur la distribution limite du terme maximum d'une serie aleatoire. Ann. Math., 44, 423-453.

[10] Goldie, C. (1980). Slow variation with remainder. Pre-print, University of Sussex, Brighton, U.K.

[11] Goldie, C. and Smith, R. (1984). Slow variation with remainder: A survey of the theory and its applications. Pre-print, Colorado State University, Ft. Collins, CO 80523, USA.

[12] Haan, L. de (1970). On Regular Variation and its Application to the Weak Convergence of Sample Extremes. Math. Centre Tract 32, Amsterdam.

[13] Hall, W. J. and Wellner, J. A. (1979). The rate of convergence in law of the maximum of an exponential sample. Statistica Neerlandica, 33, 151-154.

[14] Smith, R. (1982). Uniform rates of convergence in extreme value theory. Adv. Appl. Prob., 14, 543-565.

Original paper received: 25.03.85
Final paper received: 17.03.86

Paper recommended by D.C. Boes

PROBABILITY AND STATISTICS
Essays in Honor of Franklin A. Graybill
J.N. Srivastava (Editor)

BEST LINEAR UNBIASED ESTIMATION IN MIXED MODELS OF THE ANALYSIS OF VARIANCE

S. R. Searle[1/]

Biometrics Unit, Cornell University,
Ithaca, New York 14853
U.S.A.

A broad definition is given of balanced data in mixed models. For all such models it is shown that the BLUE (best linear unbiased estimator) of an estimable function of the fixed effects is the same as the ordinary least squares estimator (OLSE).

Running title: ESTIMATION IN MIXED MODELS

Key words and phrases: Balanced data, BLUE, Fixed effects, OLSE, Ordinary least squares, Random effects

AMS Classification: 62J10.

INTRODUCTION

a. Fixed effects models

Analysis of variance models are traditionally formulated in terms of additive main effects and additive interaction effects. For example, suppose y_{ijk} is the k'th observation on treatment i of variety j in a two-factor experiment concerned with fertilizer treatments and plant varieties. Then a usual analysis of variance model is of the form

$$y_{ijk} = \mu + \alpha_i + \beta_j + \gamma_{ij} + e_{ijk} \tag{1}$$

where μ is a general mean, α_i is the effect on the response variable due to the i'th treatment, β_j is the effect due to the j'th variety, γ_{ij} is the interaction effect between treatment i and variety j, and e_{ijk} is the residual error term defined as $e_{ijk} = y_{ijk} - E(y_{ijk})$ for

$$E(y_{ijk}) = \mu + \alpha_i + \beta_j + \gamma_{ij}$$

where E denotes expectation over repeated sampling.

Models such as (1), where estimation of (and testing of hypotheses about) parameters are the features of interest, are known as fixed effects models, and in such models the customary assumptions about variances and covariances are that each observation has the same variance and that every pair of observations has zero covariance. The dispersion matrix $\underset{\sim}{V}$ of the vector of observations $\underset{\sim}{y}$ then has the form

$$\text{var}(\underset{\sim}{y}) = \underset{\sim}{V} = \sigma^2 \underset{\sim}{I} \quad , \tag{2}$$

$\underset{\sim}{I}$ being an identity matrix and σ^2 being the variance of every observation. An assumption about $\underset{\sim}{V}$ more general than (2) is that it is simply a symmetric, positive semi-definite matrix; and in many cases that it be not just positive semi-definite but positive definite, and hence non-singular.

[1/] This paper is BU-483 in the Biometrics Unit Technical Report Series.

b. Mixed models

Variations of (1) are models where some or all of the α_i, β_j and γ_{ij} terms are assumed not to be parameters to be estimated, but are modeled as being random variables with zero means and some assumed variance-covariance structure. For example, in the no-interaction form of (1), with one observation y_{ij} on treatment i and variety j, namely

$$y_{ij} = \mu + \alpha_i + \beta_j + e_{ij} \quad , \tag{3}$$

suppose that the β_j for $j = 1, \cdots, b$, are modeled as random variables with zero mean $E(\beta_j) = 0 \; \forall \; j$. The β_j are then called random effects and, along with the random error terms e_{ij}, usually have the following variance-covariance structure attributed to them:

$$\begin{aligned} &\text{var}(\beta_j) = \sigma^2_\beta \; \forall \; j \;, && \text{cov}(\beta_j,\beta_{j'}) = 0 \; \forall \; j \neq j' \\ &\text{var}(e_{ij}) = \sigma^2_e \; \forall \; i,j, && \text{cov}(e_{ij},e_{i'j'}) = 0 \text{ except for } i=i' \text{ and } j=j' \end{aligned} \tag{4}$$

and

$$\text{cov}(\beta_j,e_{ij'}) = 0 \; \forall \; i,j,j' \quad .$$

Then with μ and the α_i in (3) being fixed effects and the β_j being random effects, (3) is known as a mixed model. And the variances σ^2_β and σ^2_e of (4) are the variance components. The structure of (4) then leads to $\underset{\sim}{V}$ having elements that are either zero, $\sigma^2_\beta + \sigma^2_e$, or σ^2_β; in general to elements that are either zero, or one of the variance components or a sum of them.

Example 1 Consider (3) and (4), where the β factor represents blocks in a randomized complete blocks experiment. Suppose there are 2 treatments and 3 blocks. Then for a zero element of a matrix being shown as a dot,

$$\underset{\sim}{V} = \text{var}\begin{bmatrix} y_{11} \\ y_{12} \\ y_{13} \\ y_{21} \\ y_{22} \\ y_{23} \end{bmatrix} = \begin{bmatrix} \sigma^2_\beta+\sigma^2_e & \cdot & \cdot & \sigma^2_\beta & \cdot & \cdot \\ \cdot & \sigma^2_\beta+\sigma^2_e & \cdot & \cdot & \sigma^2_\beta & \cdot \\ \cdot & \cdot & \sigma^2_\beta+\sigma^2_e & \cdot & \cdot & \sigma^2_\beta \\ \sigma^2_\beta & \cdot & \cdot & \sigma^2_\beta+\sigma^2_e & \cdot & \cdot \\ \cdot & \sigma^2_\beta & \cdot & \cdot & \sigma^2_\beta+\sigma^2_e & \cdot \\ \cdot & \cdot & \sigma^2_\beta & \cdot & \cdot & \sigma^2_\beta+\sigma^2_e \end{bmatrix} .$$

c. Estimation with balanced data

Section 3 gives a formal description of a wide class of balanced data which, generally speaking, are data that have equal numbers of observations in the subclasses. Model equations (3) and (1) are examples having, for each treatment-variety combination, one observation and (with $k = 1, 2, \cdots, n$) n observations, respectively. In both cases the best linear unbiased estimator (BLUE) of a treatment difference is a well known, simple function of means; i.e., when each of (1) and (3) are fixed effects models, the BLUE of $\alpha_i - \alpha_{i'}$ is

$$\text{BLUE}(\alpha_i - \alpha_{i'}) = \bar{y}_i - \bar{y}_{i'} \tag{5}$$

where $\bar{y}_i$ is the mean of all observations on treatment i. Moreover, the right-hand side of (5) is also the ordinary least squares estimator (OLSE) of $\alpha_i - \alpha_{i'}$. Hence, for these examples

$$\text{BLUE}(\alpha_i - \alpha_{i'}) = \text{OLSE}(\alpha_i - \alpha_{i'}) \quad . \tag{6}$$

For balanced data, it is salient to note that although (5) is true when (1) and (3) are fixed effects models, it is also true when (1) and (3) are mixed models with αs fixed. This generalizes (6) to the extent that for any estimable function of fixed effects in a mixed model with balanced data, BLUE = OLSE. The utility of this result is that although a BLUE is a desirable estimator, its direct derivation generally involves inverting $\underset{\sim}{V}$, which can be tedious; in contrast, with balanced data, the OLSE is often easily derived as a simple function of observed means. Moreover, the equality BLUE = OLSE for balanced data is broad in scope. For example, (5) is true not only for (1) being a fixed effects model or a mixed model with βs, or γs, or βs and γs taken as random effects, but also for (1) extended by the addition of other random effects: for example, in the model $y_{ijk\ell m} = \mu + \alpha_i + \beta_j + \gamma_{ij} + \theta_k + \tau_\ell + \delta_{jk} + e_{ijk\ell m}$ with μ and αs being fixed effects and all other effects being random, (5) and (6) are still true. We proceed to establish the generalization of (6) for any mixed model with balanced data. To do so we first describe a general mixed model and then give a broad definition of balanced data.

A GENERAL MIXED MODEL

a. Description

Recognizing the dichotomy of fixed and random effects in a mixed model, we write the model equation for a vector of observations $\underset{\sim}{y}$ as

$$\underset{\sim}{y} = \underset{\sim}{X}\underset{\sim}{\beta} + \underset{\sim}{Z}\underset{\sim}{u} \tag{7}$$

where $\underset{\sim}{\beta}$ is a vector of fixed effects and $\underset{\sim}{u}$ is a vector of random effects, including error terms. The matrices and vectors of (7) are partitioned as

$$\begin{aligned} \underset{\sim}{X} &= [\underset{\sim}{X}_1 \;\; \underset{\sim}{X}_2 \;\; \cdots \;\; \underset{\sim}{X}_d \;\; \cdots \;\; \underset{\sim}{X}_f] & \text{and} \quad \underset{\sim}{Z} &= [\underset{\sim}{Z}_1 \;\; \underset{\sim}{Z}_2 \;\; \cdots \;\; \underset{\sim}{Z}_q \;\; \cdots \;\; \underset{\sim}{Z}_r] \\ \underset{\sim}{\beta} &= [\underset{\sim}{\beta}_1' \;\; \underset{\sim}{\beta}_2' \;\; \cdots \;\; \underset{\sim}{\beta}_d' \;\; \cdots \;\; \underset{\sim}{\beta}_f']' & \underset{\sim}{u} &= [\underset{\sim}{u}_1' \;\; \underset{\sim}{u}_2' \;\; \cdots \;\; \underset{\sim}{u}_q' \;\; \cdots \;\; \underset{\sim}{u}_r']' \; . \end{aligned} \tag{8}$$

Each $\underset{\sim}{\beta}_d$ for $d = 1, 2, \cdots, f$ has as its elements the h_d effects corresponding to the h_d levels of the d'th fixed effect (main effect or interaction) factor, and $\underset{\sim}{X}_d$ is the incidence matrix corresponding to $\underset{\sim}{\beta}_d$. Similarly, $\underset{\sim}{u}_q$ (of p_q elements) and $\underset{\sim}{Z}_q$ for $q = 1, 2, \cdots, r-1$ are defined for the random effect (main effect or interaction) factors analogously to $\underset{\sim}{\beta}_d$ and $\underset{\sim}{X}_d$ for fixed effect factors. For $q = r$, we define $\underset{\sim}{u}_r = \underset{\sim}{e}$, the vector of error terms, and accordingly $\underset{\sim}{Z}_r = \underset{\sim}{I}_N$ where N is the total number of observations, and $p_r = N$.

Example 2 Using (3) and (4) as the model for a randomized complete blocks experiment for a treatments in b blocks, μ and $[\alpha_1 \cdots \alpha_a]'$ would be $\underset{\sim}{\beta}_1$ and $\underset{\sim}{\beta}_2$ of (8), respectively, and $[\beta_1 \cdots \beta_b]'$ and the e_{ij}-terms of (3) would be $\underset{\sim}{u}_1$ and $\underset{\sim}{u}_2$ of (8), respectively.

The variance and covariance properties of (4) generalized to $\underset{\sim}{u}$ are

$$\begin{aligned} \text{var}(\underset{\sim}{u}_q) &= \sigma_q^2 I_{p_q} & \text{for} \quad q &= 1, 2, \cdots, r \\ \text{and} \quad \text{cov}(\underset{\sim}{u}_q, \underset{\sim}{u}_{q'}') &= \underset{\sim}{0}_{p_q \times p_{q'}} & \text{for} \quad q &\neq q' = 1, 2, \cdots, r \; . \end{aligned} \tag{9}$$

Hence from (7) the variance-covariance matrix of $\underset{\sim}{y}$ is

$$\underset{\sim}{V} = \text{var}(\underset{\sim}{y}) = \text{var}(\underset{\sim}{Z}\underset{\sim}{u}) = \sum_{q=1}^{r} \sigma_q^2 \underset{\sim}{Z}_q \underset{\sim}{Z}_q' \quad . \tag{10}$$

Thus (7) through (10) constitute description of a general mixed model.

b. Estimation

The OLSE estimator of an estimable function $\lambda'X\beta$ of the parameters in β in the model (7) will be denoted by OLSE($\lambda'X\beta$) and is, as is well-known,

$$\text{OLSE}(\lambda'X\beta) = \lambda'X(X'X)^{-}X'y \tag{11}$$

where $(X'X)^{-}$ is any matrix satisfying $X'X(X'X)^{-}X'X = X'X$. Similarly the BLUE of that same estimable $\lambda'X\beta$ is

$$\text{BLUE}(\lambda'X\beta) = \lambda'X(X'V^{-1}X)^{-}X'V^{-1}y \ , \tag{12}$$

where V is assumed to be positive definite.

In fixed effects models, $V = \sigma^2 I$ as in (2), whereupon (12) very simply reduces to (11), as is well known. An extension to $V = [(1-\rho)I + \rho J]\sigma^2$ is given by McElroy (1967) and, in complete generality, Zyskind (1967) has shown that these two estimators are equal, if and only if

$$VX = XQ \qquad \text{for some } Q \ . \tag{13}$$

Graybill (1976, p. 209) also has this result, restricted to X of full column rank. We use (13) to show for a broad definition of balanced data that for mixed models of the form (7) through (10) the BLUE of an estimable function of the fixed effects parameters is the same as the OLSE.

BALANCED DATA

We deal with data categorized by a number of factors, each of which is either a main effects factor (including the possibility of nested main effects factors), or an interaction factor representing the interaction of two or more main effects factors. Suppose there are m main effects factors, with the t'th one having N_t levels, for $t = 1, 2, \cdots, m$. Then the k'th observation in the "cell" defined by the i_t'th level (for $i_t = 1, \cdots, N_t$) of the t'th main effect for $t = 1, \cdots, m$, where there are $n_{i_1 i_2 \cdots i_t \cdots i_m}$ such observations, is $y_{i_1 i_2 \cdots i_t \cdots i_m k}$ for $k = 1, 2, \cdots, n_{i_1 i_2 \cdots i_t \cdots i_m}$. On defining $i = [i_1\ i_2\ \cdots\ i_m]$, a typical observation can then be denoted as y_{ik} for $k = 1, 2, \cdots, n_i$. Furthermore, the total number of observations is

$$N = p_r = \sum_{i=1'_m}^{i=N'} n_i \qquad \text{for } N' = [N_1\ \ N_2\ \cdots\ N_t\ \cdots\ N_m] \ .$$

($1'_m$ is a row vector of m unities.)

A tight, rigorous, formal and complete definition of balanced data is elusive. Development of such a definition would, as Cornfield and Tukey (1956) write, involve "··· systematic algebra [which] can take us deep into the forest of notation. But the detailed manipulation will, sooner or later, blot out any understanding we may have started with." Nevertheless, one formulation of a model that yields a wide class of balanced data situations is as follows. It is similar to that used by Smith and Hocking (1978), Searle and Henderson (1979), Seifert (1979), Khuri (1981) and Anderson *et al.* (1984).

The balanced data models we consider are those that have $n_i = n\ \forall\ i$. They also have each X_d and each Z_q of (8) being a Kronecker product (KP, for brevity) of m + 1 matrices, each of which is either an identity matrix, I, or a 1-vector; i.e.,

$$\text{each } X_d \text{ and each } Z_q \text{ is a KP of m+1 matrices that are each } I \text{ or } 1 \ . \tag{14}$$

The occurrence of the I-matrices and 1-vectors in these KPs is as follows.

First, corresponding to the scalar parameter μ in the model is X_1 which is 1_N, and so every matrix in its KP is a 1:

$$X_1 = 1_N = 1_{N_1} * 1_{N_2} * \cdots * 1_{N_t} * \cdots * 1_{N_m} * 1_n = \left(\overset{t=m}{\underset{t=1}{*}} 1_{N_t}\right) * I_n ,$$

where * represents the operation of Kronecker multiplication. Second, corresponding to $u_r = e$ is Z_N, and so each of the m + 1 matrices in the KP that is $Z_r = I_N$ is an identity matrix:

$$Z_r = I_N = I_{N_1} * I_{N_2} * \cdots * I_{N_t} * \cdots * I_{N_m} * I_n = \left(\overset{t=m}{\underset{t=1}{*}} I_{N_t}\right) * I_n .$$

Third, in the KP for each X_d and Z_q (other than X_1 and Z_r), the t'th matrix corresponds to the t'th main effects factor and is I_{N_t} when that factor is part of the definition of the factor corresponding to X_d or Z_q; otherwise it is 1_{N_t}. This is for $t = 1, \cdots, m$; and for all X_d and Z_q, other than Z_r, the (m+1)'th matrix in the KP is 1_n.

The phrase "part of the definition" demands explanation. It is exemplified in the 2-factor model (1), wherein the two main effects factors are each part of the definition of the interaction factor. Similarly, if nested within an α-factor there is a β-factor, then the α-factor is part of the definition of that β-factor. (See also, comments B and C which follow the examples.)

Each h_d and p_q (number of levels in the d'th fixed factor and the q'th random factor, respectively) in the balanced data we have defined is the product of the numbers of columns in the I and 1 terms in the KP (14) that is X_d and Z_q. Hence h_d is the product of the N_t values for the main effects factors that are part of the definition of the d'th fixed effect factor; p_q is a similar product for the q'th random effects factor.

Examples We give four examples that are each in terms of those of the following vectors that are appropriate: $\alpha = [\alpha_1, \cdots, \alpha_a]'$, $\beta = [\beta_1, \cdots, \beta_b]'$ or $\beta_+ = [\beta_{11} \cdots \beta_{1b}\ \beta_{21} \cdots \beta_{2b} \cdots \beta_{a1} \cdots \beta_{ab}]'$, $\gamma = [\gamma_{11} \cdots \gamma_{1b}\ \gamma_{21} \cdots \gamma_{2b} \cdots \gamma_{a1} \cdots \gamma_{ab}]'$, and e, the vector of error terms, the same order as y. Determination of which KPs are X-matrices and which are Z-matrices is governed by which factors are defined as fixed effects and which are random. This is illustrated for only example (iii).

(i) One-way classification: $y_{ij} = \mu + \alpha_i + e_{ij}$ with $i=1,\cdots,a$ and $j=1,\cdots,n$.

$$y = (1_a * 1_n)\mu + (I_a * 1_n)\alpha + (I_a * I_n)e . \tag{15}$$

(ii) Two-way crossed classification, no interaction, and one observation per cell: $y_{ij} = \mu + \alpha_i + \beta_j + e_{ij}$ for $i=1,\cdots,a$ and $j=1,\cdots,b$.

$$y = (1_a * 1_b)\mu + (I_a * 1_b)\alpha + (1_a * I_b)\beta + (I_a * I_b)e . \tag{16}$$

(iii) Two-way crossed classification, with interaction and n observations per cell: $y_{ijk} = \mu + \alpha_i + \beta_j + \gamma_{ij} + e_{ijk}$ with $i=1,\cdots,a$, $j=1,\cdots,b$ and and $k=1,\cdots,n$

$$\begin{aligned} y = {} & (1_a * 1_b * 1_n)\mu + (I_a * 1_b * 1_n)\alpha + (1_a * I_b * 1_n)\beta \\ & + (I_a * I_b * 1_n)\gamma + (I_a * I_b * I_n)e . \end{aligned} \tag{17}$$

Suppose in (17) that elements of β and γ were taken to be random effects. Then the terms of (8) for the general mixed model would have the following values:

$m=3,\quad f=2$ with $h_1 = N_1 \quad = 1$ and $\underset{\sim}{X}_1 = \underset{\sim}{1}_a * \underset{\sim}{1}_b * \underset{\sim}{1}_n$ for $\beta_1 = \mu$,

and $h_2 = N_2 \quad = a$ and $\underset{\sim}{X}_2 = \underset{\sim}{I}_a * \underset{\sim}{1}_b * \underset{\sim}{1}_n$ for $\underset{\sim}{\beta}_2 = \underset{\sim}{\alpha}$;

$r=3$ with $p_1 = N_3 \quad = b$ and $\underset{\sim}{Z}_1 = \underset{\sim}{1}_a * \underset{\sim}{I}_b * \underset{\sim}{1}_n$ for $\underset{\sim}{u}_1 = \underset{\sim}{\beta}$,

$p_2 = N_2N_3 \quad = ab$ and $\underset{\sim}{Z}_2 = \underset{\sim}{I}_a * \underset{\sim}{I}_b * \underset{\sim}{1}_n$ for $\underset{\sim}{u}_2 = \underset{\sim}{\gamma}$,

and $p_3 = N_2N_3n = abn$ and $\underset{\sim}{Z}_3 = \underset{\sim}{I}_a * \underset{\sim}{I}_b * \underset{\sim}{I}_n$ for $\underset{\sim}{u}_3 = \underset{\sim}{e}$.

(iv) Two-way nested classification: $y_{ij} = \mu + \alpha_i + \beta_{ij} + e_{ijk}$ for $i=1,\cdots,a$, $j=1,\cdots,b$ and $k=1,\cdots,n$.

$$\underset{\sim}{y} = (\underset{\sim}{1}_a * \underset{\sim}{1}_b * \underset{\sim}{1}_n)\mu + (\underset{\sim}{I}_a * \underset{\sim}{1}_b * \underset{\sim}{1}_n)\underset{\sim}{\alpha} + (\underset{\sim}{I}_a * \underset{\sim}{I}_b * \underset{\sim}{1}_n)\underset{\sim}{\beta}_+ + (\underset{\sim}{I}_a * \underset{\sim}{I}_b * \underset{\sim}{I}_n)\underset{\sim}{e} \quad . \tag{18}$$

Comments on the examples. Several comments are in order. (A) In every case $\underset{\sim}{X}_1$ for μ is $\underset{\sim}{1}$, a KP of $\underset{\sim}{1}$-vectors; and $\underset{\sim}{Z}_r$ for $\underset{\sim}{e}$ is $\underset{\sim}{I}$, a KP of identity matrices. (B) In every case the KP that is the coefficient of $\underset{\sim}{\alpha}$ has only one identity matrix in it, namely $\underset{\sim}{I}_a$. This is so because, obviously, the definition of $\underset{\sim}{\alpha}$ involves only $\underset{\sim}{\alpha}$. The same is true of the coefficient of $\underset{\sim}{\beta}$ in (16) and (17). (C) In contrast, the KP that is the coefficient of $\underset{\sim}{\beta}_+$ in (18) has two identity matrices, $\underset{\sim}{I}_a$ and $\underset{\sim}{I}_b$. This is because $\underset{\sim}{\beta}_+$ has elements that represent the nesting of the β-factor within the α-factor. Thus the α-factor is involved in the definition of $\underset{\sim}{\beta}_+$ and so the coefficient of $\underset{\sim}{\beta}_+$ contains $\underset{\sim}{I}_a$ and $\underset{\sim}{I}_b$. Hence the coefficient of $\underset{\sim}{\beta}_+$ in (18) is the same as that of $\underset{\sim}{\gamma}$, the interaction term, in (17). Judged solely by their coefficients, $\underset{\sim}{\beta}_+$ and $\underset{\sim}{\gamma}$ might therefore appear to be the same. What makes $\underset{\sim}{\gamma}$ an interaction term is that both main effect factors that go into defining it are also present on their own in (17), whereas with $\underset{\sim}{\beta}_+$ only one factor that goes into defining it is present on its own in (18), and so $\underset{\sim}{\beta}_+$ represents nesting. In other words, a factor that looks like an interaction factor is such when all of its associated main effects factors are present in the model; otherwise it is a nested factor. (D) Equation (16) is a special case of (17) with $\underset{\sim}{\gamma}$ omitted and $n=1$ and hence, for example, $\underset{\sim}{1}_a * \underset{\sim}{1}_b * \underset{\sim}{1}_n = \underset{\sim}{1}_a * \underset{\sim}{1}_b * 1 = \underset{\sim}{1}_a * \underset{\sim}{1}_b$.

A final observation that concerns $\underset{\sim}{V} = \sum_{q=1}^{r} \sigma^2_q \underset{\sim}{Z}_q \underset{\sim}{Z}'_q$ of (10) is based on the general result that $(\underset{\sim}{A} * \underset{\sim}{B})(\underset{\sim}{P} * \underset{\sim}{Q}) = \underset{\sim}{AP} * \underset{\sim}{BQ}$, given the necessary conformability requirements. Then, for $\underset{\sim}{1}_n \underset{\sim}{1}'_n = \underset{\sim}{J}_n$ being a square matrix of order n with every element unity, we have from (14) that every $\underset{\sim}{Z}_q \underset{\sim}{Z}'_q$ is a KP of $\underset{\sim}{I}$ and $\underset{\sim}{J}$ matrices. Hence we rewrite (10) as

$$\underset{\sim}{V} = \sum_{q=1}^{r} \sigma^2_q \text{ (the KP of } \underset{\sim}{I} \text{ and } \underset{\sim}{J} \text{ matrices that is } \underset{\sim}{Z}_q \underset{\sim}{Z}'_q) \quad . \tag{19}$$

ESTIMATION FROM BALANCED DATA

We now show for mixed models as specified in (7) – (10), with balanced data as defined in the preceding section, that the BLUE of (12) equals the OLSE of (11). We do this by showing that (13) is satisfied for $\underset{\sim}{V}$ of (19) and $\underset{\sim}{X} = \{\underset{\sim}{X}_d\}$, $d=1,\cdots,f$ of (8), with $\underset{\sim}{X}_d$ being a KP of $\underset{\sim}{I}$-matrices and $\underset{\sim}{1}$-vectors, as in (14).

Writing $\underset{\sim}{W}_q$ for $\underset{\sim}{Z}_q \underset{\sim}{Z}'_q$ of (19) we have

$$\underset{\sim}{W}_q = \underset{\sim}{Z}_q \underset{\sim}{Z}'_q = (\underset{\sim}{W}_{q1} * \underset{\sim}{W}_{q2} * \cdots * \underset{\sim}{W}_{qt} * \cdots * \underset{\sim}{W}_{q,m+1}) = \mathop{*}_{t=1}^{m+1} \underset{\sim}{W}_{qt} \quad , \tag{20}$$

where, from (19) each $\mathbf{W}_{qt}$ is either an $\mathbf{I}$ or a $\mathbf{J}$ matrix. Similarly, applying (14) to (8) gives

$$\mathbf{X} = [\mathbf{X}_1 \ \ \mathbf{X}_2 \ \cdots \ \mathbf{X}_d \ \cdots \ \mathbf{X}_f] \quad \text{with} \quad \mathbf{X}_d = \mathop{*}_{t=1}^{m+1} \mathbf{X}_{dt} \ , \tag{21}$$

where each $\mathbf{X}_{dt}$ is either $\mathbf{I}_{N_t}$ or $\mathbf{1}_{N_t}$. Then from (19) and (21)

$$\mathbf{VX} = \left\{ \sum_{q=1}^{r} \sigma^2_q \mathbf{Z}_q \mathbf{Z}'_q \mathbf{X}_d \right\}_{d=1}^{d=f}$$

where, by the curly braces notation we mean that $\mathbf{VX}$ is partitioned into a row of f sub-matrices. Thus

$$\mathbf{VX} = \left\{ \sum_{q=1}^{r} \sigma^2_q \mathbf{W}_q \mathbf{X}_d \right\}_{d=1}^{d=f} \tag{22}$$

$$= \left\{ \sum_{q=1}^{r} \sigma^2_q \mathop{*}_{t=1}^{m+1} \mathbf{W}_{qt} \mathbf{X}_{dt} \right\}_{d=1}^{d=f} . \tag{23}$$

Now in (20), $\mathbf{W}_{qt}$ is either $\mathbf{I}$ or $\mathbf{J}$, and in (21) each $\mathbf{X}_{dt}$ is either $\mathbf{I}$ or $\mathbf{1}$, all of order N_t. Therefore the four possible values of the product $\mathbf{W}_{qt}\mathbf{X}_{dt}$, together with the definition of a matrix $\mathbf{M}_{qdt}$ such that $\mathbf{W}_{qt}\mathbf{X}_{dt} = \mathbf{X}_{dt}\mathbf{M}_{qdt}$ in each case, are as follows:

$\mathbf{W}_{qt}$	$\mathbf{X}_{dt}$	$\mathbf{W}_{qt}\mathbf{X}_{dt} = \mathbf{X}_{dt}\mathbf{M}_{qdt}$	$\mathbf{M}_{qdt}$
$\mathbf{I}$	$\mathbf{I}$	$\mathbf{I} = \mathbf{II}$	$\mathbf{I}$
$\mathbf{I}$	$\mathbf{1}$	$\mathbf{1} = \mathbf{1}1$	1
$\mathbf{J}$	$\mathbf{I}$	$\mathbf{J} = \mathbf{IJ}$	$\mathbf{J}$
$\mathbf{J}$	$\mathbf{1}$	$N_t\mathbf{1} = \mathbf{1}N_t$	N_t

Therefore from (23)

$$\mathbf{VX} = \left\{ \sum_{q=1}^{r} \sigma^2_q \mathop{*}_{t=1}^{m+1} \mathbf{X}_{dt} \mathbf{M}_{qdt} \right\}_{d=1}^{d=f} \tag{24}$$

$$= \left\{ \sum_{q=1}^{r} \sigma^2_q \mathbf{X}_d \mathbf{M}_{qd} \right\}_{d=1}^{d=f} \tag{25}$$

for

$$\mathbf{M}_{qd} = \mathbf{M}_{qd1} * \mathbf{M}_{qd2} * \cdots * \mathbf{M}_{qdt} * \cdots * \mathbf{M}_{q,d,m+1} \ . \tag{26}$$

Derivation both of (23) from (22) and of (25) from (24) is based both on $\mathbf{X}_d$ and $\mathbf{M}_q$ each being a KP, and on the product rule for a KP quoted earlier.

The conformability requirements of the regular products in (24) might seem to be lacking because, from the preceding table, two forms of $\mathbf{M}_{qdt}$ are scalars. However, both regular and Kronecker products of matrices do exist when one or more of the matrices is a scalar; e.g., for scalar θ, both $\mathbf{A}\theta$ and $(\mathbf{A} * \mathbf{B})(\theta * \mathbf{L}) = \mathbf{A}\theta * \mathbf{BL}$ exist. Therefore (25) exists. Hence, on writing

$$\mathbf{Q} = \text{diag}\left\{ \sum_{q=1}^{r} \sigma^2_q \mathbf{M}_{qd} \right\}_{d=1}^{d=f} ,$$

the block diagonal matrix of matrices $\sum_{q=1}^{r} \sigma^2_q \mathbf{M}_{qd}$, for d=1,2,$\cdots$,f, we get from (25)

$$\mathbf{VX} = [\mathbf{X}_1 \cdots \mathbf{X}_d \cdots \mathbf{X}_f]\begin{bmatrix} \sum_{q=1}^{r} \sigma_q^2 \mathbf{M}_{q1} & & & & \mathbf{0} \\ & \ddots & & & \\ & & \sum_{q=1}^{r} \sigma_q^2 \mathbf{M}_{qd} & & \\ & & & \ddots & \\ \mathbf{0} & & & & \sum_{q=1}^{r} \sigma_q^2 \mathbf{M}_{qf} \end{bmatrix} \tag{27}$$

$$= \mathbf{XQ} \quad .$$

Thus Zyskind's condition of (13) is satisfied. Hence, with balanced data as here defined, the BLUE of an estimable function of the fixed effects in any mixed model is the same as the OLSE.

A final note: each sum $\sum_{q=1}^{r} \sigma_q^2 \mathbf{M}_{qd}$ in (27) does exist because, as a result of (26), the order of $\mathbf{M}_{qd}$ is the product of the orders of $\mathbf{M}_{qdt}$ for $t = 1, \cdots, m+1$; and (from the Table) each $\mathbf{M}_{qdt}$ is square of order either N_t or 1. Furthermore, that order is N_t only when $\mathbf{X}_{dt} = \mathbf{I}$; and this is so only when the t'th main effects factor is involved in defining the d'th fixed effects factor. Hence the order of $\mathbf{M}_{qd}$ is the product of such N_t values, and this is h_d; thus $\mathbf{M}_{qd}$ has order h_d for all q and so $\sum_{q=1}^{r} \sigma_q^2 \mathbf{M}_{qd}$ exists.

Example Suppose in (1) and (17) that the βs and γs are random effects. Then

$$\mathbf{X} = [\mathbf{1}_a * \mathbf{1}_b * \mathbf{1}_n \quad \mathbf{I}_a * \mathbf{1}_b * \mathbf{1}_n]$$

and

$$\mathbf{V} = \sigma_\beta^2(\mathbf{J}_a * \mathbf{I}_b * \mathbf{J}_n) + \sigma_\gamma^2(\mathbf{I}_a * \mathbf{I}_b * \mathbf{J}_n) + \sigma_e^2(\mathbf{I}_a * \mathbf{I}_b * \mathbf{I}_n) \quad .$$

Hence in $\mathbf{VX}$ the first sub-matrix is

$$\begin{aligned} \mathbf{V}(\mathbf{1}_a * \mathbf{1}_b * \mathbf{1}_n) &= \sigma_\beta^2(a\mathbf{1}_a * \mathbf{1}_b * n\mathbf{1}_n) + \sigma_\gamma^2(\mathbf{1}_a * \mathbf{1}_b * n\mathbf{1}_n) + \sigma_e^2(\mathbf{1}_a * \mathbf{1}_b * \mathbf{1}_n) \\ &= (\mathbf{1}_a * \mathbf{1}_b * \mathbf{1}_n)[\sigma_\beta^2(a * 1 * n) + \sigma_\gamma^2(1 * 1 * n) + \sigma_e^2(1 * 1 * 1)] \quad . \end{aligned} \tag{28}$$

Similarly, the second sub-matrix of $\mathbf{VX}$ is

$$\begin{aligned} \mathbf{V}(\mathbf{I}_a * \mathbf{1}_b * \mathbf{1}_n) &= \sigma_\beta^2(\mathbf{J}_a * \mathbf{1}_b * n\mathbf{1}_n) + \sigma_\gamma^2(\mathbf{I}_a * \mathbf{1}_b * n\mathbf{1}_n) + \sigma_e^2(\mathbf{I}_a * \mathbf{1}_b * \mathbf{1}_n) \\ &= (\mathbf{I}_a * \mathbf{1}_b * \mathbf{1}_n)[\sigma_\beta^2(\mathbf{J}_a * 1 * n) + \sigma_\gamma^2(\mathbf{I}_a * 1 * n) + \sigma_e^2(\mathbf{I}_a * 1 * 1)] \quad . \end{aligned} \tag{29}$$

Hence

$$\mathbf{VX} = [\mathbf{1}_a * \mathbf{1}_b * \mathbf{1}_n \quad \mathbf{I}_a * \mathbf{1}_b * \mathbf{1}_n]\begin{bmatrix} \mathbf{M}_1 & \mathbf{0} \\ \mathbf{0} & \mathbf{M}_2 \end{bmatrix} = \mathbf{X}\begin{bmatrix} \mathbf{M}_1 & \mathbf{0} \\ \mathbf{0} & \mathbf{M}_2 \end{bmatrix}$$

for $\mathbf{M}_1$ and $\mathbf{M}_2$ being the matrices in square braces in (28) and (29), respectively, namely

$$\mathbf{M}_1 = an\sigma_\beta^2 + n\sigma_\gamma^2 + \sigma_e^2 \quad \text{and} \quad \mathbf{M}_2 = n\sigma_\beta^2\mathbf{J}_a + n\sigma_\gamma^2\mathbf{I}_a + \sigma_e^2\mathbf{I}_a \quad .$$

REFERENCES

[1] Anderson, R. D., Henderson, H. V., Pukelsheim, F. and Searle, S. R. Best estimation of variance components from balanced data with arbitrary kurtosis. Mathematische Operationsforschung und Statistik. Series Statistik 15 (1984) 163-176.

[2] Cornfield, J. and Tukey, J. W. Average values of mean squares in factorials. Ann. Math. Statist. 27 (1956) 907-949.

[3] Graybill, F. A. Theory and Application of the Linear Model, (Duxbury, North Scituate, Massachusetts, 1976).

[4] Khuri, A. I. Anova sums of squares for balanced mixed-effects models revisited. Technical Report No. 162, University of Florida (1981).

[5] McElroy, F. W. A necessary and sufficient condition that ordinary least squares estimators be best linear unbiased. J. American Statistical Association 62 (1967) 1302-1304.

[6] Searle, S. R. and Henderson, H. V. Dispersion matrices in variance components models. J. American Statistical Association 74 (1979) 465-470.

[7] Seifert, B. Optimal testing for fixed effects in general balanced mixed classification models. Mathematische Operationsforschung und Statistik. Series Statistics 10 (1979) 237-255.

[8] Smith, D. W. and Hocking, R. R. Maximum likelihood analysis of the mixed model: the balanced case. Communications in Statistics – Theory and Methods A7 (1978) 1253-1266.

[9] Zyskind, G. On canonical forms, non-negative covariance matrices and best and simple least squares linear estimators in linear models. Ann. Math. Statistics 38 (1967) 1092-1109.

Original paper received: 18.01.85
Final paper received: 01.07.86

Paper recommended by D.C. Bowden

PROBABILITY AND STATISTICS
Essays in Honor of Franklin A. Graybill
J.N. Srivastava (Editor)
Elsevier Science Publishers B.V. (North-Holland), 1988

HIGH RESOLUTION FREQUENCY ANALYSIS OF ALMOST PERIODIC PROCESSES

M. M. Siddiqui and Chien Chun Wang
Department of Statistics, Colorado State University
Fort Collins, Colorado 80523 U.S.A.

Abstract

We consider the problem of estimating the hidden periodicities in a stochastic process $x(t)$, $-\infty < t < \infty$, such that its mean function $\mu(t) = Ex(t)$ is an almost periodic function

$$\Sigma(\alpha_j \cos 2\Pi f_j t + \beta_j \sin 2\Pi f_j t)$$

and the error process $e(t) = x(t) - \mu(t)$ is second order stationary with an absolutely continuous spectral distribution function. The number of frequencies K, the frequencies f_1, f_2, ..., f_K, and the coefficients α_j, β_j, $j = 1, 2, \ldots, K$, are all assumed unknown. The method we propose has two desirable properties of high resolution and consistency. These two properties are illustrated by an analysis of several time series such as the time series of the earth's orbital parameters, Gaussian white noise, magnitude of a variable star and annual lynx trappings.

1. Introduction

Almost periodic processes are all around us. The routine of our lives is governed by the two periodic motions of the earth, around its axis and around the sun. On the time scale of human lives these two motions appear exactly periodic. However, there are other long-term periodicities of the earth's motion which modulate or perturb these two basic periodicities. These perturbations are not immediately perceptible to our senses but their existence can be deduced from the influence of the gravitational forces of the sun, the moon, and of the other planets upon the earth. Among other examples of almost periodic processes are the propagation of heat, light and sound waves, and the atmospheric and oceanic tides. In a mathematical treatment of such physical phenomena the almost periodic function representations are obtained as bounded and continuous solutions of linear differential equations with constant coefficients, or of the partial differential "wave" equations.

What we are concerned with in this article is, however, not a deductive theory but an inferential approach: Given a sample of a random signal $x(t)$, identify any line frequencies present in the data as precisely as possible. The sample may be a continuous record of $x(t)$ over some time interval, $a \leq t \leq a + 2T$, $-\infty < a < \infty$, $T > 0$, or a sequence of systematic observations $x_t = x(a + th)$, $t = 1, 2, \ldots, n$, where $-\infty < a < \infty$, $h > 0$ and $n \geq 1$.

Our emphasis will be on high resolution of line frequencies rather than on estimating the total spectral power within a frequency band. The multi-step procedure of identifying, isolating, and estimating the line frequencies, which will be presented in this paper, provides detailed information on the structure of the signal x that would be missed by a method of spectral density estimation which uses "smoothing" in time or frequency domain.

The organization of the paper is as follows. Almost periodic functions and their properties will be reviewed in Sections 2 through 4, almost periodic random processes will be introduced in Section 5, and their spectral form will be given therein. Our procedure of high resolution frequency analysis will be described in Section 6. In Sections 7 through 10 we will present the results of analysis of several data sets to illustrate the two properties of high resolution and consistency of our procedure.

2. Almost Periodic Functions

Let $x(t)$, $-\infty < t < \infty$, be a complex valued function of a real variable t, to be called "time". We will refer to x as signal, to $|x(t)|^2$ its instantaneous power and to $|x(t)|^2dt$ its differential energy. If the total energy

$$\int_{-\infty}^{\infty}|x(t)|^2dt$$

is finite, then the signal fades out as $|t| \to \infty$. Such signals will be called "transient". On the other hand, if the total energy is infinite but the average power, $\|x\|^2$, defined by

$$\|x\|^2 = \lim_{T\to\infty} \frac{1}{2T} \int_{-T}^{T}|x(t)|^2dt \tag{2.1}$$

is finite and positive then x will be called "persistent". Such a signal does not fade out. For each f, the function

$$\Psi(t;f) = e^{2\pi itf}, \quad i = \sqrt{-1}$$

is a persistent signal. Note that $\Psi(t;0) \equiv 1$, and for $f\neq 0$, $\Psi(t;f)$ is a "pure" periodic function with a single frequency f and period f^{-1}.

If x and y are two persistent signals, define the <u>scalar</u> <u>product</u> of x and y as

$$(x,y) = \lim_{T\to\infty} \frac{1}{2T} \int_{-T}^{T}x(t)\,\bar{y}(t)dt \tag{2.2}$$

where, if z is a complex number, $\bar{z}$ is the complex conjugate of z. We note that

$$(\Psi(\cdot;f_1),\ \Psi(\cdot;f_2)) = \begin{cases}0, & \text{if } f_1 \neq f_2,\\ 1, & \text{if } f_1 = f_2,\end{cases} \tag{2.3}$$

so that the uncountably infinte set of functions

$$P = \{\Psi(t;f) = e^{2\pi ift},\ -\infty < f < \infty\}$$

is an orthonormal set of functions with respect to the scalar product (2.2).

Now consider the linear manifold s(P) spanned by P. That is to say, the class of all trigonometric polynomials

$$p(t) = \sum_{k=1}^{n} a_k e^{2\pi if_k t}, \tag{2.4}$$

where n = 1, 2, ..., a_k are non-zero complex numbers, and f_k are real numbers such that $f_i \neq f_j$ if $i \neq j$. These conditions are imposed so that zero terms are omitted and that a_k are identifiable. We note that p(t) is continuous and is bounded in absolute value by $\sum_{k=1}^{n}|a_k|$, and that $\|p\|^2 = \sum_{k=1}^{n}|a_k|^2$ is a finite positive number. Therefore, every element in s(P) is a uniformly continuous and bounded persistent signal. If $n \geq 2$ and some two frequencies, say f_1 and f_2, are relatively irrational, then p(t) is not periodic. For example, if n = 2, $f_1 = 1$,

$f_2 = \sqrt{2}$, $a_1 = a_2 = 1$, so that $p(t) = e^{2\pi it} + e^{2\pi i\sqrt{2}t}$, then $p(0) = 2$ and there does not exist any $t \neq 0$ for which $p(t) = 2$. However, given $\epsilon > 0$, there exist (infinitely) many numbers t_ϵ such that $|p(t+t_\epsilon) - p(t)| < \epsilon$, for $-\infty < t < \infty$, so that each t_ϵ is an "ϵ-almost period" of $p(t)$. (Clearly, $t_\epsilon \to \infty$ as $\epsilon \to 0$.) We now close $s(P)$ in the uniform topology of continuous functions to obtain $\overline{s}(P)$.

That is to say, if $x \in \overline{s}(P)$ and $\epsilon > 0$, there exist a non-negative integer $n = n(\epsilon)$, frequencies $f_1, \ldots, f_n$, and coefficients $a_1, \ldots, a_n$ such that

$$|x(t) - \sum_{k=1}^{n} a_k e^{2\pi i f_k t}| < \epsilon, \text{ for } -\infty < t < \infty.$$

The class of functions $\overline{s}(P)$ will be referred to as the class of almost periodic (AP) functions. Obviously, AP functions are bounded and continuous on the real line and are closed under addition, multiplication, scalar multiplication, complex conjugation and uniform limits.

A very important property of AP functions is the following characterization theorem.

Theorem 2.1. A complex valued function $x(t)$, $-\infty < t < \infty$, is AP if and only if for each $\epsilon > 0$ there exists a number $\ell_\epsilon > 0$ such that any interval of length ℓ_ϵ of the real line contains at least one point with abscissa η such that $|x(t+\eta) - x(t)| < \epsilon$, $-\infty < t < \infty$.

The number ℓ_ϵ, which depends on ϵ, is called an ϵ-translation number of the AP function x.

3. Fourier Series of AP Functions

Let x be AP. It was noted earlier that for each $\epsilon > 0$ there exist frequencies $f_1, \ldots, f_n$ and coefficients $a_1, \ldots, a_n$ such that

$$|x(t) - \sum_{k=1}^{n} a_k e^{2\pi i f_k t}| < \epsilon, \ -\infty < t < \infty.$$

The problem of discovering these frequencies and their coefficients will now be solved.

For each f, $-\infty < f < \infty$, define

$$a(x;f) = (x, \Psi(\cdot;f)) = \lim_{T\to\infty} \frac{1}{2T} \int_{-T}^{T} x(t)e^{-2\pi ift}dt. \tag{3.1}$$

Let $n \geq 1$, and $f_1, \ldots, f_n$ be real numbers such that $f_i \neq f_j$ if $i \neq j$, $i, j = 1, \ldots, n$, and $c_1, \ldots, c_n$ be complex numbers. Put

$$S(c_1, \ldots, c_n) = \|x - \sum_{k=1}^{n} c_k \Psi(\cdot;f_k)\|^2.$$

The following theorems can be proved.

Theorem 3.1. The function $S(c_1, \ldots, c_n)$ assumes a minimum value for $c_k = a(x; f_k)$, $k = 1, \ldots, n$. We have

$$\sum_{k=1}^{n} |a(x;f_k)|^2 \leq \|x\|^2, \tag{3.2}$$

with equality if and only if $x(t) = \Sigma_{k=1}^{n} a_k \Psi(t; f_k)$.

Theorem 3.2. The set $F_x = \{f: a(x,f) \neq 0\}$ is countable.

Let $F_x = \{f_1, f_2, \ldots\}$. The numbers $f_1, f_2, \ldots$, are called the Fourier frequencies (or eigenfrequencies) of the AP function x, and $\{a(x;f_k)\}$ are the Fourier coefficients, and

$$x(t) \sim \sum_{k=1}^{\infty} a(x;f_k) e^{2\pi i f_k t}$$

the Fourier series. Here ~ symbolizes the Fourier series associated with the function $x(t)$.

Theorem 3.3. If $x(t) \sim \Sigma_k a_k e^{2\pi i f_k t}$ and b is a real number, then

(i) $\bar{x}(t) \sim \Sigma_k \bar{a}_k e^{-2\pi i f_k t}$,

(ii) $x(t+b) \sim \Sigma_k (a_k e^{2\pi i f_k b}) e^{2\pi i f_k t}$,

(iii) $e^{2\pi i b t} x(t) \sim \Sigma_k a_k e^{2\pi i (f_k + b) t}$.

Theorem 3.4. If x is AP, then

$$\sum_{f \in F_x} |a(x;f)|^2 = \|x\|^2.$$

Theorem 3.5. Two distinct AP functions have distinct Fourier series.

Theorem 3.6. Let x be AP and $F_x = \{f_1, f_2, \ldots\}$. Given $\epsilon > 0$ there exists an n_ϵ such that $n > n_\epsilon$ implies

$$\|x(t) - \sum_{k=1}^{n} a(x;f_k) e^{2\pi i f_k t}\|^2 < \epsilon.$$

Theorem 3.7. If the Fourier series of an AP function is uniformly convergent then the sum of the series is the given function.

Corollary. If $\Sigma_k |a(x;f_k)| < \infty$, where $F_x = \{f_1, f_2, \ldots\}$, then

$$x(t) = \Sigma_k a(x;f_k) e^{2\pi i f_k t}.$$

Corollary. There exist (infinitely many) AP functions whose Fourier frequencies are the arbitrarily prescribed numbers $f_1, f_2, \ldots$.

For a more detailed treatment of AP functions the reader may refer to a reference book on the subject such as Corduneanu (1968).

4. Almost Periodic Sequences

A sequence of complex numbers x_t, $t = \ldots, -1, 0, 1, \ldots$, is called AP if to any $\epsilon > 0$ there exist an integer N such that among any N consecutive integers there exist an integer p with the property $|x_{t+p} - x_t| < \epsilon$, $t = 0, \pm1, \ldots$.

For AP sequences x_t and y_t, in place of (2.2), we define the scalar product as

$$(x,y) = \lim_{T\to\infty} \frac{1}{2T} \Sigma_{t=-T}^{T} x_t \bar{y}_t.$$

If, in the discussion of Section 2, we interpret t and T to be integers, and replace integrals by summations, then all the theorems of Section 3 remain valid, except that the Fourier frequencies of an AP sequence x_t remain confined to the unit interval $-1/2 \leq f \leq 1/2$.

The following theorem connects AP sequences with AP functions.

Theorem 4.1. A necessary and sufficient condition for a sequence $\{x_n\}$ to be AP is the existence of an AP function $x(t)$ such that $x_n = x(n)$.

Corollary. If $x(t)$ is AP and $h > 0$ then $x(nh)$, $n = 0, \pm1, \ldots$, is an AP sequence.

5. Almost Periodic Random Processes

Let $\{x(t,\omega), -\infty < t < \infty\}$ be a random process, $\omega\epsilon\Omega$, where $(\Omega, B, P,)$ is an appropriate probability space. To construct an AP process we can assume that the mean function $\mu(t)$ is an AP function, and $e(t,\omega) = x(t,\omega) - \mu(t)$ is a second order stationary process with an absolutely continuous spectral distribution function. In another approach one assumes that $x(t,\omega)$ is a stationary process with a constant mean μ, and covariance function $E[x(t+s) - \mu][x(s) - \mu] = c(t) = c_1(t) + c_2(t)$, where $c_1(t)$ is an AP function, and $c_2(t) \to 0$ as $|t| \to \infty$.

In general, a record of some aspect of socio-economic or geophysical phenomenon cannot be considered as a sample function from a stationary process. For example, let $x(t,\omega)$, $t = 1, 2, \ldots$, denote the hourly median values of the critical frequency, f_0F_2, of the F_2 layer of the ionosphere at Denver, Colorado. We would expect a diurnal, a seasonal and an eleven year periodicity in a record spanning a few decades, the last mentioned cycle corresponding to the average period of sunspot numbers. From the nature of the series we expect that the contribution of the eleven year periodicity to the variance of the sample record will be very large compared to the contributions of the other two periodicities, or of any stationary noise component.

Now, if we look at the time series of monthly or annual mean sunspot numbers themselves, we find that there is no satisfactory theory which would provide a model for periodic behavior of the sun. The so-called eleven year cycle is nothing more than a statistical average over a few centuries. The most appropriate setting for such a series is to model it as an AP random signal, and then try to identify line frequencies "hidden" in its record of monthly or annual mean values.

Let $x(t)$, $-\infty < t < \infty$, be a continuous-time real random process such that $\mu(t) = Ex(t)$ is an AP function and $e(t) = x(t) - \mu(t)$ is a second order stationary process with $Ee(t) = 0$, $Ee(t+s)e(t) = \sigma^2 \rho(s)$, and $\int_{-\infty}^{\infty}|\rho(s)|ds < \infty$. We can then write, with $K \leq \infty$, and $f_1, f_2, \ldots$, all positive,

$$\mu(t) = \gamma_0 + \frac{1}{2}\Sigma_{j=1}^{K} \{\gamma_j e^{2\pi i f_j t} + \bar{\gamma}_j e^{-2\pi i f_j t}\} = \int_{-\infty}^{\infty} e^{2\pi i f t} dZ_1(f), \tag{5.1}$$

where $Z_1(f)$ is defined by its differential increments: $dZ_1(0) = \gamma_0$, $dZ_1(f_j) = \frac{1}{2}\gamma_j$, $dZ_1(-f_j) = \frac{1}{2}\bar{\gamma}_j$, $j = 1, \ldots, K$, and $dZ_1(f) = 0$ for all other values of f. We assume that, for some $\delta > 0$,

$$\inf_{m\neq n}|f_m - f_n| > \delta. \tag{5.2}$$

If $K < \infty$, this is always true, and if $K = \infty$ this condition means that the sequence $\{f_j\}$ does not have a finite limit point. Equation (5.1) gives us the spectral representation of the mean function $\mu(t)$. From the spectral representation of a second order stationary process (see Doob 1953, chapter XI, Theorem 4.1) we also have

$$e(t) = \int_{-\infty}^{\infty} e^{2\pi i t f} dZ_2(f), \tag{5.3}$$

where $Z_2(f)$ is a complex valued stochastic process with orthogonal increments and

$$EdZ_2(f) = 0,\ E|dZ_2(f)|^2 = g_e(f)df,$$

$$g_e(f) = \sigma^2 \int_{-\infty}^{\infty} \rho(t)e^{2\pi i f t}dt.$$

Combining (5.1) and (5.3) we obtain

$$x(t) = \int_{-\infty}^{\infty} e^{2\pi i f t}dZ(f),\ Z(f) = Z_1(f) + Z_2(f). \tag{5.4}$$

Equation (5.4) gives us the spectral representation of $x(t)$.

6. Estimation of Frequencies

The problem of discovering periodicities in a time series goes back to Lagrange (1736-1813), and the representation of a function, square integrable over a line segment, in a series of trigonometric functions was formulated by Fourier (1768-1830). Since then Fourier analysis has played a very significant role in the study of physical processes. If a sequence of observations is free from gross errors or disturbances and if the Fourier analysis of these data produces spectral lines, these lines are considered "real" in a physical sense even though their number may be large and their spacings small. Astronomers and physicists, especially astrophysicists, were never bothered by the question of "spurious" frequencies in their spectra. However, when the Fourier analysis was applied to economic time series, such as the time series of wheat prices, then the question of the significance and reality of the so called business cycles became important. In order to resolve this issue Fisher (1929) obtained his test of significance for the maximum of the periodogram ordinates. Later on, with a flurry of activity in the field of stationary sequences and processes, the interest of statisticians shifted from discovering line frequencies toward constructing consistent estimates of the (power) spectral density function of stationary processes. Techniques such as Fourier transform of sample autocorrelation function, smoothing, windowing and tapering, and, more recently, fitting of a high order autoregressive scheme in order to maximize the entropy, were introduced and used widely. This subject is too vast to be surveyed or even referenced here. An interested reader may refer to some recent books on time series analysis such as that of Bloomfield (1976) or Brillinger (1981).

In the present paper we are mainly interested in the spectral analysis of almost periodic processes. Because an almost periodic process is non-stationary and because its spectrum is decomposable into a sequence of line frequencies, the above mentioned techniques of estimating the power spectral density function of a stationary process are not suitable, in fact quite misleading, for the analysis of almost periodic processes. Our approach is, therefore, necessarily different from the approaches mentioned above and it has much in common with the "classical" approach of astronomers and astrophysicists. We would like to identify and separate as many line frequencies as possible which belong to the set of characteristic frequencies of the AP process under investigation. For our purposes it is much more desirable to estimate the increment $\Delta Z(f)$ of the cumulative amplitude process $Z(f)$ than to estimate the average power $E|\Delta Z(f)|^2$ in order to identify and separate the characteristic line frequencies of the AP process.

Before we describe our method of estimation, we need the following theorem. Let $f > 0$ and $\Delta f > 0$ be such that $f \pm 1/2\ \Delta f$ are continuity points of $Z_1(f)$ (and hence of $Z(f)$). Define

$$\begin{aligned}\Delta Z(f) &= Z(f + 1/2\ \Delta f) - Z(f - 1/2\ \Delta f)\\ &= \Delta Z_1(f) + \Delta Z_2(f)\end{aligned}$$

as the increment of the amplitude process $Z(f)$.

<u>Theorem 6.1</u>. (Doob 1953, p. 527). If

$$x(t) = \int_{-\infty}^{\infty} e^{2\pi i t f} dZ(f),$$

and $f \pm 1/2\ \Delta f$ are continuity points of $Z(f)$, then

$$\Delta Z(f) = \underset{T\to\infty}{\ell.\text{i.m.}}\ \Delta Z_T(f), \tag{6.2}$$

where

$$\Delta Z_T(f) = \int_{-T}^{T} e^{-2\pi i f t}\ \frac{\sin \pi\ \Delta f\ t}{\pi t}\ x(t)dt,\ -\infty < f < \infty. \tag{6.3}$$

A similar theorem (Doob 1953, pp. 481-2) is proved for discrete-time processes. For discrete-time version we write

$$\Delta Z_N(f) = \Sigma_{t=-N}^{N} e^{-2\pi i f t}\ \frac{\sin \pi\ \Delta f\ t}{\pi t}\ x(t),\ -1/2 \leq f \leq 1/2, \tag{6.4}$$

where, for $t = 0$, the term in the summation is $x(0)\ \Delta f$.

We suppose that a continuous record of the process is available which has been read at discrete epochs equally spaced in time, or else a discrete-time record is available. Let the time span covered by the record be $2T$ units, and let $2N + 1$ equi-spaced observations, denoted as $x(t)$, $t = -N, \ldots, -1, 0, 1, \ldots, N$, be taken from it covering the entire period of length $2T$. In what follows as $T \to \infty$ so does N.

Theorem 6.1 provides us with a method of consistently estimating the line frequencies f_j from this sample. Let Δf be small enough so that the interval $(f - 1/2\ \Delta f,\ f + 1/2\ \Delta f)$ includes at most one line frequency f_j of the process $x(t)$ for any $f > 0$. Then

$$\Delta Z(f) = \begin{cases} g_e(f)\,\Delta f & \text{, if the interval does not include any } f_j, \\ \gamma_j/2 + g_e(f_j)\Delta f, & \text{if it includes } f_j . \end{cases} \qquad (6.5)$$

If $\Delta f \to 0$, $\Delta Z(f) \to 0$ and $|\Delta Z(f)|/\Delta f \to g_e(f)$, for $f \neq f_j$, $j = 1, \ldots, K$; and $\Delta Z(f) \to \gamma_j/2$, $|\Delta Z(f)|/\Delta f \to \infty$, if $f = f_j$ for some $j = 1, \ldots, K$. Thus, if Δf is small, $|\Delta Z(f)|$ attains its local maximum of $O(1)$ at $f = f_j$, $j = 1, \ldots, K$. All other local maxima will be of $O(\Delta f)$.

Our analysis procedures are described in detail in Siddiqui and Wang [1984]. Here we only give a brief description of the steps:

Step 1. Do a "crude" scan of the positive frequency range $0 \leq f \leq 1/2$ with $\Delta f = 0.005$ to identify the frequencies where $|\Delta Z_N(f)|$ has a local maximum of $O(1)$. Denote these frequencies as $\hat{f}_1, \hat{f}_2, \ldots, \hat{f}_M$ according to the descending order of the magnitudes of the corresponding amplitudes.

Step 2. Proceed to a more refined scan with $\Delta f = 0.0001$ in the neighborhoods of $\hat{f}_1, \ldots, \hat{f}_M$ to pin point the hidden frequencies and to obtain a regression model of these frequencies.

Step 3. Apply Fisher's test [1929] and retain the significant frequencies $\hat{f}_1, \ldots, \hat{f}_L$.

Step 4. Use least-squares method to fit a regression equation on $\hat{f}_1, \ldots, \hat{f}_L$ and obtain residual series $\hat{e}(t) = x(t) - \hat{x}(t)$, $t = -N, \ldots, -1, 0, 1, \ldots, N$.

The residual series $\hat{e}(t)$ is now subjected to steps 1 to 4, to obtain any additional "real" frequencies which might have been missed in the first cycle.

7. Application to Earth's Orbital Elements

One way of evaluating the performance of our proposed method of high resolution frequency analysis (HRFA) is to apply it to some time series where the signal has a large number of frequencies, some of them being very close to each other. Berger's (1977) representation of the orbital elements of the earth provides us with three excellent almost periodic (AP) functions on which to test our method of HRFA. Climatologists and geophysicists are very much interested in the relationship between the solar radiation reaching at the top of the atmosphere (insolation) at a particular latitude and season and the earth's climate, especially during the Pleistocene. A note of caution is in order here. Berger's 1977 representations of the earth's orbital elements, eccentricity, obliquity and precession, which will be given below, were chosen only for the sake of illustration. A better representation is provided in Berger (1978) in which he has published more than forty terms of the trigonometric series, ordered according to decreasing amplitude, for each of the three orbital elements mentioned above.

In order to compute insolation at any time instant τ of the year t, $0 \leq \tau < 1$, at a latitude ϕ, $-90^\circ \leq \phi \leq 90^\circ$, we need to know the value of the earth's three orbital elements for the year t: (i) the eccentricity $e = a^{-1}(a^2 - b^2)^{1/2}$, where a is the length of the semi-major axis and b is the length of the semi-minor axis of the elliptical orbit of the earth (ecliptic), (ii) the obliquity ϵ, which is the angle between the earth's rotational axis and the normal to the ecliptic plane, and (iii) the precessional parameter $e \sin \omega$, where e is the eccentricity as defined in (i), and ω is the longitude of the perihelion as measured from the moving vernal equinox.

The AP function representation of these elements as given by Berger (1977) are as follows:

$$e - e_0 = \Sigma E_i(\cos \lambda_i t + \phi_i), \tag{7.1}$$

$$\epsilon - \bar{\epsilon} = \Sigma A_i \cos(\gamma_i t + \delta_i), \tag{7.2}$$

$$e \sin \omega = \Sigma P_i \sin(\alpha_i t + \beta_i), \tag{7.3}$$

where $e_0 = 0.0287069$, $\bar{\epsilon} = 23.320556$, and the epoch t = 0 refers to 1950. (In a later paper (1978) Berger gives more than 40 terms for each of the three series; but more than 99 percent variation is contained in the terms reproduced here.) Numerical values of some dominant terms of each of the three series are given in Tables 7.1, 7.2, and 7.3, respectively.

Table 7.1. Main terms in the series expansion of the eccentricity e

	Amplitude E_i	Mean rate λ_i (" per yr)	Period (yr)	Phase ϕ_i (deg)
1	0.01102940	3.138886	412,881	165.2
2	-0.00873296	13.650058	94,945	279.7
3	-0.00749255	10.511172	123,297	114.5
4	0.00672394	13.013341	99,590	291.6
5	0.00581229	9.874455	131,248	126.4
6	-0.00470066	0.636717	2,035,441	348.1
7	-0.00254464	12.639528	102,535	250.8
8	0.00231485	0.991874	1,306,618	58.6
9	-0.00221955	9.500642	136,412	85.6
10	0.00201868	2.147012	603,630	106.6
11	-0.00172371	0.373813	3,466,974	40.8
12	-0.00166112	12.658184	102,384	221.1
13	0.00131342	12.021467	107,807	233.0

Table 7.2. Main terms in the series expansion of the obliquity ϵ

	Amplitude A_i	Mean rate γ_i (" per yr)	Period (yr)	Phase δ_i (deg)
1	-2,462.2	31.609974	41,000	251.9
2	- 857.3	32.620504	39,730	280.8
3	- 629.3	24.172203	53,615	128.3
4	- 414.3	31.983787	40,521	292.7
5	- 311.8	44.828336	28,910	15.4
6	308.9	30.973257	41,843	263.8
7	- 162.5	43.668246	29,678	308.4
8	- 116.1	32.246691	40,190	240.0
9	101.1	30.599444	42,354	223.0
10	- 67.7	42.681324	30,365	268.8

Table 7.3. Main terms in the series expansion of the precession e sin ω

	Amplitude P_i	Mean rate α_i (" per yr)	Period (yr)	Phase β_i (deg)
1	0.0186080	54.646478	23,716	32.0
2	0.0162752	57.785364	22,428	197.2
3	-0.0130066	68.296536	18,976	311.7
4	0.0098883	67.659819	19,155	323.6
5	-0.0033670	67.286006	19,261	282.8
6	0.0033308	55.638352	23,293	90.6
7	-0.0023540	68.670349	18,873	352.5
8	0.0014002	76.656031	16,907	131.8
9	0.0010070	56.798442	22,818	157.5

Several comments are in order here. In the representation of e sin ω (Table 7.3) we notice that there are four frequencies with periods between 22,400 and 23,800 years, four frequencies with periods from 18,800 to 19,300 years, and one frequency with period 16,900 years. The power = $(\text{amplitude})^2$ of 16,907 periodicity is less than 0.006 times the power in periodicity 23,716.

In the representation of the obliquity ε (Table 7.2), six periods are squeezed between 39,700 and 42,400 years, one is 53,600 years, and the remaining three are bunched in a neighborhood of 29,000 years. The total power in the latter three periodicities is less than 2 percent of the total power in the 41,000-year group of periodicities.

The series representation of the eccentricity e is very complex. There are five frequencies with periods in the neighborhood of 100,000 years, three have periodicities around 130,000 years, and the remaining frequencies have very long periods (more than 400,000 years).

7.1 Analysis

Series of length 1,255 were generated by using Berger's equations with sampling interval fixed at 2,000 years, and for t = -1, -2, ..., -1255, thus spanning the past 2,510,000 years. (These series are graphed in Figure 7.1, in which the unit of time is 1000 years.) Each series was divided into five sections of length 251 so that 2N + 1 = 251, and N = 125. These sections are separated in Figure 7.1 by vertical dashed lines. A visual comparison of various sections is enough to show that different parts of any AP function, with a large number of frequencies of comparable power, look very different from each other.

Each section of each series was analyzed by our method of HRFA. The results of the analysis are given in Tables 7.4, 7.5 and 7.6. The frequencies being close together are grouped in several groups for easy comparison. The relative ranks for each frequency are given in the parentheses beside the period. The percentage of total variance explained by each frequency and the sum of the percentages of variance explained are also given.

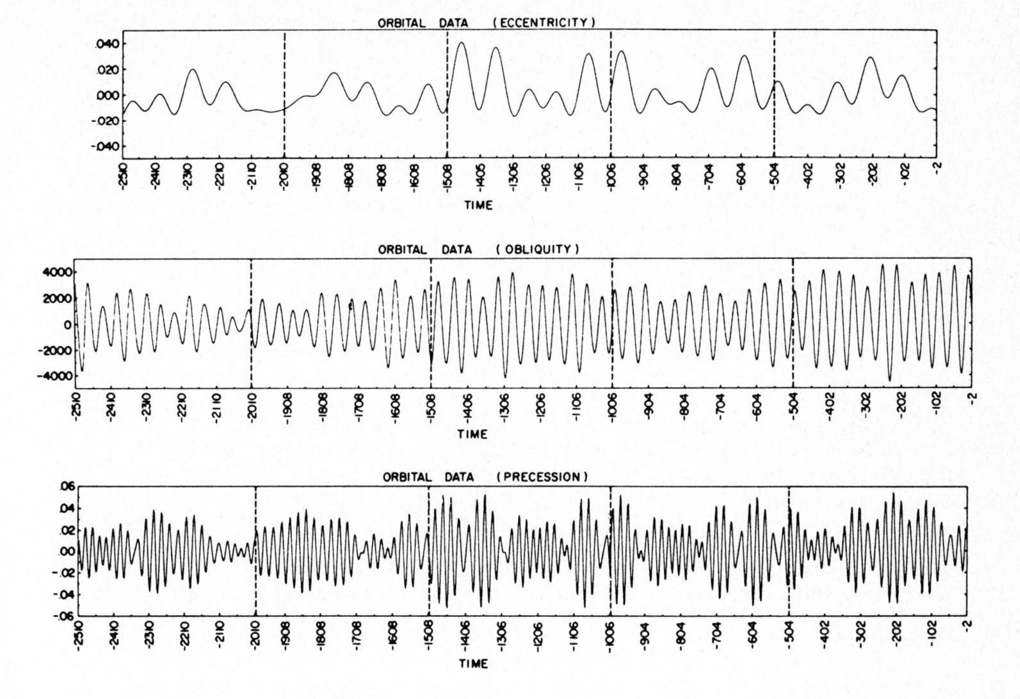
ORBITAL DATA (ECCENTRICITY)
.040
.020
.000
-.020
-.040
TIME
ORBITAL DATA (OBLIQUITY)
4000
2000
0
-2000
-4000
TIME
ORBITAL DATA (PRECESSION)
.06
.04
.02
.00
-.02
-.04
-.06
TIME

Figure 7.1.

Table 7.4 True periods (in 1000 years), estimated periods of the eccentricity e and the percentage of variance explained by each period (numbers in parentheses show relative ranks of the amplitudes)

		Section 1		Section 2		Section 3		Section 4		Section 5	
True Periods	% Variance	Estimated Periods	% Variance	Estimated Periods	% Variance	Estimated Periods	% Variance	Estimated Periods	% Variance	Estimated Periods	% Variance
412.885(1)	31.73	377.358(1)	63.40	363.636(3)	16.25	408.163(3)	15.47	425.532(1)	52.59	408.163(1)	42.16
94.945(2)	19.89	86.580(4)	3.43	94.340(2)	27.00	88.106(8)	0.01	95.694(2)	28.38	86.957(4)	5.72
99.590(4)	11.79					96.618(1)	37.99				
102.535(7)	1.69	101.523(2)	23.46								
102.384(2)	0.72										
107.807(13)	0.45			109.890(7)	0.03	112.994(5)	0.97			105.820(2)	35.88
123.297(3)	14.64	126.582(3)	7.58								
131.248(5)	8.81			128.205(1)	42.68	131.579(2)	28.16	129.870(6)	1.31		
136.412(9)	1.28									143.885(3)	14.53
2,035.441(6)	5.76										
1,306.618(8)	1.40										
603.630(10)	1.06										
3,466.974(11)	0.77										
spurious f		151.515(6)	0.69	155.039(4)	2.20	161.290(7)	0.55	153.846(3)	12.40		
		235.294(5)	1.32	235.294(6)	0.35	273.973(6)	0.89	202.020(4)	3.57	243.902(5)	1.71
		74.671(7)	0.11	74.074(8)	0.00			83.333(5)	1.75		
				740.741(5)	1.35	833.333(4)	4.10				
Sum of %			99.99		89.96		88.23		100.00		100.00

Table 7.5. True periods (in 1000 years), estimated periods of the obliquity ε and the percentage of variance explained by each period (numbers in parentheses show relative ranks of amplitudes).

		Section 1		Section 2		Section 3		Section 4		Section 5	
True Periods	% Variance	Estimated Periods	% Variance	Estimated Periods	% Variance	Estimated Periods	% Variance	Estimated Periods	% Variance	Estimated Periods	% Variance
41.000(1)	79.64	40.650(1)	91.63	41.580(1)	88.41	40.816(1)	92.17	40.900(1)	85.28	40.984(1)	79.76
39.730(2)	9.65	37.665(7)	0.19	37.175(6)	0.24	37.807(6)	0.34	38.536(3)	4.26	38.685(3)	5.48
40.521(4)	2.25										
41.843(6)	1.25										
40.190(8)	0.18										
42.354(9)	0.13	43.103(9)	0.03	44.843(5)	0.42	43.763(9)	0.04	42.918(5)	0.98	43.011(5)	1.40
53.615(3)	5.20	51.680(2)	2.24	52.493(2)	4.38	51.948(2)	3.02	52.632(2)	5.51	53.476(2)	9.79
		55.866(4)	0.64	57.803(4)	0.44	56.022(5)	0.50	57.471(6)	0.75		
		48.426(6)	0.25			48.662(7)	0.26				
28.910(5)	1.28	27.739(8)	0.07	28.653(3)	0.78	27.816(8)	0.04	28.694(4)	1.30		
29.678(7)	0.35	28.736(3)	1.49			28.563(3)	1.25			29.028(4)	3.57
30.365(10)	0.06	30.166(5)	0.28			30.030(4)	0.56				
Sum of %			96.83		94.67		98.18		98.08		100.00

Table 7.6. True periods (in 1000 years), estimated periods of the precession term $e \sin \tilde{\omega}$ and the percentage of variance explained by each period (numbers in parentheses show relative ranks of amplitudes).

		Section 1		Section 2		Section 3		Section 4		Section 5	
True Periods	% Variance	Estimated Periods	% Variance	Estimated Periods	% Variance	Estimated Periods	% Variance	Estimated Periods	% Variance	Estimated Periods	% Variance
23.716(1)	38.32	23.529(1)	49.23	23.810(3)	23.95	23.753(2)	33.36	23.669(1)	53.32	23.474(1)	49.16
22.428(2)	29.32	22.396(2)	30.21	22.346(2)	24.11	22.321(3)	19.22	22.346(2)	33.59	22.173(2)	28.73
23.293(6)	1.22			23.015(4)	0.77	23.068(4)	1.93			24.600(4)	2.34
22.818(9)	0.10			21.692(5)	0.23	21.739(8)	0.10				
						24.540(9)	0.07				
18.976(3)	18.72	18.940(3)	19.61			18.605(5)	1.37	18.709(4)	2.30		
19.155(4)	10.81			19.011(1)	40.10	19.157(1)	33.94	19.724(5)	0.33	19.139(3)	19.79
19.261(5)	1.24			19.550(6)	0.22	19.763(6)	0.27	19.268(3)	10.14		
18.873(7)	0.60			18.570(8)	0.04						
16.907(8)	0.21			16.992(7)	0.18	17.050(7)	0.20	16.779(6)	0.32		
Sum of %			99.05		89.60		90.46		100.00		100.00

From Table 7.4 we see that at least one frequency from each of the first three groups of frequencies of eccentricity e has been estimated from each section. We were not able to obtain the very low frequencies because their periods are beyond our data length. Some spurious frequencies also passed the Fisher's test which are tabulated at the bottom of the table. Some of them have very small power so, although they are statistically significant, they can be ignored practically. The "spurious" frequencies are most probably subharmonics or "combination tones" of other frequencies. For example, the periodicities 151515, 155039, and 153846 years in sections 1, 2, and 4, respectively, are the fourth subharmonic of the true periodicity 603630 years.

There is little point in trying to explain these spurious frequencies which appear in our HRFA as in most cases they are practically insignificant.

Because of the complex nature of this particular series and the shortness of data length, we do not have as good a resolution as for the other two series. Even so, we get the same frequencies in different sections, i.e. we have the consistency among different sections of the data.

There are three major groups of frequencies for obliquity. Table 7.5 shows that in every section our analysis obtains at least one and many times more than one frequency from each group. Some sections have higher resolution than others, but in each case we obtain all the major true frequencies of the process, so that different sections are mutually consistent in their frequency analysis. Note that we do not have any spurious frequency for this series.

In Table 7.6 we present the HRFA result for the precession parameter e sin ω. The true frequencies fall in three groups, two major and one minor. The minor group consists of a single frequency with period 16,900 years which explained only 0.21% of the variance. Each of the five sections was analyzed with HRFA. At least one frequency from the first two major groups was obtained for each section, and for sections 2 and 4 even the periodicity in the neighborhood of 16,900 years was significant. Actually, in our step 2 analysis of refined scanning of sections 1 and 5, the periodicity in the neighborhood of 16,900 was present, but due to its low power it did not pass the Fisher's test. Note that the results of the 5 sections are mutually consistent. The analysis did not produce any spurious frequency in any of the sections.

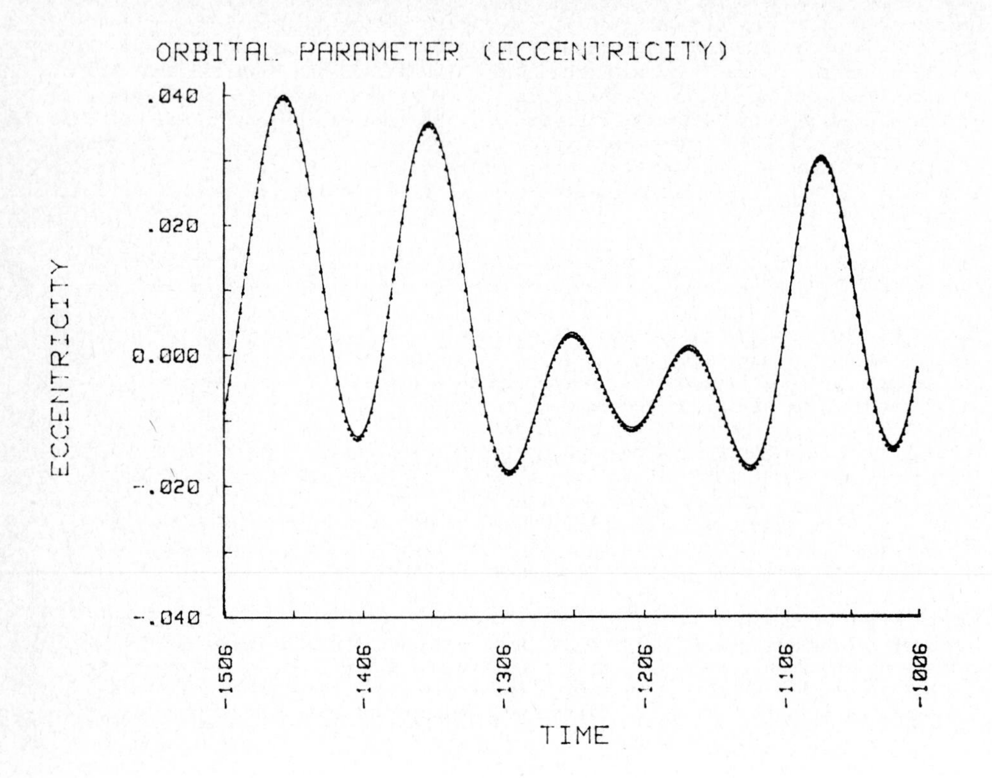
ORBITAL PARAMETER (ECCENTRICITY)
.040
.020
0.000
-.020
-.040
ECCENTRICITY
-1506
-1406
-1306
-1206
-1106
-1006
TIME

Figure 7.2.

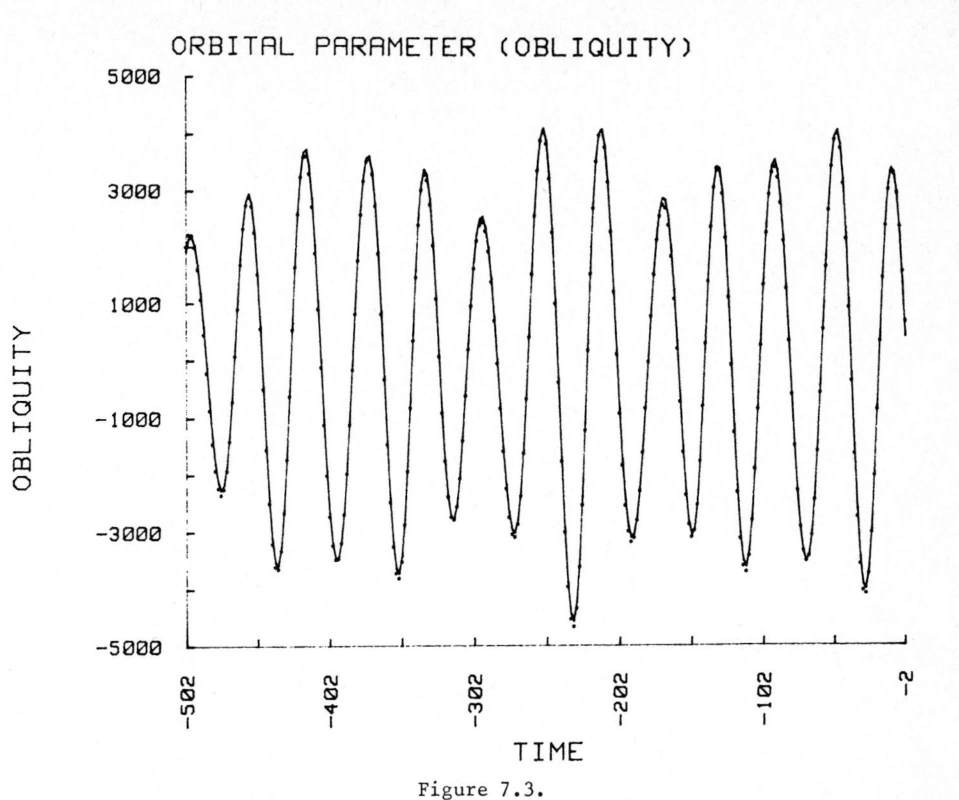

Figure 7.3.

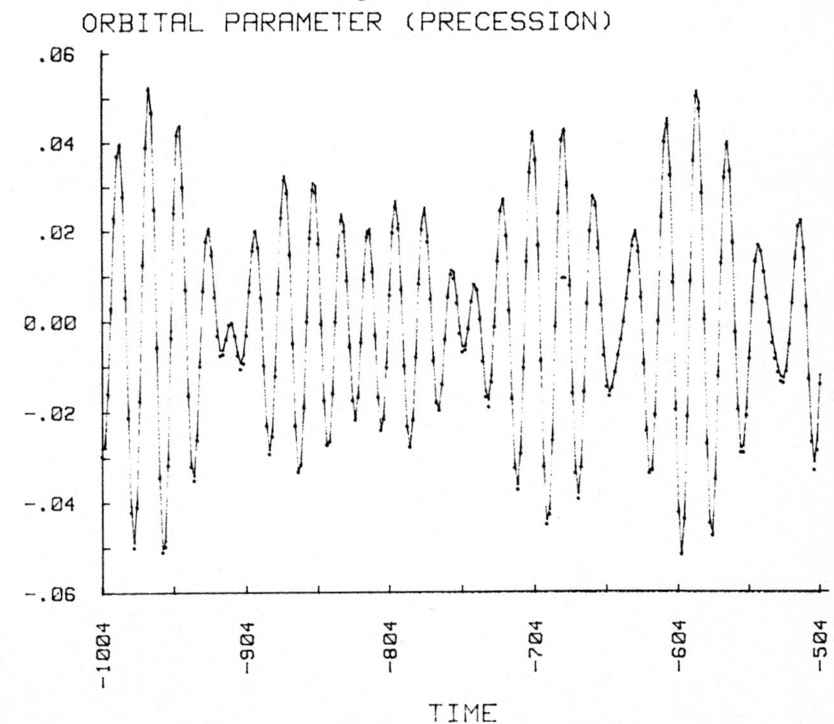

Figure 7.4.

In order to see how faithfully the original series is reproduced by our fitted curve

$$\hat{x}(t) = \hat{\mu} + \Sigma_{i=1}^{L} (\hat{\alpha}_j \cos 2\pi\hat{f}_j t + \hat{\beta}_j \sin 2\pi\hat{f}_j t),$$

we randomly picked one section from the five sections of each of the three series. The sections so chosen were as follows: Section 3 for eccentricity, Section 1 for obliquity, and Section 2 for precession. Graphs of the original series (solid lines) and the fitted series (dotted lines) are given in Figures 7.2, 7.3, and 7.4. Let the mean square of the original series be

$$\text{MST} = T^{-1} \sum_{t=1}^{T} x^2(t)$$

and the mean square of the residuals be

$$\text{MSE} = T^{-1} \sum_{t=1}^{T} (x(t) - \hat{x}(t))^2.$$

For the three series their values are as follows:

Eccentricity Section 3:

$$\text{MST} = 2.82367 \times 10^{-7}$$

$$\text{MSE} = 2.750005 \times 10^{-11}$$

Obliquity Section 1:

$$\text{MST} = 5.96150 \times 10^{6}$$

$$\text{MSE} = 2.82485 \times 10^{3}$$

Precession Section 2:

$$\text{MST} = 5.28585 \times 10^{-4}$$

$$\text{MSE} = 3.08008 \times 10^{-7}.$$

We, therefore, conclude that, at least for noise-free time series, our method of HRFA is highly accurate. It estimates the true frequencies and their amplitudes and phases so accurately that the original series are faithfully reproduced with more than 99 percent accuracy.

In order to see whether our method of HRFA can pick up the low frequency present in the time series of eccentricity by increasing the sample size we further analyzed data length covering 501 equi-spaced values spanning 1,002,000 years and also sample size of length 999 spanning 1,998,000 years. The results of these two analyses are presented in Table 7.7.

Table 7.7. True periods (in 1000 years), estimated periods of the eccentricity e and the percentage of variance explained by each frequency for data length 501 and 999 (number in parentheses show relative ranks of the amplitude).

		501 data points		999 data points	
True Periods	% Variance	Estimated Periods	% Variance	Estimated Periods	% Variance
412.885 (1)	31.73	392.157 (1)	37.79	408.163 (1)	33.73
94.945 (2)	19.89	95.694 (2)	29.64	95.238 (2)	27.10
99.590 (4)	11.79	100.000 (8)	0.31	100.000 (4)	5.13
102.535 (7)	1.69				
102.384 (12)	0.72			104.712 (5)	3.05
107.807 (13)	0.45	113.636 (5)	1.44	108.108 (10)	0.46
		89.686 (9)	0.15		
123.297 (3)	14.64	125.000 (3)	27.57	125.786 (3)	25.01
131.248 (5)	8.81				
136.412 (9)	1.28	137.931 (7)	0.56	135.135 (6)	1.93
				116.279 (9)	0.56
2,035.441 (6)	5.76			1,818.182 (7)	1.86
1,306.618 (8)	1.40				
603.630 (10)	1.06			606.061 (8)	0.85
3,466.974 (11)	0.77				
spurious f		800.000 (4)	1.86	800.000 (11)	0.32
		298.507 (6)	0.57		
		212.766 (10)	0.06		
		170.940 (11)	0.05		
sum of %			100.00		100.00

We offer the following remarks: the HRFA of 501 values, although did not produce any new low frequencies as compared to sections of length 251, the estimates are much closer to true frequencies. For HRFA of 999 data points we do pick up two additional low frequencies with periods 1,818,182 and 606,061 years as well as very much improved estimates of the other frequencies. This shows that our method is consistent in the statistical sense also. And it also possesses the property of high resolution as claimed. That is to say, as data length increases we pick up more and more true frequencies. Note also that the spurious frequencies generated by our analysis have very negligible power.

8. Gaussian White Noise

In order to see whether or not our method of HRFA picks up line frequencies which do not exist in a time series, we generated five Gaussian white noise series, each consisting of 251 values, and applied the procedures described in section 6. For these five series the values of Fisher's statistics g (= the ratio of the largest power to the total power) are as follows:

series	1	2	3	4	5
g	0.069	0.040	0.041	0.045	0.041

For N=125 (2N + 1 = 251), $\alpha = 0.05$, we have $g_\alpha = 0.1249$. Therefore, none of the g values passed Fisher's test at 5 percent level of significance; that is to say, no line frequency was picked up when there was none.

9. Magnitude of a Variable Star

Whittaker and Robinson (1946) give the magnitude (i.e. a measure of the brightness) of a variable star at midnight on 600 successive nights. They say "These magnitudes were obtained by reading off from a curve on which all the observations of the star's brightness were plotted: they have been reduced to a scale suitable for periodogram analysis." (p. 340). The periodogram analysis gives two preiods of 29 days and 24 days and a regression fit of

$$17 + 10 \sin(2\pi(t+3)/29) + 7 \sin(2\pi(t+1)/24).$$

We analyzed the last 599 observations using HRFA as described in Section 6. The results of analysis are given in Table 9.1. Some of the frequencies are close together and hence they are grouped. However, we can see that there are actually only two major frequencies, one with a period of 29.07 days explains 62% of the variance and the other with a period of 23.98 days explains 31% of the variance. These two frequencies alone explain 93.4% of the variance. The remaining eight frequencies explain only 0.395% of the total variance, so they can be ignored, even though they are statistically significant. The mean of the series is 17.10, and the amplitude of the two major frequencies are 10.0128 and 7.0885, respectively.

Table 9.1. Estimated frequencies, regression coefficients, amplitudes and percentage of variance explained by each frequency of variable star data. The ranks of the frequencies f_j are given in the parentheses. ($\bar{x}$ = 17.100, S^2 = 80.571).

Freq. (c/days) f_j	Period (days) P_j	Regression Coefficients $\hat{\alpha}_j$	$\hat{\beta}_j$	Amplitude $\lvert\hat{\gamma}_j\rvert$	% of Variance
0.0344 (1)	29.07	8.5453	5.2185	10.0128	62.222
0.0356 (3)	28.09	-.5177	.1715	0.5454	0.185
0.0333 (4)	30.03	-.0217	.5143	0.5148	0.164
0.0372 (8)	26.88	-.1001	.0474	0.1108	0.008
0.0316 (9)	31.65	-.0030	.1065	0.1066	0.007
0.0417 (2)	23.98	-.4496	7.0742	4.4996	31.182
0.1380 (5)	7.25	.0232	.1375	0.1394	0.012
0.0690 (6)	14.49	.1276	-.0171	0.1287	0.010
0.4584 (7)	2.18	-.0180	.1212	0.1226	0.009
0.1034 (10)	9.67	-.0398	.0925	0.1007	0.006
Sum of %					93.805

10. Annual Lynx Trappings

The time series of annual lynx trappings consists of the annual record of the numbers of Canadian lynx trapped in the Mackenzie River district of Northwest Canada for the period of 1821-1934 (114 data points). Because of the asymmetry of the original data series, Moran (1953) applied a $\log_{10}$ transformation to reduce the degree of asymmetry. The original data series and plots for both the original series and the $\log_{10}$ transformed data series can be found in Campbell and Walker (1977).

Applications of our method of HRFA to the data in years 1821 + t, t = 1, ..., 113 (last 113 values) and their $\log_{10}$ transform are given in Tables 10.1 and 10.2 respectively. Comparing the two tables we notice that frequencies f_1, f_3 and f_4 in Table 10.2 are the same as frequencies f_1, f_2 and f_3 in Table 10.1. We may therefore surmise that .1040, .0910, and .0253 (with periods 9.61, 10.99, and 39.53) most probably are the characteristic frequencies in the lynx data, and that the other statistically significant frequencies are perhaps a combination of these three. For example, $f_4 = 2f_1$ in Table 10.1. We have not investigated this point any further.

Table 10.1. Estimated frequencies, regression coefficients, amplitudes, and percentage of variance explained by each frequency of annual lynx trapping data. The ranks of the frequencies f_j are given in the parentheses. ($\overline{x}$ = 1549.25 S^2 = 252284.95)

Freq. (c/yr.)	Period (yr.)	Regression Coefficients		Amplitude	% of
f_j	P_j	$\hat{\alpha}_j$	$\hat{\beta}_j$	$\|\hat{\gamma}_j\|$	Variance
.1038 (1)	9.634	-744.49	-1473.01	1650.46	54.47
.0255 (2)	39.216	345.98	509.96	616.25	7.59
.0925 (3)	10.811	9.28	-499.15	499.24	4.98
.2076 (4)	4.817	-437.86	213.96	487.34	4.75
.1315 (5)	7.605	390.70	-211.71	444.37	3.95
.0503 (6)	19.881	75.10	401.68	408.64	3.34
sum of %					79.08

Table 10.2. Estimated frequencies, regression coefficients, amplitudes, and percentage of variance explained by each frequency of $\log_{10}$ transformed annual lynx trapping data. The ranks of the frequencies f_j are given in the parentheses. ($\overline{x}$ = 2.9078, S^2 = 0.3126)

Freq. (c/yr.)	Period (yr.)	Regression Coefficients		Amplitude	% of
f_j	P_j	$\hat{\alpha}_j$	$\hat{\beta}_j$	$\|\hat{\gamma}_j\|$	Variance
.1040 (1)	9.615	-.3013	-.5382	0.6168	61.40
.0384 (2)	26.042	.0042	.1984	0.1985	6.36
.0253 (3)	39.526	.0920	.1627	0.1870	5.64
.0910 (4)	10.989	-.1219	-.1298	0.1780	5.12
.0692 (5)	14.451	-.0319	.1416	0.1452	3.40
.1143 (6)	8.749	.0940	.0609	0.1120	2.02
.0162 (7)	61.728	-.0570	-.0815	0.0994	1.59
.1960 (8)	5.102	-.0652	-.0706	0.0961	1.49
sum of %					87.02

Campbell and Walker (1977) have represented $\log_{10}$ transformations of the data as a superposition of a pure sine wave and a second-order autoregressive process. Their fit to the data, for 1821 + t, t = 0, 1, ..., 113, is

$$X_t = 2.9036 + 0.6313 \cos\{(2\pi/9.63)(t + 3.19)\} + U_t \qquad (10.1)$$

$$U_t = 0.9717\, U_{t-1} - 0.2654\, U_{t-2} + \epsilon_t \qquad (10.2)$$

with $\overline{\epsilon} = 0$ and $S^2_{\epsilon} = 0.4198$.

We notice that in our analysis $|\hat{\gamma}_1| = 0.6168$ which is about the same as the amplitude 0.6313 in (10.1).

Damsleth and Spjotvoll (1982) used stepwise estimation procedure to the $\log_{10}$ transform of the data and obtained significant periods as: 9.58, 41.84, 11.03, 21.74, 15.36, 5.11, 7.13, 4.84. We notice that several frequencies including the three characteristic frequencies mentioned above found be HRFA almost coincide with the frequencies they have listed.

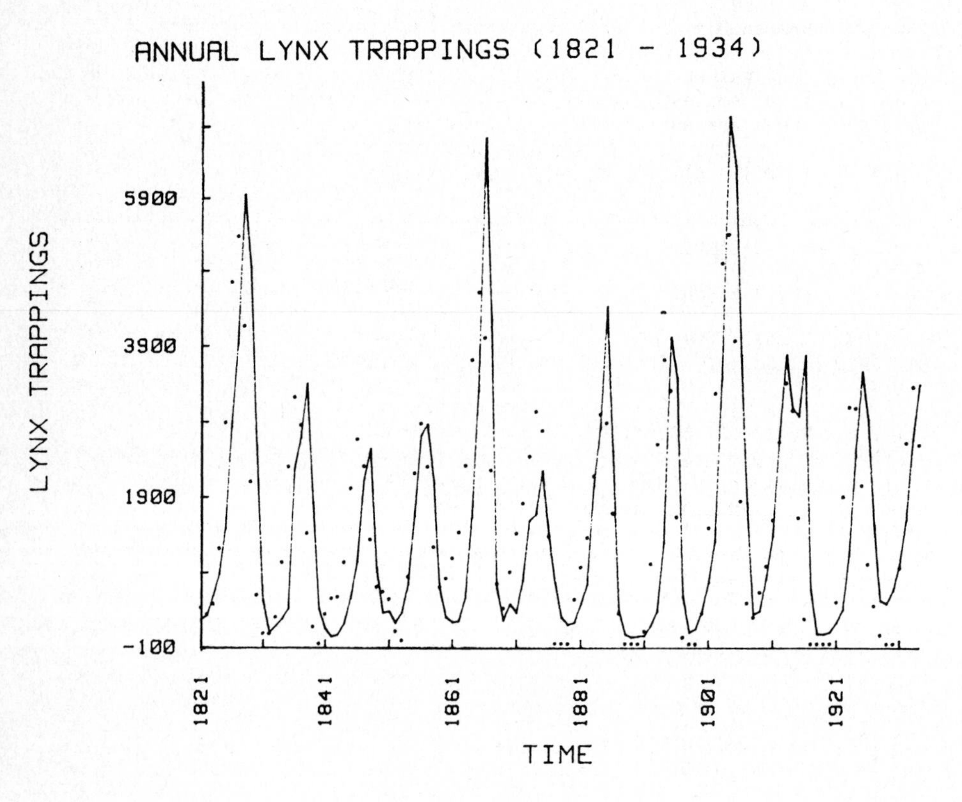

Figure 10.1

In Figure 10.1 we graph the original series (solid line) for lynx data together with the fitted series (dotted line). The fitted and the original series are slightly out of phase. The values of MST and MSE for this series are as follows:

$$\text{MST} = 2.50052 \times 10^6$$

$$\text{MSE} = 5.03104 \times 10^5$$

11. Acknowledgments

The work of Siddiqui was partially supported by the National Science Foundation Grant No. ATM-8204751 and by the National Oceanic and Atmospheric Administration Grant No. NOAA-NA81AA-D-00039. The work of Wang was supported entirely by the NOAA grant.

12. References

Berger, A. L. (1977). Support for the astronomical theory of climatic change. Nature (London) 269, 44-45.

Berger, A. L. (1978). Long-term variations of daily insolation and quaternary climatic changes. J. Atmospheric Sci. 35, 2362-2367.

Bloomfield, P. (1976). Fourier Analysis of Time Series: An Introduction. Wiley and Sons, New York.

Brillinger, D. R. (1981). Time Series Data Analysis and Theory, Expanded Edition. Holden-Day, San Francisco.

Campbell, M. J. and Walker, A. M. (1977). A survey of statistical work on the Mackenzie river series of annual Canadian lynx trappings for the years 1821-1934 and a new analysis. J. R. Statist. Soc. A, 140, Part 4, 411-431.

Corduneanu, C. (1968). Almost Periodic Functions. Wiley and Sons, New York.

Damsleth, E. and Spjotvoll, E. (1982). Estimation of trigonometric components in time series. J. Amer. Statist. Assoc. 77, 381-387.

Doob, J. L. (1953). Stochastic Processes. Wiley and Sons, New York.

Fisher, R. A. (1929). Test of significance in harmonic analysis. Proceedings of the Royal Society, Series A. 125, 54-59.

Moran, P. A. P. (1953). The statistical analysis of the Canadian lynx cycle: I. Austr. J. Zool., 1, 163-173.

Shimshoni, M. (1971). On Fisher's test of significance in harmonic analysis. Geophysical J. Roy. Astro. Soc. 23, 373-377.

Siddiqui, M. M. and Wang, C. C. (1984). High-resolution frequency analysis of geological time series. J. Geophys. Res. 89, D5, 7195-7201.

Whittaker, E. T. and Robinson, C. (1946). The Calculus of Observations, 4 Ed. Blackie and Son, London.

Original paper received: 04.03.85
Final paper received: 01.08.87

Paper recommended by D.C. Boes

PROBABILITY AND STATISTICS
Essays in Honor of Franklin A. Graybill
J.N. Srivastava (Editor)

ON A GENERAL THEORY OF SAMPLING USING EXPERIMENTAL DESIGN CONCEPTS II. RELATION WITH ARRAYS.*

JAYA SRIVASTAVA
COLORADO STATE UNIVERSITY

ABSTRACT

In this paper, we explore the relationship between Sampling Theory and Combinatorial Arrays, which constitute an integral part of the theory of factorial designs, and which also play an important role in other branches of statistical experimental design. It is pointed out how Simple Random Sampling (or, in general, t-design Sampling (Srivastava and Salah (1985)) are special cases of Balanced-Array Sampling introduced herein. Controlled Sampling is looked into, in this context. Some studies are made, under the general estimator introduced recently by the author (Srivastava (1985)).

1. Introduction

In the past, many authors have explored the natural relationship that exists between design theory and sampling theory. An illustrative (but not exhaustive) bibliography is presented at the end.

In this paper , we shall establish some important connections which exist between Sampling Theory and the Theory of Arrays, which arises in the theory of discrete factorial experiments.

In what follows, by an 'Array', we shall mean a matrix whose elements come from a given finite set. If this set has s symbols in it, then the corresponding array will be said to be an 'array with s symbols'. Without loss of generality, the symbols used in such an array will be the integers $\{0, 1, 2, \ldots, (s-1)\}$. The case $s = 2$ is specially important; in this case, the array will sometimes be called a (0,1) matrix.

Throughout, if K is any set, the symbol $|K|$ will denote the number of elements in K. The 'weight' of a matrix is the number of non-zero elements in the matrix. (This includes the case where the matrix is a vector, or even a scaler.) Also, for any real matrix Q (including the situtation where Q is a vector or scaler), the symbol Q^- will denote the Moore-inverse of Q.

We now consider traditional Sampling Theory. A population will be denoted by U. We shall assume that U consists of N units (all of which are distinct). Thus, $|U| = N$. Let 2^U denote the class of all distinct subsets of U, including the empty set. Obviously, we have $|2^U| = 2^N$. Let $\omega \in 2^U$; then ω will be called a 'sample'. We shall assume that all elements of ω are distinct, in other words, ω will contain $|\omega|$ <u>distinct</u> elements. Let Ω_U denote the $N \times 2^N$ (0,1)-matrix whose columns constitute the set of the 2^N possible distinct (0,1)-vectors, each containing N elements. For example, for $N = 3$, Ω_U may be expressed as

$$(1.1) \qquad \Omega_U = \begin{bmatrix} 0 & 1 & 0 & 0 & 1 & 1 & 0 & 1 \\ 0 & 0 & 1 & 0 & 1 & 0 & 1 & 1 \\ 0 & 0 & 0 & 1 & 0 & 1 & 1 & 1 \end{bmatrix}$$

We shall write U as

$$(1.2) \qquad U \equiv \{1, 2, \ldots, N\}.$$

In other words, the units in U will be labelled by the integers $(1, 2, \ldots, N)$. The rows of Ω_U will be considered to be in correspondence with the units in U, so that the ith row of U $(i = 1, \ldots, N)$ will correspond to the unit i. Consider now any given column of Ω_U; the set of coordinates in which this particular column has the symbol 1 constitute a subset of U, and hence a member of 2^U. Thus, every possible column of Ω_U corresponds to a unique sample ω, and vice

*This research supported by AFOSR grant #830080.

versa. For example, the column #2 of Ω_U at (1.1) corresponds to the sample (2), and the sample (2, 3) corresponds to column #6 of Ω_U. (Throughout, we shall assume that the columns of Ω_U are numbered respectively from 0 to 2^N-1.) Notice that the column of Ω_U whose weight is zero corresponds to the empty sample, and the sample consisting of the whole population U corresponds to the column of Ω_U whose weight is N. Without loss of generality, the columns in Ω_U will be arranged in the following ("standard") order:

(i) a column of smaller weight will precede a column of larger weight,
(ii) if $\underline{\alpha}$ ($\equiv(\alpha_1,\ldots,\alpha_N)'$) and $\underline{\beta}$ ($\equiv(\beta_1,\ldots,\beta_N)'$) are columns of the same weight, then $\underline{\alpha}$ precedes $\underline{\beta}$ if there is a $j \in (1,\ldots,N)$ such that $\alpha_1 = \alpha_2 = \ldots = \alpha_j = \beta_1 = \ldots \beta_{j-1} = 1$, and $\beta_j = 0$. Notice that the columns of Ω_U at (1.1) satisfy this 'standard' ordering.

Notice that the columns of Ω_U are also indexed by the members of 2^U. Hence, a general element of Ω_U can be expressed by the symbol $a_{i\omega}$, where $i \in U$, and $\omega \in 2^U$. Notice that for all permissiable i and ω, we have

(1.3) $$a_{i\omega} = 1, \text{ if } i \in \omega$$
$$= 0, \text{ if } i \notin \omega.$$

For $n \in \{0, 1,\ldots,N\}$, the set of all the $\binom{N}{n}$ distinct samples from U of size n, will be denoted by $\{U: n\}$; notice that this set corresponds to the columns of Ω_U of weight n.

2. Arrays and Sampling Measures.

Throughout, by a 'Sampling Technique' (ST), we shall mean any method of drawing a sample ω out of a population U. For example, any method of drawing a 'simple random sample without replacement' (SRSWOR) is a ST. Similarly, all physical techniques using which we draw particular kinds of samples, such as a stratified sample, a systematic sample, or a sample of clusters (including subsampling within clusters), etc., constitute examples of ST's. Other ST's include, for example, the Murthy, or Madow, or Rao-Hartley-Cochran techniques of drawing samples. Most books on sampling (an illustrative list of which is included at the end) contain discussion of vast numbers of ST's.

We shall occasionally use the phrase 'Physical Sampling Technique' (PST). By a 'PST', we shall mean any physical process by which a sample of units is drawn from a (real) population. For example, suppose we need to draw a sample of 5 students from a class of 30 students. One PST way could be as follows. Number the students from 1 to 30. Using a random number generator and computer, generate one random number between 1 and 30. Next excluding the random number drawn, draw another number (between 1 and 30), and so on, until five numbers are drawn. Identify the five students. Notice that this PST executes the ST which corresponds to SRSWOR. However, PST's do not necessarily have to correspond to the ST's which are usually discussed in the books. For example, in the above one, we could ask the students to randomly sit in a classroom in 5 rows, and then randomly pick one student <u>by eye</u> from each row. This is a certainly a PST, but since selection by eye is likely to be biased, this PST does not necessarily correspond to the ST known as SRSWOR.

Now, consider any PST, and any $\omega \in U$; this PST would induce a certain probability $p(\omega)$ such that if the PST is executed then $p(\omega)$ is the probability that the sample ω will get selected. Notice that

(2.1) $$\sum_{\omega} p(\omega) = 1,$$

where in (2.1), the sum runs over all $\omega \in 2^U$. It is clear that, in general,

$p(\omega)$ may or may not be known for all $\omega \in 2^U$. Let

(2.2) $$p(\cdot) = \{p(\omega) | \omega \in 2^U\}.$$

Notice that any given PST would give rise to a unique probability distribution $p(\cdot)$, but that the same $p(\cdot)$ may be induced by more than one (generally, an infinite number of) PST. Given any PST, any probabilistic or statistical statements made on the basis of the data from the sample resulting from the PST, would require knowledge of the distribution $p(\cdot)$ induced by the PST. In what follows in this paper, we shall assume that any PST under consideration is such that the corresponding probability distribution $p(\cdot)$ is known. For any PST, the distribution $p(\cdot)$ induced by it will sometimes be referred to as its 'Sampling Measure'.

Consider Ω_U with its columns arranged in the standard order; we shall denote by $\omega(j)(j = 0, 1, \ldots, 2^N-1)$ the sample correspoinding to the jth column of Ω_U. Given any PST, $p(\cdot)$ will sometimes be indicated by the $(2^N \times 1)$ vector $\underline{p}$, where

(2.3) $$\underline{p}' = (p(\omega(0)), p(\omega(1)), \ldots, p(\omega(2^N-1)))$$

It should be noted that in many ST's units are drawn one by one, so that along with each unit comes also the extra information as to the draw number in which the unit got drawn. Similarly, there are ST's (such as 'simple random sampling with replacement' (SRSWR)) in which, along with each unit drawn, there comes information as to the number of times the unit got drawn into the sample. In most cases, such extraneous information has nothing to do with the value of the vector $(y_1, \ldots, y_N)$, where y_i $(i = 1, \ldots, N)$ denotes the value of the variable of interest (say, y) on the unit i. We shall restrict our attention to situations where this is indeed the case, and, consequently, we shall confine ourselves to procedures which do not depend upon such information.

Now, consider any arbitrary, but fixed, PST, say S. Corresponding to this PST, we shall sometimes consider the $(N + 1)\times 2^N$ matrix $II_U(p)$, where

(2.4) $$II_U(p) = \begin{bmatrix} \Omega_U \\ \cdots\cdots \\ \underline{p}' \end{bmatrix},$$

where $\underline{p}$ is the sampling measure induced by the PST, as defined earlier. The interpretation of $II_U(p)$ is obvious: it presents each possible sample, along with the probability of obtaining that sample under the given PST. Next, assume momentarily that for all $j(j = 0, 1, \ldots, 2^N-1)$, the value of $p(\omega(j))$ is a rational number equal to, say, υ_j/υ, where υ and υ_j are nonnegative integers, with $\upsilon > 0$. For convenience, we shall assume that υ is the smallest integer, such that this holds. Let

(2.5) $$\eta = \{j | j = 0, 1, \ldots, 2^N-1;\ p(\omega(j)) > 0\}.$$

Also, let $\Delta_U(p)$ be a $(N \times \mu)$ (0, 1)-matrix (whose columns, like those of Ω_U, are arranged in standard order), in which (for all $j \in \eta$), the jth column of Ω_U occurs υ_j times. The number of columns in $\Delta_U(p)$ will then be given by μ where

(2.6) $$\mu = \sum_{j=0}^{2^N-1} \upsilon_j = \sum_{j \in \eta} \upsilon_j.$$

Notice that the columns of $\Delta_U(p)$ are not necessarily distinct. The matrix $\Delta_U(p)$ is called the 'Sampling Array' corresponding to the PST S.

Now, suppose we randomly select one column of $\Delta_U(p)$, such that each column of $\Delta_U(p)$ is given the same chance μ^{-1} of being selected. Notice that this is equivalent to selecting a column from Ω_U such that the jth ($j = 0, 1, \ldots, 2^N-1$) column of Ω_U has the chance $p(\omega(j))$ of being selected. Consider the PST, say S^*, which consists of considering $\Omega_U(p)$, and selecting one column of $\Omega_U(p)$ with probability μ^{-1}. Then, it is clear that the PST's S and S^* are equivalent in the sense that both give rise to the same sampling measure $p(\cdot)$. Indeed, we call S^*, the 'array representation' of S. Also, S^* is called an 'Array-PST' (abbreviation, APST). Notice that $\Delta_U(p)$ is the sampling array of both S and S^*.

In the above discussion, we have assumed that S is such that all elements of $\underline{p}$ are rational numbers. In case this is not true, notice that we can approximate $\underline{p}$ arbitrarily closely by a vector (say) $\underline{p}^*$, such that all elements of $\underline{p}^*$ are rational numbers. Next we can construct a matrix $\Delta_U(p^*)$ by following exactly the same procedure as is used for $\Delta_U(p)$, except that the vector $\underline{p}^*$ is used instead of $\underline{p}$. Let S^* be the PST corresponding to $\Delta_U(p^*)$. Then, obviously S and S^* are almost equivalent in the sense that the corresponding vectors $\underline{p}$ and $\underline{p}^*$ are very close to each other. We record this in

Theorem 2.1. Let S be any PST. Then, there exists an APST S^* which is arbitrarily close to S with respect to the Sampling measure. (More precisely, suppose S induces the probability distribution vector $\underline{p}$. Then, given any $\epsilon > 0$ however small, we can find an APST S^* with probability distribution vector $\underline{p}^*$, such that $(\underline{p} - \underline{p}^*)'(\underline{p} - \underline{p}^*) < \epsilon$.)

The above theorem establishes a fundamental connection between Sampling Theory and Combinatorial Arrays. Considered over GF(2) (the finite field with the two elements 0 and 1), the columns of Ω_U constitute a vector space, and correspond one-one with the points of EG(n, 2), the finite Euclidean Geometry of n dimensions based on GF(2). On the other hand, such arrays arise naturally in Statistical Design Theory. Indeed, the columns of the array Ω_U represents the 2^N distinct treatment-combinations in a 2^N factorial experiment (where the N factors correspond to the N rows of Ω_U). Thus, the above theorem connects Sampling Theory with other rich and well-cultivated fields.

Throughout the rest of this paper, we shall assume that we are working with a PST S whose sampling measure is $p(\cdot)$ (with probability distribution vector $\underline{p}$), and whose sampling array is $\Delta_U(p)$. Also, throughout, the variable of interest will be assumed to be y, with value y_i for the ith unit. We shall generally be interested in estimating the population total Y given by

$$Y = \sum_{i=1}^{N} y_i. \tag{2.7}$$

Also, we define $\underline{y}$ by

$$\underline{y} = (y_1, \ldots, y_N)'. \tag{2.8}$$

3. Estimation

We discuss some aspects of estimation of Y needed in the sequel. Consider the general linear estimator of Y given by $\hat{Y}_G$ (G for 'general') where

$$(3.1) \qquad \hat{Y}_G = \sum_{i\epsilon\omega} c_{i\omega} y_i = \sum_{i=1}^{N} c_{i\omega} a_{i\omega} y_i$$

where ω is any given sample, and the $c_{i\omega}$ are known real numbers which may depend on both i and ω, for all $(i \in U, \omega \in 2^U)$. Define $\phi_{ic}(i \in U)$ by

$$(3.2) \qquad \phi_{ic} = \sum_{\omega} c_{i\omega} a_{i\omega} p(\omega),$$

where, throughout this paper, the notation $\sum_{\omega}$ will denote sum over all $\omega \in 2^U$.

We record the obvious and well known result given by

Theorem 3.1. The estimator $\hat{Y}_G$ gives an unbiased estimate of Y if and only if we have

$$(3.3) \qquad \phi_{ic} = 1, \text{ for all } i \in U.$$

Proof: We have $E(\hat{Y}_G) = \sum_{\omega}[\sum_{i=1}^{N} y_i c_{i\omega} a_{i\omega}]p(\omega) = \sum_{i=1}^{N} y_i \phi_{ic} \equiv \sum_{i=1}^{N} y_i$, for all $\underline{y}$, leading to the result.

Throughout, for any positive integers m and n, the symbol J_{mn} will denote a $(m \times n)$ matrix every element of which equals 1. For all $i \neq j$, $(i, j \in U)$, define

$$(3.4) \qquad \phi^0_{iic} = \sum_{\omega} c^2_{i\omega} a_{i\omega} p(\omega), \quad \phi^0_{ijc} = \sum_{\omega} c_{i\omega} c_{j\omega} a_{i\omega} a_{j\omega} p(\omega)$$

$$\underline{\phi}_c = (\phi_{11c}, \ldots, \phi_{Nc})', \quad \Phi^o = ((\phi^0_{ijc})),$$

so that for all i, j $\in$ U, the (i, j) element of the $(N \times N)$ matrix Φ^0 is ϕ^0_{ijc}, and $\underline{\phi}_c$ is a $(N \times 1)$ vector with ith element ϕ_{ic}, for $i \in U$.

Theorem 3.2. Consider $\hat{Y}_G$ as an estimator of Y. The mean square of $\hat{Y}_G$, denoted by $MSE(\hat{Y}_G)$ is given by

$$(3.5) \qquad MSE(\hat{Y}_G) = \underline{y}'\Phi_c\underline{y},$$

where Φ_c is an $(N \times N)$ matrix (which is known, and is positive semi definite, and is) given by

$$(3.6) \qquad \Phi_c = \Phi^0_c - (J_{N1}\underline{\phi}'_c + \underline{\phi}_c J_{1N}) + J_{NN} = \sum_{\omega} p(\omega)\underline{u}_{c\omega}\underline{u}'_{c\omega},$$

where $\underline{u}_{c\omega}$ is a $(N \times 1)$ vector defined by

$$(3.7) \qquad \underline{u}_{c\omega} = (c_{1\omega}a_{1\omega} - 1, c_{2\omega}a_{2\omega} - 1, \ldots, c_{N\omega}a_{N\omega} - 1).$$

Proof: We have, $MSE(\hat{Y}_G) = E(\hat{Y}_G - Y)^2 = E(\hat{Y}^2_G) - 2YE(\hat{Y}_G) + Y^2$. Now,

$$E(\hat{Y}_G) = \underline{y}'\underline{\phi}_c, \quad Y = \underline{y}'J_{N1}, \text{ and}$$

$$E(\hat{Y}^2_G) = \sum_{\omega}\{\sum_{i=1}^{N}\sum_{j=1}^{N} c_{i\omega}a_{i\omega}a_{j\omega}y_i y_j\}p(\omega)$$

$$= \sum_{i=1}^{N} y^2_i\{\sum_{\omega} c^2_{i\omega}a_{i\omega}p(\omega)\} + \sum_{i\neq j} y_i y_j\{\sum_{\omega} c_{i\omega}c_{j\omega}a_{i\omega}a_{j\omega}p(\omega)\}$$

$$= \underline{y}'\Phi^o_c\underline{y},$$

which gives the first equation in (3.6). Also, $\hat{Y}_G - Y = \sum_{i=1}^{N}(c_{i\omega}a_{i\omega} - 1)y_i =$

$\underline{y}'\underline{u}_{c\omega}$, so that $E(\hat{Y}_G - Y)^2 = \underline{y}'\{E(\underline{u}_{c\omega}\underline{u}'_{c\omega})\}\ \underline{y}$, leading to the second part of (3.6).

Remark 3.1. Notice that while the vector $\underline{y}$ is (an) unknown (constant), the matrix Φ_c (which is also a constant) is known once the PST (and hence the sampling measure $p(\cdot)$) is specified, and the estimator $\hat{Y}_G$ is selected. Now, $\hat{Y}_G$ gets specified once we select the $(N\times 2^N)$ matrix Ω_{Uc} where

(3.8) $$\Omega_{Uc} = ((c_{i\omega})),$$

so that we may consider the rows of Ω_{Uc} to be indexed by the elements of U, and the columns by the elements of 2^U. Thus, the nature of Φ_c is governed by the choice of $p(\cdot)$ and Ω_{Uc}, both of which are at our disposal. The matrix Φ_c could be considered as a connecting link between Sampling Theory and Design of Experiements. In the latter field, if there are υ parameters, one often deals with a $(\upsilon\times\upsilon)$ "information matrix" (say, M). In some cases, M has a "simple structure" of the form $(\alpha I_\upsilon + \beta J_{\upsilon\upsilon})$, where (for all q) I_q denotes the $(q\times q)$ identity matrix, and α and β are real numbers. (This happens, for example, in the theory of 'Universal Optimality' of Kiefer (1975).) In other cases, M may be more general and may belong to the linear associated algebra of a (uni- or multi-dimensional) partially balanced association scheme (introduced in Bose and Srivastava (1964)). Analogous to M, one may require Φ_c to have such structure. The simple case when Φ_c is of the form $(\alpha I + \beta J)$ will be considered in this paper; it will be seen that some special cases of this simple case are already very common in Sampling Theory. Requiring Φ_c to belong to more complex algebras may open up useful new fields; this will, however, be taken up elsewhere.

We now consider a special Ω_{Uc} matrix. This arises from the estimator (of Y) given by $\hat{Y}_{Sr1}$ discussed by the author in Srivastava (1985). (Indeed, in Srivastava (1985), an extremely general theory of estimation was introduced, and $\hat{Y}_{Sr1}$, was obtained as a very special case of the same.) To consider $\hat{Y}_{Sr1}$, we introduce a function $r(\omega)$, for all $\omega \in 2^U$, where for any ω, the value of $r(\omega)$ is some finite real number. Now, let

(3.9) $$\pi_r(i) = \sum_\omega r(\omega)p(\omega)a_{i\omega}, \text{ for } i \in U,$$

(3.10) $$c_{i\omega} = r(\omega)/\pi_r(i), \text{ for all } i \in U,\ \omega \in 2^U,$$

assuming that

(3.11) $$\pi_t(i) \neq 0, \text{ for } i \in U.$$

Then, we have

(3.12) $$\hat{Y}_G = \hat{Y}_{Sr1} \text{ (under (3.9) - (3.11))}.$$

Let $\hat{Y}_{HT}$ be the Horvitz - Thompson estimator of Y. Then, we notice that

(3.13) $$\hat{Y}_{Sr1} = \hat{Y}_{HT} \text{ (if } r(\omega) = 1, \text{ for all } \omega \in 2^U).$$

In Srivastava (1985), formulae for the variance of the general estimator introduced therein (and hence also of $\hat{Y}_{Sr1}$), were presented, as also the formulae for sample-based estimators of such variance expressions. Any reader interested in such formulae for any special cases of $\hat{Y}_{Sr1}$ discussed in this paper, should refer to Srivastava (1985).

Definition 3.1. The estimator $\hat{Y}_G$ is said to be '<u>location invariant</u>' if and only if we have

(3.14) $\hat{Y}_G$ (given that $\underline{y} = \underline{y}^*$) $= -y_0 N + Y_G$ (given that $\underline{y} = \underline{y}^* + y_0 J_{1N}$),

for all real numbers y_0.

Now, $\hat{Y}_G = \sum_{i=1}^{N} c_{i\omega} a_{i\omega} y_i$. Hence, if y_i is replaced by $y_i + y_0$ (for all $i \in U$), the value of $\hat{Y}_G$ will increase by $y_0(\sum_{i=1}^{N} c_{i\omega} a_{i\omega})$. We obtain

Theorem 3.2. A necessary and sufficient conditoin that $\hat{Y}_G$ is location invariant is that, for all $\omega \in 2^U$, we have

$$\sum_{i=1}^{N} c_{i\omega} a_{i\omega} = N. \tag{3.15}$$

Corollary 3.1 The estimator $\hat{Y}_{Sr1}$ is location invariant if and only if the function $r(\cdot)$ is such that, for all $\omega \in 2^U$, we ahve

$$r(\omega) \sum_{i=1}^{N} \frac{a_{i\omega}}{\pi_r(i)} = N \tag{3.16}$$

Corollary 3.2 Let π_i $(i \in U)$ be the probability that the unit i is included in the sample, so that

$$\pi_i = \sum_{\omega} p(\omega) a_{i\omega}. \tag{3.17}$$

Then a necessary and sufficient condition that $\hat{Y}_{HT}$ is location-invariant is that we have

$$\sum_{i=1}^{N} a_{i\omega} \pi_i^{-1} = N, \text{ for all } \omega \in 2^U, \tag{3.18}$$

where the π_i are all, clearly, positive.

Proof: Follows from Corollary 3.1.

4. Balanced-Array Sampling

For every finite integer $s > 1$, let σ_s be the set $\{0, 1, 2, \ldots, s-1\}$. By an 'array over σ_s' we shall mean an array whose elements come from the set σ_s. If $K(a \times b)$ is any array over σ_{s1} and $\underline{k}(a \times 1)$ is any column vector over σ_s, then the symbol $\lambda(\underline{k}, K)$ denotes the number of times $\underline{k}$ occurs as a column of K. For any positive integer g, let Ψ_g denote the group of permutations over g symbols. If $\psi \in \Psi_a$, and $j \in \sigma_a$, then $\psi(j)$ is the image of j under the permutation ψ. Also, if $\underline{k} = (k_1, \ldots, k_a)$, then we define $\psi(\underline{k}) = (\psi(k_1), \ldots, \psi(k_a))$.

Definition 4.1. Consider $K(a \times b)$ as above. Let $t(\leq a)$ be a positive integer. Then K is said to be an <u>balanced array</u> (B-array or BA) of <u>strength</u> t and s symbols, if the following holds: For every $(t \times b)$ subarray K_0 in K, and <u>every</u> $(t \times 1)$ vector $\underline{k}_0$ over σ_s, and also for <u>every</u> permutation $\psi \in \Psi_t$, we have

$$\lambda(\underline{k}_0, K_0) = \lambda(\psi_0(\underline{k}_0), K_0). \tag{4.1}$$

Furthermore, a balanced array K of strength t, is said to be an <u>orthogonal array</u>

(OA) of strength t, if for every K_0 and $\underline{k}_0$ as above, we have

$$(4.2)\qquad \lambda(\underline{k}_0, K_0) = b \cdot s^{-t}.$$

The terms 'orthogonal array' and 'balanced array' were introduced respectively in Rao (1946) and Srivastava (1972), although versions of these structures were studied before these respective dates. Both play a very important role in the theory of factorial experiments. The subject is also deeply connected with coding theory. Below, we recall some simple properties of such arrays. Some of these are needed in the sequel, and others are presented for the sake of completeness to enable the interested reader to utilize the results of such later sections.

Theorem 4.1. Let $K(a \times b)$ and $K^*(a^* \times b^*)$ be respectively any balanced and orthogonal arrays of strengh t, where $t \leq a$ and $t \leq a^*$, and where a, b, a^*, and b^* are positive integers.

(i) Every OA is a BA, but every BA is not an OA. (Indeed, very few BA's are OA's.)

(ii) If $a_0 < a$, and $a_0^* < a^*$, and $K_0(a_0 \times b)$ and $K_0^*(a_0^* \times b^*)$ are subarrays respectively of K and K^*, then K_0 and K_0^* are respectively BA and OA of strength t_0 and t_0^*, where $t_0 = \min(a_0, t)$, and $t_0^* = \min(a_0^*, t)$.

(iii) The arrays K and K^* are also of strength $(t - 1)$.

(iv) Let $\psi_1 \in \Psi_s$, and let $\psi_1(K)$ be obtained from K by replacing any element k in K by $\psi_1(k)$. Define $\psi_1(K^*)$ similarly. Then $\psi_1(K)$ and $\psi_1(K^*)$ are respectively BA and OA. An example of a K and a $\psi_1(K)$ is given by

$$(4.3)\qquad s = 3, \psi_1 = \begin{bmatrix} 0 & 1 & 2 \\ 2 & 0 & 1 \end{bmatrix}, \ K = \begin{bmatrix} 0 & 0 & 1 & 2 & 2 & 1 \\ 1 & 2 & 0 & 1 & 0 & 2 \\ 2 & 1 & 2 & 0 & 1 & 0 \end{bmatrix}, \ \psi_1(K) = \begin{bmatrix} 2 & 2 & 0 & 1 & 1 & 0 \\ 0 & 1 & 2 & 0 & 2 & 1 \\ 1 & 0 & 1 & 2 & 0 & 2 \end{bmatrix}.$$

(v) SupPose $s > 1$. Suppose in the array K (or K^*), the symbol $(s - 1)$ is replaced everywhere by α (where $\alpha \in \sigma_{s-1}$, and α is fixed), let K_1 (or K_1^*) be the array so obtained. Then K_1 and K_1^* are both BA. (Interestingly, K_1^* is not an OA, but in certain cases it is possible that K_1 is an OA.) For example, with s = 3, $\alpha = 1$, the array K_1 obtained from the K displayed in (4.3) is a BA given by

$$(4.4)\qquad K_1 = \begin{bmatrix} 0 & 0 & 1 & 1 & 1 & 1 \\ 1 & 1 & 0 & 1 & 0 & 1 \\ 1 & 1 & 1 & 0 & 1 & 0 \end{bmatrix}.$$

(vi) Let $a_0, a_1, \ldots, a_{s-1}$ be nonnegative integers such that

$$(4.5)\qquad a = a_0 + a_1 + \ldots + a_{s-1}; \ \nu(a; a_0, a_{11}, \ldots, a_{s-1}) = \frac{a!}{a_0! a_1! \ldots a_{s-1}!}.$$

Let $K(a \times b)$ be the matrix with $b = \nu(a; a_0, a_1, \ldots, a_{s-1})$ obtained by taking all possible distinct column vectors of size $(a \times 1)$ in which the symbol j occurs a_j times, for $j = 0, 1, \ldots, s-1$. Then K is a BA over σ_s.

(vii) If $K_1, \ldots, K_r$ are arrays, such that K_i $(i = 1, \ldots, r)$ is a BA over σ_{s_i} of size $(a \times b_i)$, where $s_1, \ldots, s_t$, and $b_1, \ldots, b_r$ are any positive integers, and if

$$(4.6)\qquad K = [K_1 : K_2 : \ldots : K_r],$$

then K is a BA over σ_s, where

(4.7) $$s = \max(s_1, \ldots, s_r).$$

(viii) Suppose that s is either a prime number or a power of a prime number, so that GF(s) (the finite field with s elements) exists. Let H(axh) be a matrix over GF(s) such that no set of t columns of H is linearly dependent. Let $b = s^h$, and let K(axb) be the matrix whose set of columns constitute the set of possible distinct column vectors in the column space of H. Then K is an OA with s symbols (the symbols being elements of GF(s)). (Notice that for t = 2, such a matrix H can be easily written down with $a = (s^h - 1)/(s-1)$, by taking as rows of H all distinct nonzero row vectors over GF(s), such that the first nonzero element in any such row vector equals 1.)

(ix) Suppose K(axb) is a BA of strength t over σ_2. Then there exist nonnegative real numbers $\mu_{t0}, \mu_{t1}, \ldots, \mu_t$ such that for <u>every</u> (txb) subarray K_2 in K, and for <u>every</u> (tx1) vector $\underline{x}$ over σ_2, we have

(4.9) $$\lambda(\underline{x}, K_2) = \mu_{tg}, \text{ where } g = wt(\underline{x}),$$

where $wt(\underline{x})$ is the 'weight of $\underline{x}$', as defined in the Introduction. Also, then we have

(4.10) $$\sum_{g=0}^{t} \mu_{tg}\binom{t}{g} = b.$$

In connection with result (viii) above, it should be remarked that for general t and s, coding theory offers a variety of matrices H. (Examples are the BCH codes, Srivastava codes, etc. (Berlekamp (1968).) There, H is called the "parity check matrix", and the space orthogonal to the column space of H is called the set of code words.

Now, we connect the B-arrays with sampling. Let $\underline{i} = (i_1, \ldots, i_k) \epsilon \{U:k\}$, where k is any positive integer. Define $\pi_r(\underline{i})$ by

(4.11) $$\pi_r(\underline{i}) = \sum_{\omega \underline{i}} r(\omega)p(\omega),$$

where, through the paper, the notation $\sum_{\omega \underline{i}}$ will mean that the sum is running over all $\omega \epsilon 2^U$ such that $i_1, i_2, \ldots, i_k \in \omega$. We shall also write

(4.12) $$\pi_r(\underline{i}) \equiv \pi(\underline{i}) \equiv \pi(i_1, \ldots, i_k), \text{ if } r(\omega) = 1, \text{ for } \omega \in 2^U.$$

Furthermore, when k = 1 or 2, $i_1 = i < i_2 = j$, and i, j ϵ U, we shall use the customary notation

(4.13) $$\pi(i) \equiv \pi_i, \ \pi(i, j) = \pi_{ij}.$$

<u>Definition 4.2.</u> Let k be a nonnegative integer. Let T be any PST, and let p(•) be the corresponding sampling measure. Then p(•) is said to correspond to "<u>Balanced Array Sampling with strength k</u>" if and only if there exist nonnegative real numbers $\theta_1, \ldots, \theta_k$ such that

(4.14) $$\pi(i_1, \ldots, i_g) = \theta_g; \text{ for } g = 0, 1, \ldots, k; \ (i_1, \ldots, i_g) \in \{U:g\},$$

where the cases (k = 0, g = 0) are included only for notational convenience and correspond merely to putting no condition of the form (4.14) on p(•).

When $k \geq 2$, the phrase "BA sampling with strength k" will be abbreviated to "BA Sampling".

<u>Theorem 4.2.</u> If T is such that p(•) corresponds to BA sampling with strength k, then there exists a APST T^* with sampling measure $p^*(\cdot)$, such that the sampling array of T^* is $\Omega_U(p^*)$, where $\Omega_U(p^*)$ is a B-array of strength k, and such that p^* is arbitrarily close to p.

<u>Proof:</u> By Theorem 2.1, we can select a p^* which is arbitrarily close to p, such that $p^*(\omega)$ are rational numbers for all $\omega \in 2^U$. Suppose $p^*(\omega(j)) = \upsilon_j^*/\upsilon^*$, for $j \in (0, 1, \ldots, 2^N - 1)$, where υ^* is a positive integer, and the υ_j^* are nonnegative integers such that (for simplicity) the highest common factor of the υ_j^* equals 1. Let $\Omega_U(p^*)$ be the sampling array corresponding to the measure p^*. Let $\mu^* = \sum_j \upsilon_j^*$, so that $\Omega_U(p^*)$ is $(N \times M^*)$. Let K be the $(k \times \mu^*)$ submatrix obtained by taking any k rows (say $i_1, \ldots, i_k$) of $\Omega_U(p^*)$. Then, we have $\lambda(J_{k1}, K) = \Sigma^* \upsilon_j^*$, where Σ^* runs over all j such that $i_1, \ldots, i_k \in \omega(j)$. Since $p(\cdot)$ and hence (approximately) $p^*(\cdot)$, correspond to BA sampling with strength k, it follows that the quantity $\Sigma^* \upsilon_j^*$, and hence $L(J_{k1}, K)$ is (approximately) constant for every $(k \times \mu^*)$ submatrix K of $\Omega_U(p^*)$. From the classical theory of B-arrays (for example, Srivastava (1972)), it then follows that $\Omega_U(p^*)$ is (approximately) a B-array of strength k. This completes the proof.

<u>Theorem 4.3.</u> Suppose $\Omega^*(N \times \mu)$ is a B-array of strength k, and suppose that Ω^* is used as a APST. Then it will give rise to balanced array sampling with strength k.

<u>Proof:</u> Obvious.

We now study the properties of $\hat{Y}_{HT}$ under BA sampling.

Throughout the paper, for any non-empty sample ω, the symbol $\bar{y}_\omega$ will denote the sample mean. Also, let $\bar{Y}$ be the population mean, and S^2 (= $(N - 1)^{-1} \sum_{i=1}^{N} (y_i - \bar{Y})^2$) the population variance.

<u>Theorem 4.4.</u> Consider BA sampling with inclusion probabilities as at (4.14). Then,

$$\hat{Y}_{HT} = \theta_1^{-1} |\omega| \bar{y}_\omega \tag{4.15}$$

$$V(\hat{Y}_{HT}) = (N - 1)S^2 \cdot \frac{1}{\theta_1}(1 - \frac{\theta_2}{\theta_1}) + N\{(\frac{1}{\theta_1} - 1) - (N - 1)(1 - \frac{\theta_2}{\theta_1^2})\}\bar{Y}^2 \tag{4.16}$$

<u>Proof:</u> Equation (4.15) follows at once from (4.14) and the definition of $\hat{Y}_{HT}$. From Cochran (1977) (Equation 9A.41), we have

$$V(\hat{Y}_{HT}) = \sum_{i=1}^{N} y_i^2(\frac{1}{\pi_i} - 1) + \sum_{i \neq j} y_i y_j (\frac{\pi_{ij}}{\pi_i \pi_j} - 1). \tag{4.17}$$

Using (4.14), and the fact that $N(N - 1)S^2 = (N - 1)(\sum_{i=1}^{N} y_i^2) - (\sum_{i \neq j} y_i y_j)$, and $(N - 1)S^2 = \sum_{i=1}^{N} y_i^2 - N\bar{Y}^2$, the result (4.16) may be verified.

Define the mean δ_1 and variance δ_2 of the sample size ($|\omega|$) by

$$\delta_1 = \sum_\omega p(\omega) |\omega|, \ \delta_2 = \sum_\omega p(\omega)\{|\omega| - \delta_1\}^2. \tag{4.18}$$

Theorem 4.5. Consider BA sampling with inclusion probabilites as at (4.14). Then

$$\theta_1 = \delta_1/N, \ \theta_2 = [\delta_2 + \delta_1^2 - \delta_1]/N(N-1), \tag{4.19}$$

$$V(\hat{Y}_{HT}) = S^2 \cdot [\frac{N}{\delta_1}\{(N - \delta_1) - \frac{\delta_2}{\delta_1}\}] + \frac{N^2\delta_2}{\delta_1^2}\bar{Y}^2. \tag{4.20}$$

Proof: Since we have BA sampling, it follows that $N\theta_1 = \sum_{i=1}^{N} \pi_i = \sum_{i=1}^{N} [\sum_{\omega} p(\omega)a_{i\omega}] = \sum_{\omega} p(\omega)|\omega| = \delta_1$. Similarly, $N(N-1)\theta_2 = \sum_{i\neq j} \pi_{ij} = \sum_{i\neq j} \{\sum_{\omega} p(\omega)a_{i\omega}a_{j\omega}\} = \sum_{\omega} p(\omega)[\sum_{i\neq j} a_{i\omega}a_{j\omega}] = \sum_{\omega} p(\omega)[|\omega|)(|\omega| - 1)] = \{\sum_{\omega} p(\omega)|\omega|^2 - \delta_1] = [\delta_2 + \delta_1^2 - \delta_1]$.
This give (4.19). Substitution into (4.16) leads to (4.20).

Remark 4.2. Notice that under SRSWOR, with fixed sample size δ_1 (in case δ_1 is an integer), we would obtain $V(\hat{Y}_{HT}) = S^2N^2(1/\delta_1 - 1/N)$, which is the first term in the second expression on the r.h.s. of (4.20). Thus, relative to this situation, we shall have a decrease in variance if and only if the second term on the r.h.s. of (4.20) is negative. Clearly, this would happen if and only if we have

$$C.V.(y) > \sqrt{N}, \tag{4.21}$$

where C.V.(y) is the coefficient of variation of the variable of interest y in the population, so that

$$C.V.(y) = S/|\bar{Y}|. \tag{4.22}$$

Remark 4.3. The above reveals the potential usefulness of BA sampling. Notice that the value of $\bar{Y}$ could be influenced by us. We do not know $\bar{Y}$, but suppose we have a guess estimate $\bar{Y}_g$. We new consider the conceptual population in which the variable of interest is x, with value x_i for unit $i(\epsilon U)$, where $x_i = y_i - \bar{Y}_g$. The mean $\bar{X}$ of this population is $(\bar{Y} - \bar{Y}_g)$, while the variance is S^2. We then try to estimate $\bar{X}$ using BA sampling and $\bar{X}_{HT}$, the new estimate of Y being $\hat{\hat{Y}}$ where

$$\hat{\hat{Y}} = \hat{X}_{HT} + \bar{Y}_g \cdot N. \tag{4.23}$$

The above results show that $\hat{\hat{Y}}$ will do better than SRSWOR (with fixed sample size δ_1) provided that we have

$$\{S/|\bar{Y} - \bar{Y}_g|\} > \sqrt{N} \tag{4.24}$$

Clearly, whether or not (4.24) is satisfied would depend upon how close the given value is to the true value. It appears that (4.24) should be satisfiable in a wide variety of situations. One is the class of situations of cluster sampling, where a good estimate $\bar{Y}_g$ of $\bar{Y}$ is available, S is large (which could happen, for example, if cluster means are widely different), and the number of clusters is not too large. Notice that such cluster sampling could be easily combined with subsampling within clusters. Another class of applications will be in successive sampling, where S is relatively large, but good estimates Y_g of Y are obtainable by considering the estimates of Y on previous occasions and using Bayesian techniques.

5. Proportional Array Sampling

Consider the $\pi(i_1,\ldots,i_k)$ defined at (4.12) and (4.11).

Definition 5.1. Let k (> 1) be a positive integer. Let T be any PST, and p(•) it's sampling measure. Then p(•) is said to correspond to "Proportioanl array sampling with strength k", sometimes abbreviated as 'Proportional Sampling', if and only if, for all g such that $2 \leq g \leq k$, and all $i_1,\ldots,i_g$ such that $1 \leq i_1 < i_2 <\ldots, i_g \leq N$, we have

$$\pi(i_1,\ldots,i_g) = \pi(i_1)\pi(i_2)\ldots\pi(i_g). \tag{5.1}$$

Because of Theorem 2.1, without essential loss of generality, we shall assume that each $\pi(i)$ (for $i \in U$) is a rational number, and furthermore that

$$\pi(i) = s_i/s, \text{ for } i \in U, \tag{5.2}$$

where s is a prime number (or a power of a prime number).

We now produce a APST whose sampling measure corresponds to proportionnal sampling with strength k. Let L(N×b) be a OA of strength k with s symbols (assumed, for simplicity, to be the elements of σ_s). (Notice that this means that $b = b_0 s^k$, where b_0 is a positive integer.) Let $\underline{\ell}'_i (i = 1,\ldots,N)$ denote the ith row of L. In $\underline{\ell}'_i$ $(i = 1,\ldots,N)$, replace the $(s_i - 1)$ symbols $\{2, 3,\ldots,s_i\}$, each by 1, and (if $s > s_i + 1$) replace the $(s - 1 - s_i)$ symbols $\{s_i + 1, s_i + 2,\ldots,s - 1\}$ by 0. Let $L^*(N\times b)$ be the array so obtained.

Theorem 5.1. The array L^*, considered as a APST, leads to proportional sampling of strength k, such that (5.2) is satisfied by the probabilities π_i corresponding to L^*.

Proof: That (5.2) is satisfied by L^* is obvious in view of the fact that the proportion of 1's in the ith $(i = 1,\ldots,N)$ row of L^* is $\{1 + (s_i - 1)\}\times b_0 s^{k-1}/b_0 s^k = s_i/s$. Also, if $1 \leq i_1 < i_2 <\ldots< i_k \leq N$, and $L_0^*(k\times b)$ is the submatrix consisting of the rows $\#i_1, i_2,\ldots,i_k$ of L^*, then using the properties of OA, it can be checked that

$$\lambda(J_{k1}, L_0^*) = b_0 s_{i_1} s_{i_2},\ldots,s_{i_k}. \tag{5.3}$$

This completes the proof.

Definition. If, under proportional sampling with strength k, we further have

$$\pi(i) = \theta_1 \text{ (say), for } i = 1,2,\ldots,N, \tag{5.4}$$

then we refer to it as 'balanced proportional sampling with strength k'.

Remark 5.1. (i) If U has N units, then what is referred to in the literature as 'Binomial Sampling' is actually balanced proportional sampling of strength N. (ii) 'Balanced proportional sampling with strength k' is a special case of 'BA sampling with strength k'. (iii) As remarked earlier, BA sampling with strength 4 should be satisfactory for 'imitating' SRSWOR up to fourth moments. Thus, instead of binomial sampling, 'balanced proportional sampling with strength four' would perhaps be satisfactory for most usages.

Theorem 5.2. (a) Under proportional sampling, we have

$$V(\hat{Y}_{HT}) = \sum_{i=1}^{N} y_i^2\left(\frac{1}{\pi_i} - 1\right) \tag{5.5}$$

Under balanced proportional sampling (with, say, (5.4)), the above reduces to

$$V(\hat{Y}_{HT}) = \left(\frac{1}{\delta_1} - 1\right)\{(N - 1)S^2 + N\bar{Y}^2\}. \tag{5.6}$$

Proof: Obvious, from (4.17).

6. Weight-Balanced Sampling

Consider a population in which $y_i \geq 0$, for all $i \in U$. If $\hat{Y}_{HT}$ is used, then it is clear that if ω_1 and ω_2 are two samples of unequal size, with $\omega_1 \subset \omega_2$, then $\hat{Y}_{HT}$ will give a larger estimate for Y if ω_2 is drawn relative to when ω_1 is drawn. In other words, $\hat{Y}_{HT}$ appears to be unduly influenced by sample size.

An alternative estimator of Y, denoted by $\hat{Y}_{S2}$ will now be presented. Recall the definition of $\hat{Y}_{Sr1}$ at (3.12). The estimator $\hat{Y}_{S2}$ is a special of $\hat{Y}_{Sr1}$, such that

(6.1) $$\hat{Y}_{S2} \equiv \hat{Y}_{Sr1}, \text{ when}$$

(6.2) $$r(\omega) = |\omega|^{-1}, \text{ for all } \omega \in 2^U \text{ (such that } \omega \text{ is nonempty).}$$

Lemma 6.1. For $i \in U$, let

(6.3) $$\pi_i' = \pi_r(i), \text{ when (6.2) holds.}$$

Then, we have

(6.4) $$\pi_i' = \Sigma_\omega^0 \frac{p(\omega)a_{i\omega}}{|\omega|}$$

(6.4) $$\hat{Y}_{S2} = |\omega|^{-1} \sum_{i \in \omega} y_i/\pi_i',$$

where, throughout this paper, Σ_ω^0 means that the sum is over all $\omega \in 2^U$ such that ω is nonempty.

Proof: Follows from (3.9) and (3.12).

Let us introduce the notation

(6.5) $$\pi_j'' = \Sigma_\omega^0 \frac{p(\omega)a_{j\omega}}{|\omega|^2}; \quad \pi_{ij}'' = \Sigma_\omega^0 \frac{p(\omega)a_{i\omega}a_{j\omega}}{|\omega|^2}, \quad (i \neq j);$$

for all $i, j \in U$.

Theorem 6.1. We have

(6.6) $$E(\hat{Y}_{S2}) = Y$$

(6.7) $$V(\hat{Y}_{S2}) = \sum_{i=1}^{N} y_i^2\{\frac{\pi_i''}{(\pi_i')^2} - 1\} + \sum_{i \neq j=1}^{N} y_i y_j\{\frac{\pi_{ij}''}{\pi_i'\pi_j'} - 1\}.$$

Proof: Equation (6.6) is obvious. Also, $V(\hat{Y}_{S2}) = E(\hat{Y}_{S2})^2 - Y^2$, and

(6.8) $$E(Y_{S2}^2) = E[|\omega|^{-2} \sum_{i=1}^{N} \sum_{j=1}^{N} (y_i y_j/\pi_i'\pi_j')a_{i\omega}a_{j\omega}]$$
$$= \sum_{i=1}^{N} \sum_{j=1}^{N} (y_i y_j/\pi_i'\pi_j')\{\sum_\omega |\omega|^{-2} a_{i\omega}a_{j\omega}p(\omega)\},$$

from which (6.7) follows.

Definition 6.1. Consider a PST T with sampling measure $p(\cdot)$. Then, $p(\cdot)$ is said to correspond to 'Weight-Balanced Sampling' (WB sampling), if and only if the following two conditions hold:

(6.9) (i) The quantities $\{\pi_i''/(\pi_i')^2\}$ are constant (say, equal to β_1), for all $i \in U$;

(ii) The quantities $\{\pi_{ij}''/\pi_i'\pi_j'\}$ are constant (say, equal to β_2), for all $i, j \in U$, with $i \neq j$.

Theorem 6.1 gives

Corollary 6.1. Under (6.9), we have

$$V(\hat{Y}_{S2}) = (\beta_1 - 1)(\sum_{i=1}^{N} y_i^2) + (\beta_2 - 1)(\sum_{i\neq j=1}^{N} y_i y_j) \tag{6.10}$$

$$= (N - 1)S^2(\beta_1 - \beta_2) + N\bar{Y}^2[(\beta_1 - \beta_2) + N(\beta_2 - 1)].$$

Definition 6.2. Suppose $p(\cdot)$ is such that in addition to (6.9), we have

(6.11) The quantites π_i' are constant (say, equal to β_3), for all $i \in U$.

Then $p(\cdot)$ is said to correspond to 'Strongly Weight-Balanced Sampling' (SWB Sampling).

Lemma 6.2. We have

$$\sum_{j=1}^{N} \pi_j'' = \sum_{\omega} p(\omega)/|\omega| \equiv \beta_0, \text{ say;} \tag{6.12}$$

$$\sum_{j=1}^{N} \pi_{ij}'' = \pi_i', \text{ for all } i \in U; \tag{6.13}$$

$$\sum_{j=1, j\neq i}^{N} \pi_{ij}'' = \pi_i' - \pi_i'', \text{ for all } i \in U. \tag{6.14}$$

Proof: The l.h.s. of (6.12) equals $\sum_{j=1}^{N} \{\sum_{\omega}^{0} p(\omega)a_{j\omega}|\omega|^{-2}\} = \sum_{\omega} p(\omega)|\omega|^{-1}$. Also, the l.h.s. of (6.14) equals $\{\sum_{j=1}^{N} \pi_{ij}'' - \pi_i''\}$, and $\sum_{j=1}^{N} \pi_{ij}'' = \sum_{j=1}^{N} [\sum_{\omega}^{0} p(\omega)a_{i\omega}a_{j\omega}|\omega|^{-2}] = \sum_{\omega}^{0} p(\omega)a_{i\omega}|\omega|^{-1} = \pi_i'$.

Consider a PST T., with sampling measure $p(\cdot)$, such that $p(\omega \text{ is empty}) = 0$. Also, under T, for $1 \leq n \leq N$, let $q(n)$ denote the probability that the sample size equals n, so that $q(n)$ equals $\sum^n p(\omega)$, where $\sum^n$ denotes the sum over all ω such that $|\omega| = n$. Also consider SRSWOR, and suppose $\hat{Y}_{SRSn} (\equiv N\bar{y}_\omega)$ is the usual estimator of Y based on a sample of size n. It is well known that

$$V(\hat{Y}_{SRSn}) = N^2 S^2 (\tfrac{1}{n} - \tfrac{1}{N}) \tag{6.15}$$

Definition 6.3. For T as in the last paragraph, we define the "Weighted Average Variance corresponding to T" (WAV(T)) by

$$WAV(T) = \sum_{n=1}^{N} q(n) \cdot V(\hat{Y}_{SRSn}) = \sum_{n=1}^{N} N^2 S^2 (\tfrac{1}{n} - \tfrac{1}{N}) q(n). \tag{6.16}$$

(The interpretation of WAV(T) is clear. Under T, a sample of size n is used with probabilty $q(n)$. Suppose T is replaced by the PST T^* which first selects a sample size (such that T^* selects size n with probability $q(n)$), and then draws a SRSWOR of size n form U. Suppose from the SRSWOR (say, ω), we estimate Y using an estimator $N\bar{y}_\omega$. Then, under T^* and with an estimator such as the one mentioned, the variance of the estimator will be WAV(T).)

Lemma 6.3. We have, for T as above,

$$WAV(T) = N^2 S^2 \{\beta_0 - N^{-1}\}. \tag{6.17}$$

Proof: The result follows from (6.16), (6.12), and the definition of $q(n)$.

Theorem 6.2. Consider a PST T, whose sampling measure $p(\cdot)$ is strongly weight balanced. Suppose, from the sample obtained by using T, the estimator $\hat{Y}_{S2}$ is computed. Then

(6.18) $$V(\hat{Y}_{S2}) = WAV(T).$$

Proof: Under (6.9) and (6.11), we have (using (6.12) - (6.14))

(6.19) (i) $\beta_0 = N\beta_1\beta_3^2$

(ii) $(N - 1)\beta_2\beta_3^2 = \beta_3 - \beta_1\beta_3^2$

(iii) $N\beta_3 = \sum_{i=1}^{N} \pi_i' = \sum_{i=1}^{N} [\Sigma_\omega^0 p(\omega)a_{i\omega}/|\omega|] = \Sigma_\omega^0 p(\omega) = 1,$

so that

(6.20) $$\beta_1 = N\beta_0,\ \beta_2 = N(N - 1)^{-1}(1 - \beta_0),$$

$$(\beta_1 - \beta_2) = N^2(\beta_0 - N^{-1})(N - 1)^{-1} = -N(\beta_2 - 1).$$

Using (6.20), (6.10), and (6.17), the result (6.18) follows.

Remark 6.1. The last result shows that, in a sense, the technique T coupled with $\hat{Y}_{S2}$, "imitates" SRSWOR.

We now present an example of an APST (say T_0) which corresponds to SWB sampling. Coonsider the case when N = 5. Then the matrix $II_U^*(p)$ corresponding to this T_0 is given by

(6.21) $$\Pi_U^*(p) = \begin{bmatrix} 1&1&1&0&0&0&0&1&1&1&1&0&0&0&0 \\ 0&0&0&1&1&1&0&1&1&1&0&1&0&0&0 \\ 1&1&0&1&1&0&1&1&0&0&1&1&1&1&0 \\ 1&0&1&1&0&1&1&0&1&0&1&1&1&0&1 \\ 0&1&1&0&1&1&1&0&0&1&1&1&0&1&1 \\ a&a&a&a&a&a&b&c&c&c&d&d&f&f&f \end{bmatrix},$$

where

(6.22) $$a = \frac{612}{7650},\ b = \frac{288}{7650},\ c = \frac{774}{7650},\ d = \frac{576}{7650},\ f = \frac{72}{7650}.$$

For T_0, it turns out that, for i, j $\epsilon\{1,\ldots,5\}$ with $i \neq j$, we have

(6.23) $$\pi_i' = \frac{1}{5},\ \pi_i'' = \frac{498}{7650},\ \pi_{ij}'' = \frac{258}{7650},\ q(1) = q(5) = 0,\ q(2) = \frac{216}{7650},\ q(3) = \frac{6282}{7650},$$
$$q(4) = \frac{1152}{7650}.$$

(6.24) $$WAV(T_0) = 3.137.$$

Notice that this To avoids 16 of the 31 possible non empty samples, but does use all possible samples of size 3.

We close this section with

Theorem 6.3. If a PST T corresponds to SWB sampling, then $\hat{Y}_{S2}$ is location invariant.

Proof: Recall Corollary 3.1. Here, we have $r(\omega) = |\omega|^{-1}$, and $\pi_r(i) = N^{-1}$, for all i. Hence $r(\omega) \sum_{i=1}^{N} \{a_{i\omega}/\pi_r(i)\} = N \cdot |\omega|^{-1} \cdot |\omega| = N$, so that (3.16) is satisfied for all $i \in U$. This completes the proof.

7. Controlled Sampling.

Consider a population U, and suppose that the class 2^U of all samples is divided into two subclasses C_0 and C_1, such that we wish to avoid all samples in C_1. The class C_0 maybe called the "class of preferred samples", and C_1 the "class of undesirable samples". Such situations arise quite often, and form an important part of the theory of 'controlled sampling', which was started by Kish and further contributed to (among others by B.V. Sukhatme and his collaborators). (An illustrative list of references is included at the end.)

An example given in Avadhani and Sukhatme (1973), which arose from real life applications is this. We have N = 7, and the 7 units are arranged in a plane, as in the following figure.

Figure 1

	2		5	
1		4		7
	3		6	

We wish to take a sample of size 3. To reduce travel costs, we wish to a sample of only 'neighboring' units. In Figure 1, any two units which are adjacent to each other and are on the outer ellipse, are 'neighboring'. Also the unit in the center (i.e. #4) is considered neighboring to all units. Then the class C_0 of preferred samples is shown by the columns of the following array:

$$C_0 = \left[\begin{array}{cccccc|cccccc|cccccc|ccc} 4&4&4&4&4&4&1&2&5&7&6&3&1&2&5&7&6&3&1&2&5 \\ 1&2&5&7&6&3&2&5&7&6&3&1&4&4&4&4&4&4&4&4&4 \\ 2&5&7&6&3&1&5&7&6&3&1&2&5&7&6&3&1&2&7&6&3 \end{array}\right], \qquad (7.1)$$

where the bars are used to separate subclasses of similar samples (for convenience of inspection). Consider the PST (say, T_1) which select one of the following seven samples with probability 1/7 each: (123, 145, 167, 246, 257, 347, 356). The set of these samples forms a balanced incomplete block design (BIBD). This T_1 will imitate SRSWOR up to second moments (Srivastava and Saleh (1985)). Avadhani and Sukhatme (1973) recommended T_1 over SRSWOR (of size 3), since this T_1 avoids the undesirable class C_1 much more than the competing SRSWOR would. But as these authors pointed out, this T_1 (or any other PST obtained from T_1 by permuting the units in U) does not totally avoid C_1; indeed two of the seven samples are in C_1.

Although, the BIBD approach does not offer a complete solution, a balanced array which does avoid C_1, is possible for this case. Thus, let T_2 be a PST which selects one of 16 possible samples in the list {147, 246, 543; 125, 257, 576, 763, 631, 312;, 15, 27, 56, 73, 61, 32; 5} such that the first three samples are each assigned the probability (1/11) and the remaining ones are assigned the probability (1/22). Notice that besides using some samples in C_0, we are using subsamples of such samples.

We now consider a general formulation of the problems of controlled sampling in terms of the concepts of array sampling introduced in this paper.

Remark 7.1. The problem of controlled sampling may be regarded as follows.

(i) Decide upon C_0 and C_1.

(ii) Decide upon a fixed sample size or an expected sample size; let this be n (where n is not necessarily an integer).

(iii) Decide upon whether we want BA sampling, or WB sampling.

(iv) Suppose we want BA sampling. Then the problem is to find a BA whose columns correspond to samples in C_0; if this is not possible, try to use subsamples of these samples. If samples in C_0 do not allow the formation of a B-array such that the requirement in (ii) is met, then one would have to choose between relaxing this requirement on the one hand, and using samples from C_1 on the other. (For example, in the Avadhani-Sukhatme example, if one wants a fixed sample size 3, then the BIBD should be used. If on the other hand, samples from C_1 are to be totally avoided, then (the B-array corresponding to) T_2 will be better.)

(v) The problem in (iv) is quite open. Even more so is the case if WB sampling is decided upon.

Remark 7.2. (General Comments) The use of arrays is quite popular in discrete experimental design theory, both in factorial designs and in other areas. The attempt in this paper is to connect it with sampling theory, so that the latter may gain from the wealth of material available in the former.

One main advantage in using arrays rather than classical sampling approaches such as SRSWOR is that the former generally leads to much practical gain since using it we can practice 'controlled sampling'. In the actual practice of survey sampling, efficiency of organisation, reliability of measurements (of the variable of interest), travel and other overhead costs, are also major factors. Controlled sampling is a realistic approach to all such influential but vaguely defined factors.

References

Avadhani, M.S. and Sukhatme, B.V. (1973), "Controlled sampling with equal probabilities and without replacement", International Statistical Review, 41, No. 2, 175-182.

Bose, R.C. and Srivastava, J.N. (1964), "Analysis of Irregular Factorial Fractions". Sankhya. 226A, 117-144.

Brewer, K.R.W. and Hanif, M. (1983), "Sampling with unequal probabilities", Springer-Verlag: New York, Heidelberg, Berlin.

Cochran, W.G. (1977),, "Sampling techniques", 3rd edition, New York: Wiley.

Kiefer, J.C. (1975) Generalised Youden designs, In "A Survey of Statistical Design and Linear Models" (Ed: J. Srivastava), North Holland, Amsterdam, pages 333-353.

Kish, L. (1965), "Survey sampling", J. Wiley and Sons, Inc., New York.

Raj, D. (1972), "Sampling Theory", McGraw Hill Book Company; New York

Rao, C.R. (1946), "Hypercubes of Strength d Leading to Confounded Designs in Factorial Experiements". Bulletin of the Calcutta Mathematical Society, 38, 67-73.

Rao, J.N.K. (1963), "On three procedures of unequal probability sampling without replacement", Journal of the American Staatistical Association, 58, 202-215.

Singh, D., and Chaudhary, F.S. (1987) Theory and Analysis of Sample Survey Designs. Wiley Eastern Limited, New Delhi.

Srivastava, A.K. and Singh, D. (1981), "A sampling procedure with inclusion probabilities proportional to size", Biometrika, 68, 732-734.

Srivastava, J.N. (1972) Some general existence conditions for balanced arrays of strength t and 2 symbols. Journal of Combinatorial Theory. 13, 198-206.

Srivastava, J.N. (1985) On a general theory of sampling using experimental design concepts, I.: Estimation. Proc. Intern. Stat. Inst., 45th Session, Amsterdam, Invited Paper Vol. I, pages 10.3 (1-16).

Srivastava, J.N., and Saleh, F. (1985) On the need of t-designs in sampling theory. Utilitas Mathematica. vol 28, pages 5-17.

Sukhatme, P.V. and Sukhatme, B.V. (1970), "Sampling theory of surveys with applications", Asian Publishing House, Calculta.

Sukhatme, P.V. and Sukhatme, B.V. (1976), "Sampling theory of surveys with applications", New Delhi: Indian Society of Agricultural Statistics.

Wynn, H.P. (1976), "Optimum designs for finite populations sampling", in Statistical Decision Theory and Related Topics, eds. S.S. Gupta and D.S. Moore, New York: Academic Press.

Original paper received: 29.04.87
Final paper received: 23.06.87

Paper recommended by P.W. Mielke

PROBABILITY AND STATISTICS
Essays in Honor of Franklin A. Graybill
J.N. Srivastava (Editor)

β-Expectation Tolerance Limits for Balanced One-Way Random-Effects Model

C. Ming Wang

Computer Science Department
General Motors Research Laboratories
Warren, Michigan 48090

In this article the existing method for determining the approximate β-expectation tolerance limits of the one-way random-effects model is examined. Some algebraic properties of these tolerance limits are also derived. In addition, an iterative algorithm which can be used to obtain an 'exact' β-expectation tolerance limit of the random-effects model is presented.

- Key words and phrases: Satterthwaite approximation; iterative algorithm.
- AMS subject classification: Primary 62F25; secondary 62J10.

1. INTRODUCTION

A common problem in statistical quality control is to verify that a product meets certain specifications. Statistical tolerance limits have been widely used for this purpose. A lower β-content tolerance limits, say L, is defined such that with $100\gamma\%$ confidence, at least $100\beta\%$ of the population is above L. For the population which is assumed to be normally distributed, the usual form for L is $L = \bar{X} - kS$, where $\bar{X}$ and S are the sample mean and standard deviation respectively. Tables of factors k for various combinations of βs, γs and smaple sizes have been given by several authors, for example, see Odeh and Owen(1980). In a related work, Hall(1984), based on the Satterthwaite approximation, provided a procedure for obtaining approximate lower β-content tolerance limits for the difference or sum of two normal deviates.

In many applications, it is also often desired to determine the tolerance limits when the population consists of many batches. The appropriate model for this population is the one-way random-effects model,

$$X_{ij} = \mu + b_i + \epsilon_{ij}, \quad i = 1, 2, \ldots, I, \quad j = 1, 2, \ldots, J$$

where X_{ij} denotes the jth observation from the ith batch, μ is an unknown constant, and b_i's and ϵ_{ij}'s are independent normal random variables with zero means and variances σ_b^2 and σ_w^2 respectively. Thus, X is normally distributed with mean μ and variance $\sigma_X^2 = \sigma_b^2 + \sigma_w^2$. Let $\bar{X}$, S_1^2, and S_2^2 denote, respectively, the sample mean, the among and the within mean squares with degrees of freedom $n_1 = I - 1$ and $n_2 = I(J - 1)$, then the unbiased estimators of μ and σ_X are X and $S_X^2 = S_1^2/J + (1 - 1/J)S_2^2$.

Two types of tolerance limits have been discussed in the literature for the one-way random-effects model. Lemon(1977) derived an approximate one-sided β-content tolerance limit which has the form of $\bar{X} - kS_1$. More recently, Mee and Owen(1983) described a procedure based on the Satterthwaite approximation, for obtaining an approximate one-sided β-content tolerance limit $\bar{X} - kS_X$. Their study also showed that Lemon's procedure is too conservative. Mee(1984a) extended the results to the two-sided case. In addition, he presented a two-sided β-expectation tolerance limit, which is defined as

$$E\{ Pr_X[\bar{X} - kS_X < X < \bar{X} + kS_X \mid \bar{X}, S_X] \} = \beta,$$

for the one-way random-effects model.

Hahn(1982) presented a method for determining whether a sufficient proportion of a product meets specifications when the data are subject to measurement errors. This problem is closely related to the tolerance limits of the one-way random-effects model. See, for example, Mee(1984b) and Jaech(1984).

In this article, the procedures of Mee(1984a) for one- and two- sided β-expectation tolerance limits of the one-way random-effects model are examined. Some algebraic properties of these tolerance limits are also derived. Then in Section 3, an iterative algorithm, analogous to the one used in the Behrens-Fisher problem for obtaining the β-expectation tolerance limits, is proposed. This method gives a tolerance limit with confidence coefficient essentially equal to a prescribed value. The method is illustrated in Section 5.

2. β-EXPECTATION TOLERANCE LIMITS

A two-sided β-expectation tolerance limit is defined such that

$$E\{ Pr_X[\bar{X} - kS_X < X < \bar{X} + kS_X] \} = \beta \tag{2.1}$$

where the expectation is with respect to $\bar{X}$ and S_X. Since a two-sided β-expectation tolerance limit is equivalent to a one-sided α-expectation tolerance limit for $\alpha = (1 + \beta)/2$, the problem is to determine the value of k such that

$$E\{ Pr_X[X < \bar{X} + kS_X] \} = \alpha, \tag{2.2}$$

or, equivalently,

$$E[\Phi(\lambda Z + kS_X/\sigma_X)] = \alpha \tag{2.3}$$

where $\Phi(\cdot)$ denotes the distribution function of standard normal, Z is distributed as a standard normal, and

$$\lambda^2 = (J\rho + 1 - \rho)/(IJ),$$

and where

$$\rho = \sigma_b^2/(\sigma_b^2 + \sigma_w^2)$$

Using the result of Fleiss(1971) (also Graybill and Wang(1980)), we have

$$\begin{aligned} S_X^2/\sigma_X^2 &= \gamma U_1/n_1 + (1 - \gamma)U_2/n_2 \\ &= [\gamma(1 - W)/n_1 + (1 - \gamma)W/n_2]U, \end{aligned}$$

where $\gamma = \rho + (1 - \rho)/J$, U_1 and U_2 are independently distributed as chi-squared distributions with n_1 and n_2 degrees of freedom respectively. $U = U_1 + U_2$, i.e. U is a chi-squared random variable with $\nu = n_1 + n_2$ degrees of freedom, and W is a

Beta random deviate with parameters $(n_2/2, n_1/2)$, and U and W are independent. Thus, (2.3) can be written as

$$E[\ Pr_Y\{\ Y < k[\gamma(1 - W)/n_1 + (1 - \gamma)W/n_2]^{1/2}\ U^{1/2}\ \}\ ]$$

where the expectation is with respect to W and U, and Y is distributed as a normal with mean 0 and variance $1 + \lambda^2$ and is independent of U. Thus, $t = Y/[(1 + \lambda^2)U/\nu]^{1/2}$ is subject to the Student's t distribution with ν degrees of freedom, or we have

$$\begin{aligned} P(\rho) &= E\{\ Pr_X[\ X < \bar{X} + kS_X\ ]\ \} \\ &= E_W[Pr_t\{t < k[\nu(\gamma(1 - W)/n_1 + (1 - \gamma)W/n_2)/(1 + \lambda^2)]^{1/2}\}] \\ &= \int_0^1 T_\nu[k\{\nu(\gamma(1 - w)/n_1 + (1 - \gamma)w/n_2)/(1 + \lambda^2)\}^{1/2}]f(w)dw \qquad (2.4) \end{aligned}$$

where $T_\nu(\cdot)$ and $f(\cdot)$ are respectively the distribution function of Student's t with ν degrees of freedom and the Beta density function with parameters $(n_2/2, n_1/2)$.

Given k, the actual probability coverages of the tolerance limit, which depends on ρ, can be evaluated using (2.4). It is also of interest to investigate the behavior of $P(\rho)$ when k is constant. Since $W = (n_2F/n_1)/(1 + n_2F/n_1)$, where F is an F distributed random variable with n_2 and n_1 degrees of freedom, we have

$$\begin{aligned} P(\rho) &= E\{\ Pr_X[\ X < \bar{X} + kS_X\ ]\ \} \\ &= E_F\{\ Pr_t[\ t < kF_\rho^{1/2}\ ]\ \} \end{aligned}$$

where $F_\rho = (p_1 + q_1F)/(p_2 + q_2F)$, and $p_1 = \gamma/(1 + \gamma/I)$, $q_1 = (1 - \gamma)/(1 + \gamma/I)$, $p_2 = n_1/(n_1 + n_2)$, and $q_2 = n_2/(n_1 + n_2)$. Thus the derivative of $P(\rho)$ with respect to ρ is

$$\begin{aligned} P'(\rho) = -c_1\int_0^\infty[(1 + 1/I)F - 1]\ F^{\,n_2/2-1}(p_2 + q_2F)^{-\nu/2-1} \\ \cdot\ (1 + k^2F_\rho/\nu)^{-(\nu+1)/2}\ F_\rho^{-1/2}\ dF \end{aligned}$$

where c_1 is a positive constant. Since $p_1 > p_2$ and $q_1 < q_2$ for all ρ, F_ρ is a decreasing function of F. Furthermore, since $p_1 + q_1F = 1/(1 + 1/I)$, which is independent of ρ when $F = 1/(1 + 1/I)$, there exists a constant $b > 0$ such that $bF_\rho = b(p_1 + q_1F)/(p_2 + q_2F) > 1$ for $F < 1/(1 + 1/I)$ and $bF_\rho < 1$ for $F > 1/(1 + 1/I)$. Thus, for $0 \le \rho \le 1$, we have

$$P'(\rho) < -(c_2/c_3)\int_0^\infty[(1 + 1/I)F - 1]\ F^{\,n_2/2-1}\ (p_2 + q_2F)^{-\nu/2-1}dF$$

where c_2 is a positive constant and $c_3 = (1 + k^2/\nu)^{(\nu+1)/2}$. Since the integral in the right side of the above inequality equals $n_1\Gamma(n_1/2)\Gamma(n_2/2)/[2I\Gamma(\nu/2 + 1)]$, we find that $P'(\rho) < 0$ for $0 \le \rho \le 1$. In other word, $P(\rho)$ is a decreasing function of ρ when k is constant, or a tolerance limit of (2.1) requires that k be an increasing function of ρ.

Based on the Satterthwaite approximation to the distribution of S_X/σ_X, Mee(1984a) proposed a solution to (2.2)

$$k_M = (1 + L^2)^{1/2} t_{a,d} \tag{2.5}$$

where L is the estimate of λ and $t_{a,d}$ denotes the 100a percentile for a Student's t distribution with d degrees of freedom. If we denote the estimate of ρ by R, i.e.

$$R = \text{maximum}\{ 0, (S_1^2 - S_2^2)/[S_1^2 + (J-1)S_2^2] \},$$

then

$$L^2 = (JR + 1 - R)/(IJ),$$

and

$$d = I(I-1)J^2/[I(JR + 1 - R)^2 + (I-1)(J-1)(1-R)^2]$$

Clearly, R can be expressed as a function of ρ and W. Also, k_M is an increasing function of ρ. Values of

$$E\{ Pr_X[X < \bar{X} + k_M S_X] \} \tag{2.6}$$

are obtained by numerical integration using (2.4) (see Section 4 for more computational details) and displayed in Table 1.

Since $(I - 1) \le d \le J^2/[1/(I-1) + (J-1)/I]$, k_M gives correct asymptotic results for large I, but not for fixed I and large J. Table 1 also bears this out. Using the fact that J tends to ∞, the random quantity U_2/n_2 approaches unity in probability, the extreme value of (2.6) can be shown as

$$E[\Phi\{ [(\rho U_1/n_1 + 1 - \rho)(1 + R/I)/(1 + \rho/I)]^{1/2} t_{a,d} \}] \tag{2.7}$$

where $d = (I - 1)/R^2$. Table 2 reports the values of (2.7) for some selected values of I.

The results of Tables 1 and 2 indicate that k_M is adequate for moderate values of I. It is also apparent that a more accurate tolerance limit K_M is obtained by increasing the numbers of batches (I) as compared to increasing number of samples (J) within each batch.

3. AN ITERATION METHOD

The $P(\rho)$ in (2.4) is similar to the one in the Behrens-Fisher problem in the sense that it can be represented as an average of values of a Student's t distribution over a Beta distribution. The difference is in the argument of the t distribution. It follows that it is possible to apply various techniques used in the Behrens-Fisher problem to the β-expectation tolerence limits. The following procedure we use to determine the tolerance limit is similar to the one used by Trickett and Welch(1954) to obtain a confidence interval in the Behrens-Fisher problem.

What we attempt, as stated above, is to determine a function of R, $k(\cdot)$, such that

$$E\{ Pr_X[X < \bar{X} + k(R)S_X] \} = a$$

Table 1. Probability coverages (times 10^4) of tolerance limit (2.5)

I	J	ρ = 0.0	0.1	0.2	0.3	0.4	0.5	0.6	0.7	0.8	0.9	1.0
						$\alpha = 0.90$						
2	2	9155	9133	9106	9077	9044	9007	8968	8927	8888	8863	9000
3	3	9057	9045	9031	9014	8996	8976	8957	8939	8927	8931	9000
3	10	9016	9009	8999	8984	8966	8946	8926	8907	8897	8905	9000
3	50	9003	9000	8993	8979	8962	8941	8919	8900	8889	8898	9000
5	3	9024	9019	9013	9006	8999	8992	8984	8979	8976	8979	9000
5	10	9007	9003	9000	8995	8989	8982	8975	8969	8967	8972	9000
10	3	9007	9005	9003	9001	8999	8998	8996	8995	8994	8996	9000
10	10	9002	9001	9000	8999	8997	8996	8994	8993	8992	8994	9000
						$\alpha = 0.95$						
2	2	9610	9592	9571	9547	9517	9483	9444	9398	9347	9302	9500
3	3	9544	9533	9518	9500	9480	9457	9433	9410	9393	9394	9500
3	10	9513	9505	9492	9473	9450	9424	9396	9371	9352	9358	9500
3	50	9502	9499	9488	9470	9446	9419	9389	9362	9342	9348	9500
5	3	9519	9515	9509	9502	9494	9485	9476	9468	9464	9469	9500
5	10	9505	9502	9498	9492	9484	9474	9464	9456	9452	9459	9500
10	3	9506	9504	9503	9501	9498	9496	9494	9492	9492	9494	9500
10	10	9502	9500	9500	9498	9496	9494	9491	9490	9489	9492	9500
						$\alpha = 0.99$						
2	2	9935	9931	9925	9917	9907	9894	9877	9852	9815	9755	9900
3	3	9915	9909	9902	9892	9879	9863	9845	9823	9800	9786	9900
3	10	9905	9900	9890	9876	9858	9836	9810	9782	9756	9745	9900
5	3	9908	9906	9902	9897	9891	9884	9877	9869	9864	9866	9900
5	10	9902	9900	9897	9892	9884	9876	9867	9857	9851	9855	9900
10	3	9903	9902	9901	9899	9898	9896	9894	9892	9892	9893	9900
10	10	9901	9900	9899	9898	9896	9894	9892	9890	9889	9891	9900

Thus, we want to determine a function $k(\cdot)$ that satisfies the integral equation below for $0 \le \rho \le 1$.

$$\int_0^1 T_\nu[k(R)g(\gamma,w)]f(w)dw = \alpha$$

where $g(\gamma,w) = \{\nu[\gamma(1-w)/n_1 + (1-\gamma)w/n_2]/(1+\lambda^2)\}^{1/2}$. To find the function $k(\cdot)$ we proceed as follows. Suppose that a first approximation of k is given by k_0, a specified function of R; we write

$$k(R) = k_0(R) + k_1(R)$$

Then, expand $k(R)$ into two terms of a Taylor series about $k_0(R)$ to obtain

$$\begin{aligned} \alpha &= \int_0^1 T_\nu[k(R)g(\gamma,w)]f(w)dw \\ &\simeq \int_0^1 T_\nu[k_0(R)g(\gamma,w)]f(w)dw + \\ &\quad \int_0^1 k_1(R)g(\gamma,w)\cdot t_\nu[k_0(R)g(\gamma,w)]f(w)dw \end{aligned} \tag{3.1}$$

which we denote by the equation $Q = Q_0 + Q_1$; where $t_\nu(\cdot) = T_\nu'(\cdot)$ is the density

Table 2. Values of (2.7) (times 10^4)

ρ	0.0	0.1	0.2	0.3	0.4	0.5	0.6	0.7	0.8	0.9	1.0
I						$\alpha = 0.90$					
3	9000	8999	8992	8979	8961	8940	8918	8899	8887	8896	9000
5	9000	9000	8998	8993	8987	8980	8973	8966	8964	8969	9000
10	9000	9000	8999	8998	8997	8995	8993	8992	8992	8994	9000
						$\alpha = 0.95$					
3	9500	9498	9488	9470	9446	9418	9388	9360	9340	9345	9500
5	9500	9499	9497	9491	9482	9472	9462	9452	9448	9455	9500
10	9500	9500	9499	9498	9496	9493	9491	9489	9488	9491	9500
						$\alpha = 0.99$					
3	9900	9898	9889	9874	9855	9830	9802	9771	9742	9730	9900
5	9900	9899	9897	9891	9884	9874	9864	9854	9847	9850	9900
10	9900	9900	9899	9898	9896	9894	9891	9889	9888	9890	9900

function of Student's t with ν degrees of freedom.

If we can obtain $k_1(R)$ to satisfy (3.1), then an improved approximation $k_0(R) + k_1(R)$ to the solution might be expected. Since $k_0(R)$ is set equal to a known function, all quantities in Q_0 are known except ρ, and the integral Q_0 can be evaluated by numerical integration for any set of values of ρ, say $\rho = 0$, 0.1,..,0.9, 1.0. The integral Q_1 can not be evaluated as it stands because it contains the unknown function of R, $k_1(\cdot)$. In order to solve for $k_1(R)$, a simplification to the integral Q_1 is made as follows.

The density function f(w) has mean value at $w = n_2/(n_1 + n_2)$. It follows that for large I, $R \to \rho$ and $g(\gamma,w) \to 1$. With these results, Q_1 becomes $k_1(\rho)\cdot t_\nu[k_0(\rho)]$ and (3.1) becomes

$$\alpha \simeq \int_0^1 T_\nu[k_0(R)g(\gamma,w)]f(w)dw + k_1(\rho)\cdot t_\nu[k_0(\rho)]$$

or

$$k_1(\rho) = (\alpha - Q_0)/t_\nu[k_0(\rho)]$$

Note that Q_0 depends on ρ and $k_0(\cdot)$ is a known function. By taking sufficient number of values of ρ, the tables of $k_1(\rho)$ and hence $k(\rho)$ can be constructed. Since $k(R)$ is the same function of R as $k(\rho)$ of ρ, this is equivalent to tabulating $k(R)$.

The function $k_1(R)$ so obtained will not exactly satisfy (3.1) because of the approximation made to the integral Q_1, but may bring us closer to a solution of (3.1). We continue the iteration until $\max_\rho|\alpha - Q_0(\rho)| < \epsilon$ for a selected small value ϵ, say 0.0001.

4. COMPUTATIONS

It seems reasonable for our problem to set $k_0(R)$ to $k_M(R)$ of Mee(1984a). Thus as ρ takes on values 0, 0.1, 0.2,..., 1.0, we can evaluate Q_0 and perform the iteration. An automatic adaptive routine based on the 8-panel Newton-Cotes rule (Forsythe, Malcolm and Moler(1977)) is used to evaluate the integral and the cubic natural spline interpolation routines (subroutines ICSCCU and ICSEVU of IMSL) are used to calculate $k_1(R)$ for sample values of R.

Tables 3 and 4 contains values of k(R) for R = 0, 0.1, 0.2,...,1.0, for α = .9 and .95 and for some selected values of I and J. The value ϵ = 0.0005 was used to compute the function $k(\cdot)$ so that

$$E\{\ Pr_X[\ X < \bar{X} + k(R)S_X\ ]\ \}$$

is between .8995 and .9004 when Table 3 is used and between .9495 and .9504 when Table 4 is used. For values of R not in the table, a linear interpolation is probably adequate.

A Fortran program which produces values of k(R) is available from the author on request.

Table 3. Values of k(R) for α = 0.90

I	J (R)	0.0	0.1	0.2	0.3	0.4	0.5	0.6	0.7	0.8	0.9	1.0
4	2	1.4436	1.4681	1.5004	1.5401	1.5864	1.6379	1.6919	1.7434	1.7839	1.8022	1.8311
	3	1.3883	1.4117	1.4423	1.4801	1.5251	1.5770	1.6346	1.6953	1.7532	1.7963	1.8311
	4	1.3607	1.3837	1.4136	1.4513	1.4969	1.5504	1.6111	1.6763	1.7401	1.7896	1.8311
	5	1.3446	1.3673	1.3965	1.4338	1.4796	1.5342	1.5967	1.6648	1.7323	1.7858	1.8311
	6	1.3341	1.3565	1.3851	1.4220	1.4680	1.5233	1.5872	1.6573	1.7273	1.7833	1.8311
	7	1.3266	1.3488	1.3768	1.4135	1.4597	1.5155	1.5804	1.6519	1.7237	1.7816	1.8311
	8	1.3209	1.3430	1.3706	1.4071	1.4534	1.5097	1.5753	1.6479	1.7211	1.7803	1.8311
	9	1.3166	1.3385	1.3658	1.4021	1.4485	1.5051	1.5713	1.6448	1.7191	1.7793	1.8311
	10	1.3131	1.3348	1.3618	1.3981	1.4446	1.5015	1.5682	1.6423	1.7174	1.7785	1.8311
	∞	1.2816	1.3003	1.3262	1.3621	1.4093	1.4682	1.5383	1.6176	1.6999	1.7707	1.8311
5	2	1.4172	1.4333	1.4536	1.4780	1.5059	1.5369	1.5699	1.6032	1.6339	1.6580	1.6795
	3	1.3705	1.3872	1.4080	1.4331	1.4624	1.4959	1.5330	1.5730	1.6136	1.6504	1.6795
	4	1.3474	1.3645	1.3855	1.4111	1.4415	1.4767	1.5164	1.5598	1.6045	1.6459	1.6795
	5	1.3339	1.3511	1.3721	1.3978	1.4287	1.4650	1.5063	1.5518	1.5991	1.6433	1.6795
	6	1.3250	1.3423	1.3630	1.3888	1.4202	1.4572	1.4996	1.5465	1.5955	1.6415	1.6795
	7	1.3187	1.3359	1.3566	1.3824	1.4140	1.4516	1.4948	1.5427	1.5929	1.6403	1.6795
	8	1.3140	1.3312	1.3517	1.3776	1.4094	1.4474	1.4912	1.5398	1.5910	1.6393	1.6795
	9	1.3104	1.3275	1.3479	1.3738	1.4058	1.4441	1.4884	1.5376	1.5895	1.6386	1.6795
	10	1.3075	1.3245	1.3448	1.3707	1.4029	1.4415	1.4861	1.5359	1.5884	1.6381	1.6795
	∞	1.2816	1.2965	1.3166	1.3431	1.3764	1.4169	1.4642	1.5176	1.5746	1.6301	1.6795

Table 3. (continued)

I	J \ R	0.0	0.1	0.2	0.3	0.4	0.5	0.6	0.7	0.8	0.9	1.0
6	2	1.3978	1.4094	1.4239	1.4410	1.4605	1.4822	1.5056	1.5300	1.5540	1.5757	1.5941
	3	1.3566	1.3699	1.3859	1.4050	1.4270	1.4520	1.4797	1.5093	1.5397	1.5684	1.5941
	4	1.3387	1.3516	1.3673	1.3861	1.4083	1.4339	1.4629	1.4951	1.5298	1.5650	1.5941
	5	1.3269	1.3401	1.3561	1.3753	1.3982	1.4248	1.4552	1.4891	1.5257	1.5631	1.5941
	6	1.3192	1.3325	1.3486	1.3681	1.3915	1.4188	1.4500	1.4850	1.5229	1.5618	1.5941
	7	1.3137	1.3271	1.3432	1.3630	1.3866	1.4144	1.4463	1.4821	1.5210	1.5608	1.5941
	8	1.3096	1.3230	1.3392	1.3591	1.3830	1.4112	1.4435	1.4799	1.5195	1.5601	1.5941
	9	1.3065	1.3199	1.3361	1.3560	1.3802	1.4086	1.4414	1.4782	1.5183	1.5596	1.5941
	10	1.3039	1.3173	1.3335	1.3536	1.3779	1.4066	1.4396	1.4768	1.5174	1.5591	1.5941
	∞	1.2816	1.2940	1.3103	1.3314	1.3574	1.3885	1.4246	1.4650	1.5085	1.5526	1.5941
7	2	1.3847	1.3933	1.4037	1.4161	1.4303	1.4463	1.4638	1.4827	1.5025	1.5222	1.5392
	3	1.3482	1.3583	1.3706	1.3851	1.4020	1.4212	1.4427	1.4662	1.4913	1.5167	1.5392
	4	1.3309	1.3416	1.3545	1.3699	1.3879	1.4086	1.4319	1.4578	1.4856	1.5140	1.5392
	5	1.3207	1.3318	1.3450	1.3607	1.3794	1.4010	1.4255	1.4527	1.4821	1.5123	1.5392
	6	1.3141	1.3253	1.3386	1.3546	1.3737	1.3959	1.4212	1.4493	1.4798	1.5112	1.5392
	7	1.3093	1.3206	1.3340	1.3502	1.3696	1.3922	1.4181	1.4469	1.4781	1.5104	1.5392
	8	1.3058	1.3171	1.3305	1.3469	1.3665	1.3895	1.4157	1.4451	1.4769	1.5098	1.5392
	9	1.3031	1.3144	1.3279	1.3444	1.3642	1.3874	1.4139	1.4436	1.4759	1.5093	1.5392
	10	1.3009	1.3122	1.3257	1.3423	1.3623	1.3857	1.4125	1.4425	1.4752	1.5089	1.5392
	∞	1.2816	1.2921	1.3059	1.3233	1.3444	1.3694	1.3982	1.4305	1.4659	1.5029	1.5392
8	2	1.3718	1.3790	1.3877	1.3981	1.4099	1.4231	1.4376	1.4532	1.4695	1.4859	1.5008
	3	1.3400	1.3487	1.3591	1.3713	1.3854	1.4014	1.4193	1.4389	1.4597	1.4811	1.5008
	4	1.3249	1.3342	1.3451	1.3581	1.3732	1.3905	1.4101	1.4316	1.4548	1.4787	1.5008
	5	1.3160	1.3256	1.3368	1.3501	1.3658	1.3839	1.4045	1.4273	1.4518	1.4772	1.5008
	6	1.3102	1.3199	1.3312	1.3448	1.3609	1.3796	1.4008	1.4244	1.4498	1.4762	1.5008
	7	1.3060	1.3158	1.3272	1.3409	1.3573	1.3764	1.3981	1.4223	1.4484	1.4755	1.5008
	8	1.3029	1.3127	1.3242	1.3381	1.3547	1.3741	1.3961	1.4207	1.4473	1.4750	1.5008
	9	1.3005	1.3103	1.3218	1.3358	1.3526	1.3722	1.3946	1.4195	1.4465	1.4746	1.5008
	10	1.2986	1.3084	1.3199	1.3341	1.3510	1.3708	1.3933	1.4185	1.4458	1.4742	1.5008
	∞	1.2816	1.2908	1.3027	1.3175	1.3354	1.3563	1.3803	1.4074	1.4371	1.4687	1.5008
9	2	1.3617	1.3679	1.3755	1.3843	1.3944	1.4056	1.4179	1.4312	1.4451	1.4592	1.4724
	3	1.3337	1.3413	1.3502	1.3607	1.3728	1.3865	1.4018	1.4186	1.4365	1.4549	1.4724
	4	1.3203	1.3284	1.3379	1.3490	1.3620	1.3770	1.3937	1.4122	1.4322	1.4527	1.4724
	5	1.3124	1.3207	1.3304	1.3420	1.3555	1.3712	1.3889	1.4084	1.4295	1.4514	1.4724
	6	1.3071	1.3156	1.3255	1.3373	1.3512	1.3673	1.3856	1.4059	1.4278	1.4506	1.4724
	7	1.3034	1.3120	1.3219	1.3339	1.3481	1.3646	1.3833	1.4041	1.4265	1.4499	1.4724
	8	1.3007	1.3093	1.3193	1.3314	1.3458	1.3625	1.3815	1.4027	1.4256	1.4495	1.4724
	9	1.2985	1.3071	1.3172	1.3294	1.3440	1.3609	1.3802	1.4016	1.4249	1.4491	1.4724
	10	1.2968	1.3054	1.3155	1.3278	1.3425	1.3596	1.3791	1.4008	1.4243	1.4488	1.4724
	∞	1.2816	1.2897	1.3003	1.3132	1.3288	1.3470	1.3679	1.3912	1.4169	1.4443	1.4724
10	2	1.3537	1.3592	1.3658	1.3735	1.3823	1.3920	1.4027	1.4142	1.4263	1.4387	1.4505
	3	1.3286	1.3353	1.3432	1.3524	1.3630	1.3750	1.3883	1.4030	1.4186	1.4348	1.4505
	4	1.3165	1.3237	1.3321	1.3419	1.3533	1.3664	1.3811	1.3973	1.4148	1.4329	1.4505
	5	1.3094	1.3168	1.3254	1.3356	1.3475	1.3613	1.3768	1.3939	1.4124	1.4317	1.4505
	6	1.3047	1.3123	1.3210	1.3314	1.3436	1.3578	1.3739	1.3917	1.4109	1.4309	1.4505
	7	1.3013	1.3090	1.3178	1.3283	1.3409	1.3554	1.3718	1.3900	1.4098	1.4304	1.4505
	8	1.2988	1.3065	1.3154	1.3261	1.3388	1.3535	1.3703	1.3888	1.4089	1.4300	1.4505
	9	1.2969	1.3046	1.3135	1.3243	1.3372	1.3521	1.3691	1.3879	1.4083	1.4296	1.4505
	10	1.2953	1.3031	1.3120	1.3229	1.3359	1.3510	1.3681	1.3871	1.4078	1.4294	1.4505
	∞	1.2816	1.2889	1.2983	1.3098	1.3234	1.3394	1.3575	1.3780	1.4005	1.4249	1.4505

5. EXAMPLE

The example discussed by Lemon(1977) is used to illustrate the method proposed in Section 3. The data was concerned with static strength design allowable for a boron-reinforced epoxy matrix composite material system. Six samples from each of five independent batches were selected, i.e. I = 5 and J = 6. The summary statistics were $\bar{X}$ = 186 ksi(thousand pounds per square inch), S_1^2 = 317.521, and S_2^2 = 34.3396. To determine a 95%-expectation upper tolerance limit, we first calculate S_X = 9.04 and R = 0.5788. From Table 4 (4 iterations were needed for this particular combination of I and J) and by interpolation, we have k(0.5788) = 2.0198, and hence $\bar{X} + kS_X$ = 186 + 2.0198(9.04) = 204.2591. Also, a 90%-expectation two-sided tolerance limit is 186 ± 2.0198(9.04) = [167.7409, 204.2591].

6. CONCLUDING REMARKS

In this article, we have proposed an iteration method for obtaining β-expectation tolerance limits for a balanced one-way random-effects model. In most cases, only a few iterations are needed to produce an interval with confidence coefficients very close to the stated value.

The iteration method has also found useful in other similar problems (e.g. see Maric and Graybill(1979)). Further study is planned for the problems of determining the confidence intervals on linear combinations of variance components. At the present stage it does not seem practical to extend the method to the problems of β-content tolerance limits. The computation involved in the evaluation of the non-central t distribution function in an iterative process is prohibitive even with the speed of today's computers.

Table 4. Values of k(R) for α = 0.95

I	J \ R	0.0	0.1	0.2	0.3	0.4	0.5	0.6	0.7	0.8	0.9	1.0
4	2	1.9080	1.9492	2.0124	2.0987	2.2099	2.3456	2.4983	2.5843	2.5892	2.5291	2.6312
	3	1.8177	1.8541	1.9098	1.9859	2.0843	2.2069	2.3524	2.5090	2.5441	2.5177	2.6312
	4	1.7732	1.8068	1.8596	1.9337	2.0314	2.1548	2.3033	2.4645	2.5183	2.4897	2.6312
	5	1.7474	1.7793	1.8298	1.9020	1.9989	2.1228	2.2735	2.4384	2.5028	2.4740	2.6312
	6	1.7304	1.7613	1.8099	1.8807	1.9769	2.1013	2.2539	2.4217	2.4924	2.4639	2.6312
	7	1.7183	1.7486	1.7957	1.8655	1.9612	2.0861	2.2401	2.4102	2.4849	2.4568	2.6312
	8	1.7092	1.7391	1.7850	1.8540	1.9495	2.0747	2.2299	2.4018	2.4792	2.4515	2.6312
	9	1.7025	1.7310	1.7745	1.8439	1.9431	2.0740	2.2309	2.3810	2.4576	2.4492	2.6312
	10	1.6968	1.7252	1.7679	1.8367	1.9357	2.0668	2.2246	2.3761	2.4526	2.4488	2.6312
	∞	1.6449	1.6704	1.7105	1.7747	1.8690	1.9977	2.1620	2.3406	2.4385	2.4372	2.6312
5	2	1.8653	1.8905	1.9265	1.9734	2.0305	2.0966	2.1680	2.2362	2.2848	2.2932	2.3353
	3	1.7881	1.8131	1.8489	1.8958	1.9540	2.0230	2.1008	2.1813	2.2497	2.2804	2.3353
	4	1.7510	1.7757	1.8108	1.8576	1.9167	1.9880	2.0701	2.1569	2.2326	2.2687	2.3353
	5	1.7293	1.7537	1.7881	1.8344	1.8938	1.9666	2.0514	2.1423	2.2228	2.2619	2.3353
	6	1.7151	1.7393	1.7728	1.8188	1.8785	1.9523	2.0390	2.1327	2.2163	2.2574	2.3353
	7	1.7053	1.7293	1.7624	1.8073	1.8653	1.9375	2.0241	2.1233	2.2306	2.2417	2.3353
	8	1.6977	1.7215	1.7542	1.7990	1.8573	1.9301	2.0177	2.1183	2.2273	2.2391	2.3353
	9	1.6917	1.7154	1.7479	1.7926	1.8510	1.9243	2.0127	2.1143	2.2247	2.2371	2.3353
	10	1.6870	1.7105	1.7427	1.7874	1.8460	1.9198	2.0086	2.1112	2.2227	2.2355	2.3353
	∞	1.6449	1.6653	1.6960	1.7403	1.8002	1.8769	1.9708	2.0783	2.1850	2.2561	2.3353

Table 4. (continued)

I	J	R 0.0	0.1	0.2	0.3	0.4	0.5	0.6	0.7	0.8	0.9	1.0
6	2	1.8354	1.8527	1.8762	1.9058	1.9413	1.9822	2.0276	2.0751	2.1200	2.1531	2.1765
	3	1.7672	1.7861	1.8118	1.8446	1.8845	1.9313	1.9843	2.0412	2.0967	2.1397	2.1765
	4	1.7353	1.7546	1.7807	1.8141	1.8555	1.9048	1.9615	2.0235	2.0848	2.1330	2.1765
	5	1.7167	1.7362	1.7621	1.7957	1.8379	1.8887	1.9477	2.0127	2.0777	2.1290	2.1765
	6	1.7045	1.7240	1.7496	1.7834	1.8260	1.8779	1.9385	2.0056	2.0730	2.1264	2.1765
	7	1.6959	1.7153	1.7407	1.7745	1.8175	1.8702	1.9319	2.0005	2.0697	2.1245	2.1765
	8	1.6894	1.7087	1.7340	1.7678	1.8112	1.8644	1.9269	1.9967	2.0672	2.1230	2.1765
	9	1.6844	1.7036	1.7287	1.7626	1.8062	1.8599	1.9231	1.9937	2.0653	2.1219	2.1765
	10	1.6804	1.6995	1.7245	1.7585	1.8023	1.8563	1.9200	1.9914	2.0637	2.1210	2.1765
	∞	1.6449	1.6618	1.6864	1.7205	1.7650	1.8205	1.8869	1.9628	2.0428	2.1138	2.1765
7	2	1.8081	1.8217	1.8401	1.8632	1.8906	1.9218	1.9560	1.9916	2.0257	2.0536	2.0773
	3	1.7528	1.7673	1.7865	1.8105	1.8395	1.8736	1.9127	1.9563	2.0029	2.0479	2.0773
	4	1.7246	1.7398	1.7598	1.7850	1.8158	1.8524	1.8948	1.9426	1.9941	2.0446	2.0773
	5	1.7082	1.7236	1.7439	1.7695	1.8015	1.8395	1.8839	1.9342	1.9888	2.0425	2.0773
	6	1.6973	1.7129	1.7333	1.7595	1.7919	1.8309	1.8766	1.9286	1.9852	2.0411	2.0773
	7	1.6897	1.7053	1.7258	1.7522	1.7850	1.8247	1.8713	1.9245	1.9826	2.0402	2.0773
	8	1.6840	1.6996	1.7201	1.7467	1.7799	1.8201	1.8674	1.9215	1.9807	2.0394	2.0773
	9	1.6791	1.6952	1.7157	1.7430	1.7777	1.8199	1.8691	1.9237	1.9796	2.0290	2.0773
	10	1.6756	1.6917	1.7121	1.7395	1.7744	1.8169	1.8666	1.9217	1.9783	2.0283	2.0773
	∞	1.6449	1.6593	1.6799	1.7077	1.7432	1.7869	1.8386	1.8975	1.9607	2.0220	2.0773
8	2	1.7913	1.8016	1.8153	1.8323	1.8524	1.8756	1.9016	1.9300	1.9596	1.9881	2.0095
	3	1.7394	1.7516	1.7676	1.7875	1.8114	1.8394	1.8713	1.9067	1.9444	1.9815	2.0095
	4	1.7148	1.7277	1.7445	1.7655	1.7910	1.8211	1.8559	1.8948	1.9367	1.9782	2.0095
	5	1.7004	1.7137	1.7307	1.7523	1.7787	1.8101	1.8466	1.8876	1.9320	1.9762	2.0095
	6	1.6910	1.7043	1.7215	1.7434	1.7704	1.8027	1.8403	1.8828	1.9288	1.9748	2.0095
	7	1.6843	1.6977	1.7150	1.7371	1.7645	1.7975	1.8358	1.8793	1.9265	1.9738	2.0095
	8	1.6793	1.6927	1.7100	1.7324	1.7601	1.7935	1.8325	1.8767	1.9248	1.9731	2.0095
	9	1.6754	1.6888	1.7062	1.7287	1.7567	1.7904	1.8299	1.8747	1.9235	1.9725	2.0095
	10	1.6723	1.6857	1.7031	1.7257	1.7539	1.7879	1.8278	1.8730	1.9224	1.9721	2.0095
	∞	1.6449	1.6574	1.6751	1.6985	1.7280	1.7639	1.8062	1.8542	1.9066	1.9597	2.0095
9	2	1.7746	1.7835	1.7951	1.8095	1.8264	1.8459	1.8677	1.8914	1.9162	1.9404	1.9601
	3	1.7290	1.7395	1.7532	1.7702	1.7905	1.8141	1.8410	1.8708	1.9025	1.9342	1.9601
	4	1.7072	1.7184	1.7328	1.7508	1.7725	1.7981	1.8275	1.8603	1.8956	1.9310	1.9601
	5	1.6944	1.7059	1.7206	1.7391	1.7617	1.7885	1.8193	1.8540	1.8914	1.9291	1.9601
	6	1.6860	1.6977	1.7125	1.7313	1.7545	1.7820	1.8139	1.8498	1.8886	1.9278	1.9601
	7	1.6800	1.6918	1.7067	1.7258	1.7493	1.7774	1.8100	1.8467	1.8866	1.9269	1.9601
	8	1.6756	1.6874	1.7024	1.7216	1.7454	1.7739	1.8071	1.8445	1.8851	1.9262	1.9601
	9	1.6721	1.6839	1.6990	1.7184	1.7424	1.7712	1.8048	1.8427	1.8839	1.9257	1.9601
	10	1.6694	1.6812	1.6962	1.7158	1.7400	1.7691	1.8030	1.8413	1.8830	1.9253	1.9601
	∞	1.6449	1.6560	1.6714	1.6916	1.7167	1.7471	1.7826	1.8231	1.8678	1.9146	1.9601
10	2	1.7614	1.7691	1.7792	1.7916	1.8063	1.8230	1.8417	1.8620	1.8833	1.9044	1.9226
	3	1.7206	1.7299	1.7419	1.7567	1.7742	1.7947	1.8179	1.8436	1.8710	1.8986	1.9226
	4	1.7011	1.7110	1.7236	1.7393	1.7582	1.7804	1.8059	1.8343	1.8648	1.8957	1.9226
	5	1.6896	1.6998	1.7127	1.7289	1.7486	1.7718	1.7986	1.8286	1.8610	1.8939	1.9226
	6	1.6820	1.6924	1.7054	1.7219	1.7421	1.7661	1.7938	1.8249	1.8585	1.8927	1.9226
	7	1.6766	1.6871	1.7002	1.7169	1.7375	1.7620	1.7903	1.8222	1.8567	1.8919	1.9226
	8	1.6726	1.6831	1.6963	1.7132	1.7341	1.7589	1.7877	1.8201	1.8553	1.8912	1.9226
	9	1.6695	1.6800	1.6933	1.7103	1.7314	1.7565	1.7857	1.8186	1.8543	1.8907	1.9226
	10	1.6670	1.6775	1.6908	1.7080	1.7292	1.7546	1.7841	1.8173	1.8534	1.8903	1.9226
	∞	1.6449	1.6548	1.6686	1.6865	1.7087	1.7353	1.7664	1.8017	1.8407	1.8817	1.9226

REFERENCES

[1] Fleiss, J.L., Distribution of a Linear Combination of Independent Chi Squares, Journal of the American Statistical Association 66 (1971), 142-144.

[2] Forsythe, G.E., Malcolm, M.A., and Moler, C.B., Computer Methods for Mathematical Computatitons (New Jersey: Prentice-Hall, New Jersey, 1977).

[3] Graybill, F.A., and Wang, C.M., Confidence Intervals on Nonnegative Linear Combinations of Variances, Journal of the American Statistical Association 75 (1980), 869-873.

[4] Hahn, G.J., Removing Measurement Error in Assessing Conformance to Specifications, Journal of Quality Technology 14 (1982), 117-121.

[5] Hall, I.J., Approximate One-Sided Tolerance Limits for the Difference or Sum of Two Independent Normal Deviates, Journal of Quality Technology 16 (1984), 15-19.

[6] Jaech, J. L., Removing the Effects of Measurement Errors in Constructing Statistical Tolerance Intervals, Journal of Quality Technology 16 (1984), 69-73.

[7] Lemon, G.H., Factors for One-Sided Tolerance Limits for Balanced One-Way-ANOVA Random-Effects Model, Journal of the American Statistical Association 72 (1977), 676-680.

[8] Maric, N., and Graybill, F.A., Confidence Intervals on Common Mean of Two Normal Distributions with Unequal Variances, Communications in Statistics A8(13) (1979), 1233-1269

[9] Mee, R.W., and Owen, D.B., Improved Factors for One-Sided Tolerance Limits for Balanced One-Way ANOVA Random Model, Journal of the American Statistical Association 78 (1983), 901-905.

[10] Mee, R.W., β-Expectation and β-Content Tolerance Limits for Balanced One-Way ANOVA Random Model, Technometrics 26 (1984a), 251-254.

[11] Mee, R.W., Tolerance Limits and Bounds for Proportions Based on Data Subject to Measurement Error, Journal of Quality Technology 16 (1984b), 74-80.

[12] Odeh, R.E., and Owen, D.B., Tables for Normal Tolerance Limits, Sampling Plans, and Screening (Marcel Dekker, New York, 1980).

[13] Trickett, W.H., and Welch, B.L., On the Comparison of Two Means: Further Discussion of Iterative Methods for Calculating Tables, Biometrika 41 (1954), 361-374.

Original paper received: 11.02.85
Final paper received: 31.08.87

Paper recommended by D.C. Bowden

PROBABILITY AND STATISTICS
Essays in Honor of Franklin A. Graybill
J.N. Srivastava (Editor)
Elsevier Science Publishers B.V. (North-Holland), 1988

Linear Calibration When the Coefficient of Variation is Constant

Y-C. Yao
Department of Statistics
Colorado State University
Fort Collins, CO 80523

D. F. Vecchia
Statistical Engineering Division
National Bureau of Standards
Boulder, CO 80303

and

H. K. Iyer
Department of Statistics
Colorado State University
Fort Collins, CO 80523

Abstract

We consider point estimation of the unknowns in the calibration problem when the calibration curve is a straight line through the origin and the responses are normal with variances proportional to the square of their mean. We present four different estimators and compare their performances using the criterion of Pitman closeness. For all the practical ranges of parameter values, one of these estimators is found to dominate the others under the normality assumption. Non-normal cases are not considered in this paper.

Running title: Linear Calibration.

Key words: classical estimator; equivariant estimator; heteroscedastic errors; inverse estimator; maximum likelihood estimator; Pitman closeness.

AMS 1980 subject classification: 62J05, 62F10, 62F11.

1. INTRODUCTION

1.1 Background: In many chemical, biological, and other calibration experiments the "standard curves" are often straight lines. In gas chromatography the area under the peak for a particular compound is linearly related to its concentration. In the calibration of a pressure transducer the response (e.g., voltage) is linearly related to the pressure. In certain bioassays the response (e.g., light intensity) is linearly related to the concentration of antibodies in the sample.

The statistical literature on calibration mostly deals with the straight-line model $y = \alpha + \beta x + \epsilon$ where ϵ represents random error. The usual assumption is that the errors are homoscedastic. If u epresents the unknown true value of a test sample and z represents the response corresponding to this sample then the usual estimator (often referred to as the classical estimator) of u is

$$(z - a_{LS})/b_{LS} \tag{1.1}$$

where a_{LS} and b_{LS} are the ordinary least squares estimators of α and β, respectively. There was a flurry of activity during the late sixties and the early seventies concerning the point estimation of u. Most of the discussion centered around the controversy arising from Krutchkoff's (1967) suggestion that the regression of x on y be used to estimate u. The reader is referred to the papers by Williams (1969); Lamborn (1970); Martinelle (1970); Shukla (1972); Berkson (1969); Krutchkoff (1971); Ali and Singh (1981); Lwin and Maritz (1982); Turiel et al. (1982), and the references contained therein. For some recent work on Bayesian and nonparametric approaches to calibration the reader is referred to Hunter and Lamboy (1981) and Knafl et al. (1984). A discussion of various calibration problems that arise in applications may be found in Rosenblatt and Spiegelman (1981).

It is quite often the case that the errors are heteroscedastic. This is certainly common in the field of analytical chemistry. For example, the reader is referred to Ross and Fraser (1977); Schwartz (1979); Garden et al. (1980); Mitchell and Garden (1982); and Prudnikov and Shapkina (1984).

C. P. Cox (1971) has discussed the construction of confidence intervals for u when the error variance is proportional to x and when the error standard deviation is proportional to x when the calibration curve is a straight line through the origin. Scheffe (1973) has considered the case where the calibration curve is more general and the error variance is proportional to any known function of x and has discussed the construction of simultaneous confidence intervals for several unknowns when the same calibration curve is to be used an unlimited number of times. The point estimation problem for this case has not been examined in sufficient detail in the literature. In the case of a straight-line calibration curve the obvious estimator for u is of the same form as in (1.1) with the exception that a and b are now the weighted least-squares estimators of α and β, respectively. While the estimator in (1.1) is also the maximum likelihood estimator of u in the homoscedastic case, the obvious estimator and the maximum likelihood estimator are in general different in the heteroscedastic case.

1.2 Statement of the Problem: We shall adopt the following notation throughout. Let $x_1, x_2, \ldots, x_n$ be the values of the standards (not necessarily distinct) and $y_1, y_2, \ldots, y_n$ the corresponding responses. The true unknown value of a test sample is denoted by u and the corresponding response is denoted by z. All the observations are mutually independent and normally distributed. The unknown u is assumed to be nonzero. It is assumed that $E[y_i] = \beta x_i$, $\text{var}(y_i) = \sigma^2 x_i^2$, $i = 1, 2, \ldots, n$, $E[z] = \beta u$, $\text{var}(z) = \sigma^2 u^2$ and σ^2, β, u are unknown parameters. Often the standards are restricted to some interval [A, B] with $0 < A < B$ and each unknown test sample is run only once. If several observations are available for each test sample our results can be suitably modified without much difficulty. The model assumptions imply that the coefficient of variation of the response random variable is constant in the calibration interval. This model is quite adequate for many calibration experiments in analytical chemistry. The reader is referred to Ross and Fraser (1977, p. 130), and Mitchell and Garden (1982, p. 922) among others. We remark that many of our results can be easily extended to the straight-line model with intercept. However, for the sake of simplicity we shall only deal with the model without intercept.

Our objective is to obtain an "efficient" point estimator for u. The classical estimator of u is

$$\hat{u}_c = z/b$$

where $b = n^{-1}\Sigma(y_i/x_i)$ is the weighted least squares estimator of β based on (x_i, y_i), $1 \le i \le n$. Several other approaches are avilable for finding potentially good estimators of u. These include maximum likelihood estimation, equivariant estimation, unbiased estimation, and "inverse estimation" along the lines of Krutchkoff (1967). In section 2 we study the properties of the minimal sufficient statistics for this problem. In section 3 we derive the maximum likelihood estimator of u and discuss its properties. In section 4 we investigate the existence of best equivariant estimators. In section 5 we consider an "inverse estimator" for u. Because some of the estimators do not have finite expectation, the usual criteria of unbiasedness and minimum variance are not suitable for this problem. Therefore, we use the closeness criterion of Pitman (1937) to compare estimators. Although we give the asymptotic distributions of the various estimators and make large sample comparisons, our main interest is in the more common "small" sample situation. For this reason section 6 deals with the small sample comparison of the estimators. Finally, section 7 contains a discussion of the results and a summary. It should be emphasized that all the results of this paper are obtained under the normality assumption.

2. MINIMAL SUFFICIENT STATISTICS

In this section we exhibit the minimal sufficient statistics for the model described in section 1.2 and examine their properties. Let

$$b = \frac{1}{n} \sum_{i=1}^{n} \frac{y_i}{x_i}$$

and

$$s^2 = \frac{1}{n-1} \sum_{i=1}^{n} \left(\frac{y_i}{x_i} - b\right)^2$$

denote the UMVU estimators of β and σ^2, respectively, using only the calibration data $(x_1, y_1), (x_2, y_2), \ldots, (x_n, y_n)$. Recall that z is the observed response corresponding to an unknown u. It is easily shown that $\{b, s^2, z\}$ is minimal sufficient. We conjecture that b, s^2 and z are complete for β, σ^2 and u, however, we can only obtain a result which is weaker than completeness but stronger than bounded completeness. To this end we first define the concept of νth-moment completeness.

Definition 2.1: Let $F = \{F_\theta: \theta \in \Omega\}$ be a family of distributions indexed by θ. Then a statistic T whose cdf belongs to F is ν^{th}-moment complete if any function f(T) of T which has a finite absolute ν^{th}-moment for all $\theta \in \Omega$ must be identically zero a.e.(F_θ), $\theta \in \Omega$, whenever $E_\theta f(T)$ is identically zero in θ.

Proposition 2.1: For the model described in section 1.2,
(a) b, s^2 and z are minimal sufficient for β, σ^2 and $u \neq 0$. They are mutually independent.
(b) $\{b, s^2, z\}$ is 2nd-moment complete for β, σ^2 and $u \neq 0$. (In particular, it is boundedly complete.) Moreover, any function $g(b, s^2, z)$ with finite variance at some β, u and σ^2 is an UMVUE of its expected value. Thus z, b and s^2 are UMVUE's of βu, β and σ^2, respectively.

The proof can be found in Yao and Iyer (1986).

It follows from proposition 2.1 that the "classical estimator" $\hat{u}_c = z/b$ of u is a ratio of UMVU estimators as in the homoscedastic case. Also, it is easy to see that the distribution of $\hat{u}_c$ is asymptotically $N(u, u^2/\rho^2)$ as $n \to \infty$ where $\rho = \beta/\sigma$, so that $\hat{u}_c$ is asymptotically unbiased.

We wish to point out that the concept of νth-moment completeness is a genuine extension of the concept of completeness in the following sense. If $1 \le \mu < \nu$ then μ^{th}-moment completeness implies ν^{th}-moment completeness but the converse is not true as the following example demonstrates.

Example 2.1: Let $1 \le \mu < \nu$. Define a family $P = \{p_\theta | \theta \epsilon (0,1/2]\}$ of probability densities on the nonnegative integers by $p_\theta(0) = (\sum_{i=1}^{\infty} 2^{i/\nu}\theta^i)/s(\theta)$, $p_\theta(i) = \theta^i/s(\theta)$ for $i \ge 1$ with $s(\theta) = \sum_{i=1}^{\infty} (2^{i/\nu} + 1)\theta^i$. It is easily verified that P is ν^{th}-moment complete but not μ^{th}-moment complete.

It should be noted that 1st-moment completeness is the same as completeness while ∞^{th}-moment completeness may be defined as bounded completeness.

3. MAXIMUM LIKELIHOOD ESTIMATION

We now derive the maximum likelihood estimates for the unknown parameters. The log-likelihood function, apart from constants, is

$$\ell(\beta, \sigma^2, u) = -\frac{m}{2}\ell n\, \sigma^2 - \ell n|u| - \frac{1}{2\sigma^2}\left[\sum_{i=1}^{n}\left(\frac{y_i - \beta x_i}{x_i}\right)^2 + \left(\frac{z-\beta u}{u}\right)^2\right] \tag{3.1}$$

where $m = n + 1$.

For fixed β and u, $\ell(\beta, \sigma^2, u)$ is maximized with respect to σ^2 at

$$\hat{\sigma}^2 = \hat{\sigma}^2(\beta, u) = \frac{1}{m}\left[\sum_{i=1}^{n}\left(\frac{y_i - \beta x_i}{x_i}\right)^2 + \left(\frac{z-\beta u}{u}\right)^2\right].$$

Substituting $\hat{\sigma}^2$ for σ^2 in (3.1) we have the "reduced" log-likelihood function ℓ^* given by

$$\ell^*(\beta, u) = -\frac{m}{2}\,\ell n\, \hat{\sigma}^2(\beta, y) - \ell n|u| - \frac{m}{2} \tag{3.2}$$

which is now a function of β and u only. For a fixed value of u, ℓ^* is maximized with respect to β at

$$\hat{\beta} = \hat{\beta}(u) = \frac{nb + \frac{z}{u}}{n + 1}.$$

Substituting $\hat{\beta}$ in (3.2) the reduced log-likelihood function for u is given by

$$\ell^{**}(u) = -\frac{m}{2}\,\ell n\, \hat{\sigma}^2(\hat{\beta}, u) - \ell n|u| - \frac{m}{2}.$$

We now find that ℓ^{**} is maximized with respect to u at

$$\hat{u} = \frac{z}{\xi(b, s^2)}$$

where

$$\xi(b, s^2) = b\left[\frac{n-1}{2n} + \frac{n+1}{2n}\sqrt{1 + \frac{4(n-1)s^2}{(n+1)b^2}}\right].$$

We have thus proved,

Theorem 3.1: Let $(x_1, y_1), \ldots, (x_n, y_n)$, $n \geq 2$, be the "calibration data" and z the response to the unknown u. The unrestricted maximum likelihood estimators of β, σ^2, and u are given by

$$\hat{u} = \frac{z}{\xi(b, s^2)} \tag{3.3}$$

$$\hat{\beta} = \frac{nb+\xi(b, s^2)}{n+1}.$$

and

$$\hat{\sigma}^2 = \frac{n-1}{n+1}s^2 + \frac{n}{(n+1)^2}(b-\xi(b, s^2))^2.$$

Note that $\xi(b, s^2) > b$ with probability 1. Thus the MLE for u is a shrinked estimator, the obvious (classical) estimator being z/b. We shall denote the maximum likelihood estimator of u in (3.3) by $\hat{u}_{MLE}$. The following properties of $\hat{u}_{MLE}$ are easily established.

Proposition 3.1: (a) The distribution of $\hat{u}_{MLE}$ is asymptotically $N[u/D, u^2/(\rho D)^2]$ as $n \to \infty$, where $D = (1 + \sqrt{1+4/\rho^2})/2$, and $\rho = \beta/\sigma$ is the sensitivity of the calibration experiment. Thus, $\hat{u}_{MLE}$ is asymptotically biased.

(b) As $n \to \infty$, $P[|\hat{u}_c-u| \leq |\hat{u}_{MLE}-u|] \to \Phi(2|\rho|/(1+D^{-1}) - |\rho|) - \Phi(-|\rho|)$, where $\Phi(\cdot)$ is standard normal cdf. Hence, in large samples $\hat{u}_c$ is Pitman closer to u than $\hat{u}_{MLE}$ whenever $|\rho| \geq 1.4$.

4. EQUIVARIANT ESTIMATION

Let $G = \{g_c | \ c \neq 0\}$ be a group of transformations defined by

$$g_c: (y_1, \ldots, y_n, z) \to (y_1, \ldots, y_n, cz).$$

This induces the group $\bar{G}$ of transformations $\bar{g}_c$ on the parameter space with

$$\bar{g}_c: (\beta, \sigma^2, u) \to (\beta, \sigma^2, cu)$$

and the group $\tilde{G}$ of transformations $\tilde{g}_c$ on the space of minimal sufficient statistics with

$$\tilde{g}_c: (b, s^2, z) \to (b, s^2, cz).$$

Assuming that the loss of an estimator $f(b, s^2, z)$ of u is $L(f, u) = (f - u)^2/u^2$, $f(b, s^2, z)$ is an equivariant estimator of u if and only if $cf(b, s^2, z) = f(b, s^2, cz)$ for all $c \neq 0$. Taking $c = 1/z$ we get $f(b, s^2, z) = zf(b, s^2, 1)$. Writing $\zeta = \zeta(b, s^2) = f(b, s^2, 1)$ we see that the risk of an equivariant estimator equals

$$\begin{aligned} E[(z\zeta(b, s^2) - u)^2] &= u^{-2}E[E[(z\zeta(b, s^2) - u)^2|b, s^2]] \\ &= u^{-2}E[\zeta^2(\beta^2u^2 + \sigma^2u^2) - 2\zeta\beta u^2 + u^2] \\ &= E(\zeta^2(\beta^2 + \sigma^2) - 2\zeta\beta + 1]. \end{aligned}$$

This is minimized when $\zeta(\beta, s^2) = \beta/(\beta^2+\sigma^2)$. Thus a locally best equivariant estimator of u is

$$f(z) = \frac{z\beta}{\beta^2+\sigma^2} = \frac{z}{\beta}\left(\frac{1}{1+\sigma^2/\beta^2}\right). \tag{4.1}$$

This leads us to consider the estimator

$$\hat{u}_{BEQ} = \frac{z}{b}\left(\frac{1}{1+s^2/b^2}\right) \tag{4.2}$$

as a candidate for the estimation of u.

We observe in passing that f(z) in (4.1) is in fact a locally minimum mean-squared error estimator of u of the form $z\zeta(b, s^2)$ and hence $\hat{u}_{BEQ}$ is an "estimated" locally minimum mean-squared error estimator of u of the form $z\zeta(b, s^2)$.

The following result is easily established.

Proposition 4.1: (a) As $n \to \infty$ the distribution of $\hat{u}_{BEQ}$ is asymptotically $N[u\rho^2/(\rho^2+1), u^2\rho^2/(\rho^2+1)^2]$ where $\rho = \beta/\sigma$. Thus, $\hat{u}_{BEQ}$ is asymptotically biased.
(b) As $n \to \infty$ $P[|\hat{u}_c-u|] < |\hat{u}_{BEQ}-u|] \to \Phi(2|\rho|(\rho^2+1)/(2\rho^2+1) - |\rho|) - \Phi(-|\rho|)$.
Hence, $\hat{u}_c$ is Pitman closer to u than $\hat{u}_{BEQ}$ whenever $|\rho| \geq 1.2$.

5. INVERSE ESTIMATION

In the homoscedastic case Krutchkoff (1967) has proposed regressing x on y by ordinary least squares (OLS) and then using the resulting equation to estimate u. This procedure may be considered a candidate in the heteroscedastic case as well. Although the inverse OLS estimator does not explicitly make use of the fact that the errors are heteroscedastic, it does have the following justification.

The inverse OLS estimator of u is given by

$$\hat{u}_{IOLS} = \frac{Sxy}{Syy}\, z \tag{5.1}$$

where $S_{xy} = \sum_{i=1}^{n} x_i y_i$ and $S_{yy} = \sum_{i=1}^{n} y_i^2$. Since $E[S_{xy}] = \beta S_{xx}$ and $E[S_{yy}] = (\beta^2+\sigma^2) S_{xx}$, we note that S_{xy}/S_{yy} is a "method of moments" estimator of $\beta/(\beta^2+\sigma^2)$. So $\hat{u}_{IOLS}$ can be justified as the locally best equivariant estimator of u with $\beta/(\beta^2+\sigma^2)$ replaced by an estimate, namely, S_{xy}/S_{yy}.
Proposition 5.1: (a) Suppose the sequence of design points $x_1, x_2, \ldots, x_n, \ldots$ is such that $\sum_{i=1}^{n} x_i^4/(\sum_{i=1}^{n} x_i^2)^2 \to 0$ as $n \to \infty$. Then, the asymptotic distribution of $\hat{u}_{IOLS}$ is $N(u\rho^2/(\rho^2+1), u^2\rho^2/(\rho^2+1)^2]$ as $n \to \infty$. Hence $\hat{u}_{IOLS}$ is asymptotically biased.
(b) As $n \to \infty$, $P[|\hat{u}_c-u| < |\hat{u}_{IOLS}-u|] \to \Phi(2|\rho|(\rho^2+1)/(2\rho^2+1) - |\rho|) - \Phi(-|\rho|)$.
Hence, $\hat{u}_c$ is Pitman closer to u than $\hat{u}_{IOLS}$ whenever $|\rho| \geq 1.2$.

6. COMPARISON OF ESTIMATORS

In this section we compare the following estimators.

Classical estimator: $\hat{u}_1 = \hat{u}_c = z/b$,

MLE: $\hat{u}_2 = \hat{u}_{MLE}$ given in (3.3),

Estimated best equivariant estimator: $\hat{u}_3 = \hat{u}_{BEQ}$ given in (4.2),

Inverse estimator (OLS): $\hat{u}_4 = \hat{u}_{IOLS}$ given in (5.1).

It is easy to verify that the distributions of the first three estimators depend only on $\rho = \beta/\sigma$, the sensitivity of the calibration experiment, and u. This facilitates the comparison of these estimators. The distribution of the estimator $\hat{u}_{IOLS}$ depends on β/σ, u and the design. It is also clear that, for fixed n, $|\hat{u}_i - \hat{u}_j| \to 0$ a.s. as $\rho \to \infty$ for all $1 \le i, j \le 4$.

As is well-known, the usual criteria, namely, unbiasedness and minimum variance, are not appropriate for the problem under consideration. Instead, we adopt Pitman's closeness criterion for comparison of the performance of the four estimators. According to this criterion an estimator T_1 of u is "closer to u" than another estimator T_2 if and only if

$$P[|T_1 - u| < |T_2 - u|] > 1/2. \tag{6.1}$$

We calculate this probability for all pairs of estimators (u_i, u_j) above and for various parameter settings by numerical integration and, when this is too costly, by simulation. The probability in (6.1) is independent of u for all comparisons. It is also independent of the design when comparing estimators $\hat{u}_j$ for j = 1, 2, 3.

Let each estimator of u be written as

$$\hat{u}_j = \frac{z}{b} \Psi_j(y_1,\dots,y_n)$$

for j = 1, 2, 3, 4. We then have

$$\Psi_1(y_1,\dots,y_n) = 1$$

$$\Psi_2(y_1,\dots,y_n) = \left[\frac{n-1}{2n} + \frac{n+1}{2n}\sqrt{1 + \frac{4(n-1)s^2}{(n+1)b^2}}\right]^{-1}$$

$$\Psi_3(y_1,\dots,y_n) = \frac{b^2}{b^2+s^2}$$

and

$$\Psi_4(y_1,\dots,y_n) = b\,\frac{\Sigma x_i y_i}{\Sigma y_i^2}.$$

It is easily verified that $\Psi_3 < \Psi_2 < \Psi_1$ a.s. Note also that Ψ_1, Ψ_2, Ψ_3, are actually functions of (b, s^2), while Ψ_4 is not. If $\Psi_i < \Psi_j$, the event E_{ij}: $|\hat{u}_i - u| < |\hat{u}_j - u|$ is equivalent to the event

$$\{\epsilon^* \le \min(-\frac{\beta}{\sigma}, \frac{2b}{\sigma(\Psi_i+\Psi_j)} - \frac{\beta}{\sigma})\}\cup\{\epsilon^* > \max(-\frac{\beta}{\sigma}, \frac{2b}{\sigma(\Psi_i+\Psi_j)} - \frac{\beta}{\sigma})\}$$

where ϵ^* has a N(0, 1) distribution and is independent of $y_1, y_2, \ldots, y_n$. Hence, conditional on $\{y_1, \ldots, y_n\}$, the probability of E_{ij} is:

$$P(E_{ij} | y_1, \ldots, y_n) = \Phi(m(y_1, \ldots, y_n)) + 1 - \Phi(M(y_1, \ldots, y_n))$$

when $\Psi_i \leq \Psi_j$ and

$$P(E_{ij} | y_1, \ldots, y_n) = \Phi(M(y_1, \ldots, y_n)) - \Phi(m(y_1, \ldots, y_n))$$

when $\Psi_i > \Psi_j$, where

$$m(y_1, \ldots, y_n) = \min(-\frac{\beta}{\sigma}, \frac{2b}{\sigma(\Psi_i + \Psi_j)} - \frac{\beta}{\sigma})$$

$$M(y_1, \ldots, y_n) = \max(-\frac{\beta}{\sigma}, \frac{2b}{\sigma(\Psi_i + \Psi_j)} - \frac{\beta}{\sigma})$$

and Φ is the cdf of the standard normal distribution. Thus, the required probability is $P(E_{ij}) = E(P(E_{ij} | y_1, y_2, \ldots, y_n)]$ where the expectation is with respect to $y_1, y_2, \ldots, y_n$. $P(E_{ij})$ can be expanded in powers of $1/\rho$ to examine its behavior as $\rho \to \infty$ For instance, as $\rho \to \infty$, we get,

$$P(E_{12}) = \frac{1}{2} + \frac{C_{12}}{\rho} + o(\frac{1}{\rho}) \text{ with } C_{12} = \frac{n - 1}{\sqrt{8\pi n(n+1)}} \tag{6.2}$$

$$P(E_{13}) = \frac{1}{2} + \frac{C_{13}}{\rho} + o(\frac{1}{\rho}) \text{ with } C_{13} = \sqrt{\frac{n}{8\pi(n+1)}} \tag{6.3}$$

and,

$$P(E_{14}) = C_{14}^* + o(1) \text{ if } x_1, x_2, \ldots, x_n \text{ are not all the same}$$

$$= \frac{1}{2} + \frac{C_{14}}{\rho} + o(\frac{1}{\rho}) \text{ if } x_1 = x_2 = \ldots = x_n (\neq 0) \tag{6.4}$$

with $C_{14} = C_{12}$. The constant C_{14}^* is equal to $2E\{\Phi(W\sqrt{nv/4(n+1)})I(W > 0)\}$ where

$W \sim N(0, 1)$ and $v = \sum_{i=1}^{n} x_i^4 / (\sum_{i=1} x_i^2)^2 - (1/n)$. Calculated values of C_{14}^* for some 2-point designs are shown in table 1. These values have been used as a check of our numerical and simulation results for large values of ρ. Expansions for $P(E_{ij})$ when $i \neq 1$, $j > i$ can be carried out easily and are omitted here.

If F denotes the cdf of b, and G the cdf of s^2, we have

$$P(E_{ij}) = \int P(E_{ij} | b, s^2) dFdG \tag{6.5}$$

for $1 \leq i \neq j \leq 3$ since m and M in these cases are functions of (b, s^2). For these cases the probability in (6.5) can be evaluated by a numberical double integration for specified values of β/σ. However, to compute $P(E_{i4})$, $i = 1, 2, 3$, a multiple integration, which is expensive to carry out, is required. Therefore, we have estimated $P(E_{ij})$ by simulation in this case. This is done by computing the average of $P(E_{ij} | y_1, \ldots, y_n)$ over 4000 independent realizations of $(y_1, y_2, \ldots, y_n)$ such that y_i's are independently distributed as $N(\beta x_i, \sigma^2 x_i^2)$. The IMSL subroutines DBLIN and GGNML were used for the numerical quadrature and to generate the random normal deviates, respectively.

Table 1. Values of C^*_{14} for Various 2-Point Designs

Number of Standards	Design Points		C^*_{14}
2	$x_1=1$	$x_2=3$	0.57224
4	$x_1=x_2=1$	$x_3=x_4=3$	0.55635
12	$x_1=...=x_6=1$	$x_7=...=x_{12}=3$	0.53517
100	$x_1=...=x_{50}=1$	$x_{51},...=x_{100}=3$	0.51266
2	$x_1=1$	$x_2=2$	0.55459
4	$x_1=x_2=1$	$x_3=x_4=2$	0.54245
12	$x_1=...=x_6=1$	$x_7=...=x_{12}=2$	0.52642
100	$x_1=...=x_{50}=1$	$x_{51}=...=x_{100}=2$	0.50950
2	$x_1=3$	$x_2=4$	0.52567
4	$x_1=x_2=3$	$x_3=x_4=4$	0.51990
12	$x_1=...=x_{50}=3$	$x_7=...=x_{12}=4$	0.51235
100	$x_1=...=x_{50}=3$	$x_{51}=...=x_{100}=4$	0.50443
2	$x_1=10$	$x_2=11$	0.50873
4	$x_1=x_2=10$	$x_3=x_4=11$	0.50676
12	$x_1=...x_6=10$	$x_7=...=x_{12}=11$	0.50419
100	$x_1=...x_{50}=10$	$x_{51}=...=x_{100}=11$	0.50150

Three different sample sizes were considered. Since we are mainly interested in small sample behavior we have considered $n = 2$, 4, and 12. It should be pointed out that 2-point calibrations are very common in analytical chemistry. In each case a 2-point design with equal number of observations at each point was used. The two design points used were $x = 1$ and $x = 3$. As mentioned earlier the choice of the design points is irrelevant for our comparisons except when $i = 1, 2, 3$ and j is 4. Note also that the choice of u is irrelevant. We chose $u = 2$ in all cases. Several values of $\beta/\sigma = \rho$ (covering the range 0 through 20) were considered for computing $P(E_{ij})$ $1 \le i \ne j \le 3$. In estimating $P(E_{i4})$, $i = 1, 2, 3$, the values of ρ used were $\rho = 1, 2, 5, 10$, and 20. The numerical and simulation results are shown in table 2 and graphically displayed in figures 1(a)-(c), 2(a)-(c) and 3(a)-(c).

7. RESULTS AND SUMMARY

The numerical calculations and simulations clearly indicate that in general $\hat{u}_c$ is closer (in the sense of Pitman) to u than the other estimators when ρ is greater than about 1.5, for the range of sample sizes considered. For values of ρ between about 0.75 and 1.5, $\hat{u}_{MLE}$ is superior. For large values of ρ (say $\rho \ge 20$), all three estimators $\hat{u}_c$, $\hat{u}_{MLE}$, $\hat{u}_{BEQ}$, perform equally well and each one is clearly bettern than $\hat{u}_{IOLS}$. This is to be expected based on the expansions (6.2) through (6.4).

It is worth contrasting these conclusions with the results of various authors (e.g., Halperin (1970); Krutchkoff (1971); and Turiel et al. (1982)) who have used the Pitman closeness criterion to compare the classical and inverse estimators in the linear calibration problem with homoscedastic errors. Firstly, interpolation or extrapolation makes no difference in our case and, for the most part, the conclusions are independent of the choice of design. Also, the classical estimator here is clearly the best performer, since in almost all real situations the sensitivity ρ is certainly larger than two. However, in the homoscedastic case it has been reported that the inverse estimator is better than the classical estimator for interpolation in small sample problems (see Turiel et al. (1982)).

In summary, four candidates for the estimation of u, the unknown value of a test sample, have been considered and their performances compared using the criterion of Pitman closeness. The calibration curve was assumed to be a straight line through the origin and the response variances were assumed to be proportional to the squares of their expected values. Asymptotic results indicate that the MLE, the inverse estimator and the equivariant estimator are inferior to the classical estimator, and the same conclusion holds in small samples as well. The classical estimator is generally preferable to the others whenever the sensitivity ρ of the calibration procedure is greater than or equal to 1.4 in absolute value, which is almost always the case in practice.

Non-normal measurement errors are not considered in this paper. The referee has suggested that u be estimated by z/b_1 in the presence of outliers where $b_1 = \text{median}\{y_1/x_1, \ldots, y_n/x_n\}$ is the weighted least absolute deviations estimator of β.

Table 2

Pitman Closeness of Estimators - Probability (%) Method i Closer than Method j

n	Method i	Method j	Sensitivity, β/σ 1	2	5	10	20
2	WLS	MLE	39.86	50.81	51.54	50.79	50.40
	(Classical)	BEQ	41.66	53.88	53.10	51.60	50.80
		IOLS	45.34	53.71	56.96	57.08	57.11
	MLE	BEQ	54.71	59.11	54.78	52.42	51.21
		IOLS	52.53	56.34	57.22	57.16	57.13
	BEQ	IOLS	49.38	52.24	57.34	57.21	57.15
4	WLS	MLE	40.51	52.82	52.54	51.32	50.67
	(Classical)	BEQ	43.70	55.30	53.48	51.78	50.89
		IOLS	43.07	53.42	55.17	55.38	55.28
	MLE	BEQ	59.23	62.29	56.16	53.11	51.56
		IOLS	56.18	56.12	55.55	55.29	55.24
	BEQ	IOLS	48.64	51.85	55.05	55.27	55.24
12	WLS	MLE	42.33	54.36	53.32	51.73	50.88
	(Classical)	BEQ	45.93	56.14	53.75	51.91	50.96
		IOLS	45.01	54.95	54.26	53.80	53.74
	MLE	BEQ	61.62	64.29	57.20	53.23	51.84
		IOLS	58.79	57.97	53.68	53.54	53.60

8. ACKNOWLEDGEMENT

Part of the research of H. K. Iyer was conducted as a consultant at the National Bureau of Standards, Boulder, Colorado.

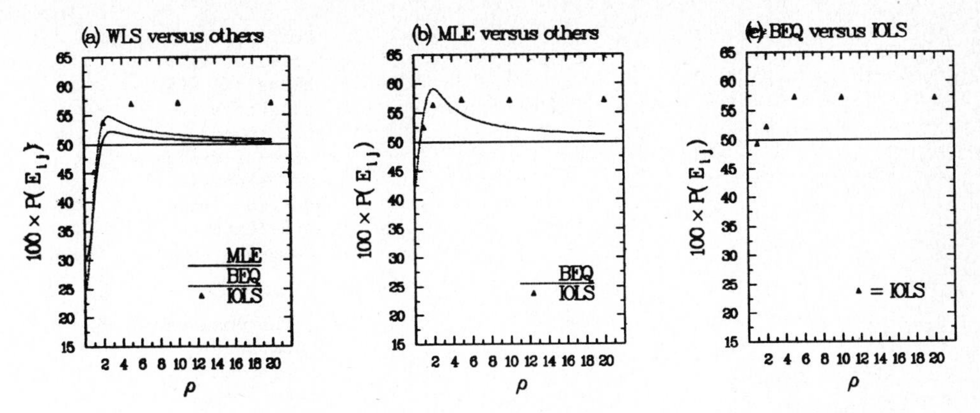

Figure 1. Pitman closeness (percent) versus $\rho = \beta/\sigma$ for n = 2 standards.

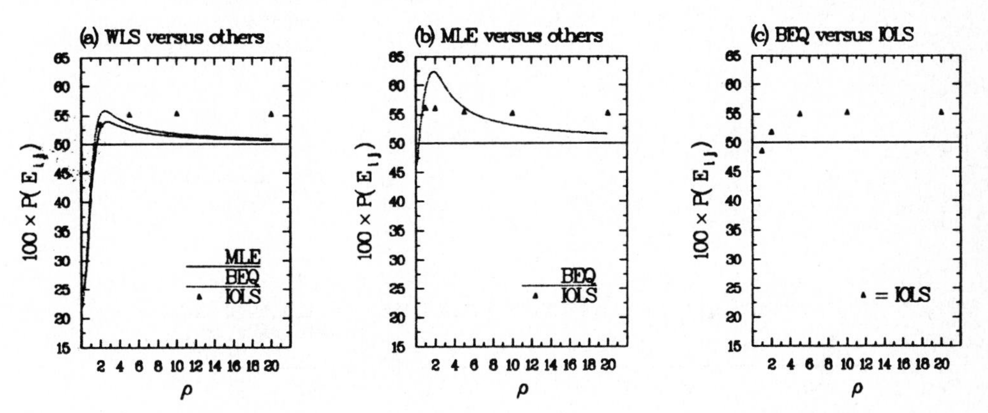

Figure 2. Pitman closeness (percent) versus $\rho = \beta/\sigma$ for n = 4 standards.

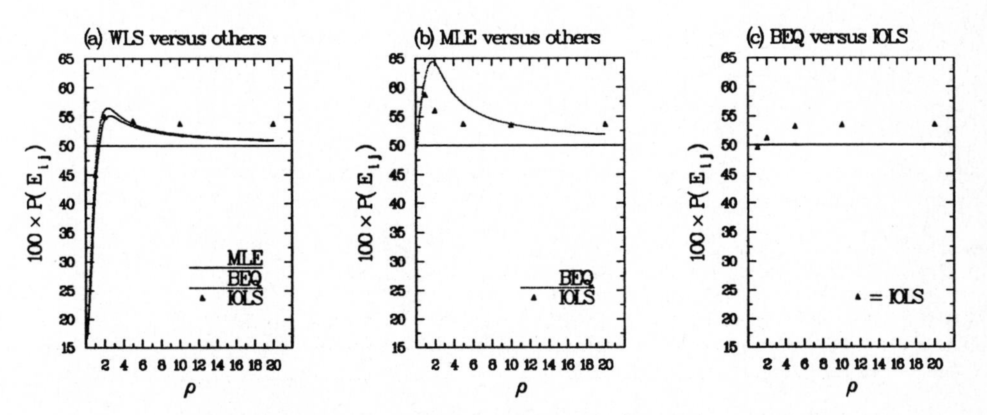

Figure 3. Pitman closeness (percent) versus $\rho = \beta/\sigma$ for n = 12 standards.

REFERENCES

Ali, M. A., and Singh, N. (1981), "An Alternative Estimator in Inverse Linear Regression," J. Statist. Comput. Simul., 14, 1-15.

Berkson, J. (1969), "Estimation of a Linear Function for a Calibration Line; Consideration of a Recent Proposal," Technometrics, 11, 649-660.

Cox, C. P. (1971), "Interval Estimation for X-Predictions from Linear Y on X Regression Lines Through the Origin," JASA, 66, 749-751.

Garden J. S., Mitchell D. G., and Mills, W. N. (1980), "Nonconstant Variance Regression Techniques for Calibration-Curve-Based Analysis," Anal. Chem., 52, 2310-2315.

Halperin, M. (1970), "On Inverse Estimation in Linear Regression," Technometrics, 12, 727-736.

Hunter, W. G., and Lamboy, W. F. (1981), "A Bayesian Analysis of the Linear Calibration Problem," Technometrics, 23, 323-328.

Knafl, G. J., Sacks, J., Spiegelman, C. H., and Ylvisaker, D. (1984), "Nonparametric Calibration," Technometrics, 26, 233-241.

Krutchkoff, R. G. (1967), "Classical and Inverse Regression Methods of Calibration," Technometrics, 9, 425-439.

Krutchkoff, R. G. (1971), "The Calibration Problem and Closeness," J. Statist. Comput. Simul., 1, 87-95.

Lamborn, K. (1970), "A Comparison of Two Methods of Estimating Calibration Lines," Technical Report 22, Stanford University.

Lwin, T., and Maritz, J. S. (1982), "An Analysis of the Linear-Calibration Controversy from the Perspective of Compound Estimation," Technometrics, 24, 235-242.

Martinelle, S. (1970), "On the Choice of Regression in Linear Calibration," comments on a paper by R. G. Krutchkoff, Technometrics, 12, 157-161.

Mitchell, D. G., and Garden, J. S. (1982), "Measuring and Maximizing Precision in Analyses Based on Use of Calibration Graphs," Talanta, 29, 921-929.

Pitman, E. J. G. (1937), "The "Closest" Estimates of Statistical Parameters," Proceedings of Cambridge Philosophical Society, 33, 212.

Prudnikov, E. D., and Shapkina, Y. S. (1984), "Random Errors in Analytical Methods," Analyst, 109, 305-307.

Rosenblatt, J. R., and Spiegelman, C. H. (1981), discussion on "A Bayesian Analysis of the Linear Calibration Problem," Technometrics, 23, 329-333.

Ross, J. W., and Fraser, M. D. (1977), "Analytical Clinical Chemistry Precision-State of the Art for Fourteen Analytes," Amer. J. Clin. Path., 68, Supplement, 130-141.

Scheffe, H. (1973), "A Statistical Theory of Calibration," The Annals of Statistics, 1, 1-37.

Schwartz, L. M, (1979), "Calibration Curves with Nonuniform Variance," Anal. Chem., 51, 723-727.

Shukla, G. K, (1972), "On the Problem of Calibration," Technometrics, 14, 547-553.

Turiel, T. P., Hahn, G. J., and Tucker, W. T. (1982), "New Simulation Results for the Calibration and Inverse Median Estimation Problems," Comm. Statist. - Simul. Computa., 11, 677-713.

Williams, E. J. (1969), "A Note on Regression Methods in Calibration," Technometrics, 11, 189-192.

Yao, Y-C., and Iyer, H. K. (1986), "A Note on the Completeness of the Normal Family with Constant Coefficient of Variation," Technical Report, Colorado State University, Fort Collins, CO.

Original paper received: 30.07.86
Final paper received: 12.08.87

Paper recommended by P.W. Mielke

PROBABILITY AND STATISTICS
Essays in Honor of Franklin A. Graybill
J.N. Srivastava (Editor)

GENERALIZATIONS FOR SECOND-ORDER AND MULTIVARIATE NORMAL LINEAR MODELS

James S. Williams
Department of Statistics, Colorado State University
Fort Collins, Colorado

and

M.K. Zaatar[1)]
Department of Mathematics,
University of Kuwait

ABSTRACT

The basic theory of Best Linear Unbiased and Maximum Likelihood Estimation of nonhomogeneous linear functions of regression parameters that are subject to linear constraints is the subject of this paper. The theory is applied to the study of estimators that are important in the methods available for growth-curve, longitudinal, profile, and repeated-measures data, all of which can be described by a common linear model.

The emphasis in our presentation is on careful statement of generalizations, new concepts, and new problems. Only methods of intermediate matrix algebra are used in order to provide a unified mathematical approach to the exposition.

This work is presented as a gesture of appreciation to Franklin Graybill who in all the years that we have known him has worked unfailingly to make the theory of statistics, particulary of linear models, accessible in full detail to his many students.

1. INTRODUCTION

Our objective in this paper is to present generalizations, and to illustrate them with one important application, of problems and their solutions for linear estimation associated with multivariate Second-Order and Normal linear models. Most of the generalizations are modest extensions of standard formulations of problems in the theory of linear models, and the relationships among solutions simply bring together various approaches that others and we have developed for the study of Best Linear Unbiased and Maximum Likelihood Estimation of the linear forms that characterize the cited models. The application with a model for longitudinal analysis is sufficiently complicated to demonstrate that the theoretical results, when properly applied, can form at least one powerful research tool.

2. SECOND-ORDER AND MULTIVARIATE NORMAL LINEAR MODELS

A vector-valued random variable $\underline{Y}$ will be either incompletely or fully characterized by

$$E(\underline{Y}) = A\underline{\beta} \text{ and } E[(\underline{Y} - A\underline{\beta})(\underline{Y} - A\underline{\beta})'] = \sigma^2 C \text{ or } \underline{Y} \sim MVN(A\underline{\beta}, \sigma^2 C)$$

where A is an $n \times p$ matrix of real constants with rank between 1 and p inclusively, σ^2 is a positive real constant, C is a positive, definite, real,

1) The first four sections of this paper are mainly needed background material for a study of inferential methods for the generalzied multivariate linear model set out here in the fifth section. These studies were completed while the second author was on sabbatical leave from the University of Kuwait to spend the academic year of 1984 - 85 at Colorado State University.

symmetric matrix, and MVN(•, •) denotes a multivariate Normal random variable with appropriate mean vector and dispersion matrix in the first and second positions. We will call these respective characterizations "Second-Order" and "Normal" ("Distributional") to denote the extent to which the distribution function of $\underline{Y}$ is specified by each.

A model for $\underline{Y}$ is a class of distribution functions that are either incompletely or fully characterized. We will let

$$(\underline{\beta}, \sigma^2) \in B \times R^+, \; B = \{\underline{b}: F\underline{b} = \underline{\Omega}_1, \; \underline{b} \in R^p\}, \tag{2.1}$$

where F is an m × p matrix of real constants, $\underline{\Omega}_1$ is a m × 1 vector of real constants, R^+ is the set of positive real numbers, and R^p is the set of p-dimensional real numbers. Then a characterization with an index for the distribution function of $\underline{Y}$ identifies a model, i.e.

$$E(\underline{Y}) = A\underline{\beta}, \; E[(\underline{Y} - A\underline{\beta})(\underline{Y} - A\underline{\beta})'] = \sigma^2 C, \; (\underline{\beta}, \sigma^2) \in B \times R^+, \tag{2.2}$$

is a Second-Order linear model for $\underline{Y}$, and

$$\underline{Y} \sim MVN(A\underline{\beta}, \sigma^2 C), \; (\underline{\beta}, \sigma^2) \in B \times R^+, \tag{2.3}$$

is a Normal linear model for $\underline{Y}$.

There is no need here or in the following theoretical development to constrain the rank of A. Indeed, in many applications rank (A) $<$ p, and in some it is either easier or more meaningful to work with a less than full-rank characterization than it is to work with one that is based on a reparameterization to introduce full rank. Therefore we will take $1 \leq \text{rank}(A) \leq p$.

On the other hand, rank (F) = m will be a characteristic of any well formulated application. Therefore, we will apply rank (F) = m to all further considerations of (2.1) - (2.3). There are then three additional rank conditions to distinguish. They are m = 0, $1 \leq m \leq p - 1$, and m = p. The first and last of these are trivial special cases, and therefore the development from here on will be limited to $1 \leq m \leq p - 1$.

There is the following real convenience of the rank conditions on F. If G is any (p - m) × p full-rank row completion of F, that is

$$H = \begin{bmatrix} F \\ \hline G \end{bmatrix}$$

is a nonsingular matrix, then $\underline{\Omega}_1$ is a constant, and $\underline{\Omega}_2$ in

$$\underline{\Omega} = \begin{bmatrix} \underline{\Omega}_1 \\ \hline \underline{\Omega}_2 \end{bmatrix} = \begin{bmatrix} F \\ \hline G \end{bmatrix} \underline{\beta} = H\underline{\beta}$$

is an arbitrary element of R^{p-m}.

Also if $H^{-1} = (J{:}K)$, say, where $FJ = I_m$, $FK = 0$, $GJ = 0$, $GK = I_{p-m}$, then $\underline{\beta}$ is nonhomogeneous in $\underline{\Omega}_2$, i.e.

$$\underline{\beta} = H^{-1}\underline{\Omega} = J\underline{\Omega}_1 + K\underline{\Omega}_2.$$

3. GENERALIZATION OF THE LINEAR ESTIMATION PROBLEM

The relationship of linear estimation to the Second-Order and Normal models set out in (2.2) and (2.3) is based on the following four definitions.

Definition 3.1: A function is said to be linear in the elements of $\underline{\beta}$ if it is a polynomial $\underline{m} + N\underline{\beta}$ in which $\underline{m}$ is a t × 1 vector of real constants, N is a t × p matrix of real constants, and $\underline{\beta}$ is a p × 1 vector that varies over the set B described in (2.1).

Definition 3.2: An estimator of the function $\underline{m} + N\underline{\beta}$, $\underline{\beta} \in B$, is said to be linear in the elements of $\underline{Y}$ if it is a polynomial $\underline{p} + Q\underline{Y}$ in which $\underline{p}$ is a $t \times 1$ vector of real constants, Q is a $t \times n$ matrix of real constants, and $\underline{Y}$ is an $n \times 1$ random vector with a distribution function that belongs to the Second-Order model described in (2.2).

Observe that the polynomials in Definition 3.1 - 3.2 are nonhomogeneous in $\underline{\beta}$ and $\underline{Y}$ if $\underline{m} + NJF\underline{\beta}$, $\underline{p} \neq \underline{0}$ and homogeneous in $\underline{\beta}$ and $\underline{Y}$, if $\underline{m} + NJF\underline{\beta} = \underline{p} = \underline{0}$. Also observe that the Normal model set out in (2.3) is a subclass of the Second-Order model that appears in (2.2). Therefore Definition 3.2 applies to Normal $\underline{Y}$ as well as Second-Order $\underline{Y}$.

Definition 3.3: A linear function $\underline{m} + N\underline{\beta}$ is said to be linearly estimable if and only if there exists a linear estimator $\underline{p} + Q\underline{Y}$ such that

$$E(\underline{p} + Q\underline{Y}) = \underline{m} + N\underline{\beta},\ (\underline{\beta}, \sigma^2) \in B \times R^+.$$

Definition 3.4: A linear estimator $\underline{p} + Q\underline{Y}$ is best (or minimum dispersion) linear unbiased for $\underline{m} + N\underline{\beta}$, $\underline{\beta} \in B$, if and only if:

1. $E(\underline{p} + Q\underline{Y}) = \underline{m} + N\underline{\beta},\ (\underline{\beta}, \sigma^2) \in B \times R^+$,

and for every $\underline{\delta} + \Delta\underline{Y}$ such that

2. $E(\underline{\delta} + \Delta\underline{Y}) = \underline{0},\ (\underline{\beta}, \sigma^2) \in B \times R^+$,

the following ordering of ranked characteristic roots obtains:

3. $ch_i[QCQ'] \leq ch_i[(Q + \Delta)C(Q + \Delta)'],\ i = 1(1)t$,

with at least one strict inequality if $P(\underline{\delta} + \Delta\underline{Y} \neq \underline{0}) > 0$.

Definition 3.4 is general because every linear unbiased estimator of $\underline{m} + N\underline{\beta}$, $\underline{\beta} \in B$, can be written in the form of $(\underline{p} + \underline{\delta}) + (Q + \Delta)\underline{Y}$ and consequently is characterized by a second-order linear model of the following form:

$$E[(\underline{p} + \underline{\delta}) + (Q + \Delta)\underline{Y}] = \underline{m} + N\underline{\beta},$$

$$E\{[(\underline{p} + \underline{\delta}) + (Q + \Delta)\underline{Y} - \underline{m} - N\underline{\beta}][(\underline{p} + \underline{\delta}) + (Q + \Delta)\underline{Y} - \underline{m} - N\underline{\beta}]'\}$$
$$= \sigma^2(Q + \Delta)C(Q + \Delta)',\ (\underline{\beta}, \sigma^2) \in B \times R^+.$$

An equivalent of Condition 3 in Definition 3.4 is that the matrix difference

$$(Q + \Delta)C(Q + \Delta)' - QCQ'$$

be nonnegative definite in general and nonzero in addition if $\Delta \neq 0$. This result is part of an extension due to Ky Fan (1949) of the Courant-Fischer min-max theorem (see Courant and Hilbert, 1931; Fischer, 1905) and can be found in the book by Bellman (1970, Second Edition; see also 1960, First Edition, for a fuller exposition of the Ky Fan results) on matrix analysis. We prefer to express Condition 3 in terms of characteristic roots because the scalar criteria that have been proposed in the statistical literature as summaries of dispersion are all individually monotone in ranked roots. We will, however, work with the matrix difference displayed here because it is so easy to handle.

The generalization of the estimation problem is threefold and clearly evident in the definitions. Firstly, any linear constraint on $\underline{\beta}$ is included in the description of the set B and therefore appears explicitly in each definition. Consideration of linear constraints is not a new idea (see for example Rao, 1945), but almost always the restriction is introduced without justification into the solution of a problem, e.g. $\underline{Y} - AJ\underline{\Omega}_1$ is used in place of $\underline{Y}$. Secondly, nonhomogeneous polynomials in $\underline{\beta}$ and $\underline{Y}$ are interesting replacements for the homogeneous forms that have been the focus of attention in the past. The latter, of course, are not always truly homogeneous because, for example, if $\underline{Y}$ is replaced by $\underline{Y} - AJ\underline{\Omega}_1$ to estimate $N\underline{\beta}$, then the Best Linear Unbiased Estimator, if one exists, is in fact homogeneous in $\underline{Y} - AJ\underline{\Omega}_1$ and therefore nonhomogeneous in $\underline{Y}$. Finally, the definition of minimum dispersion for an unbiased linear estimator allows one to account in a satisfactory way for simultaneous estimation of several scalar linear functions of $\underline{\beta}$. That is, when it is satisfied, then any scalar optimality criterion that is usually used to characterize estimators in linear multivariate analysis is also satisfied.

4. SOLUTION OF THE LINEAR ESTIMATION PROBLEM.

Sets of necessary and sufficient conditions are stated and proved for the following: i) that $\underline{p} + Q\underline{Y}$ be an unbiased estimator of $\underline{m} + N\underline{\beta}$ when (2.1) - (2.2) obtain, ii) that $\underline{p} + Q\underline{Y}$ be a Best Linear Unbiased Estimator of $\underline{m} + N\underline{\beta}$ when (2.1)-(2.2) obtain, and iii) that $\underline{p} + Q\underline{Y}$ be the Maximum Likelihood Estimator of $\underline{m} + N\underline{\beta}$ when (2.1) and (2.3) obtain. Equivalent sets of conditions are given for ii). These and the condition in (iii) are separated by a short lemma to the effect that there exists at most one Best Linear Unbiased Estimator of $\underline{m} + N\underline{\beta}$ when (2.1) - (2.2) obtain. Finally one estimator that satisfies i) is shown to satisfy both set of conditions ii) and the single set in iii) when (2.1) and (2.3) obtain. Therefore, for any linearly estimable $\underline{m} + N\underline{\beta} \in B$, there exists a unique Best Linear Unbiased Estimator, when (2.1) and (2.2) obtain, that is also the Maximum Likelihood Estimator when (2.1) and (2.3) obtain.

All statements in the following lemmas pertain only to models, functions, and criteria set out in (2.1) - (2.3) and Definitions (3.1) - (3.4). Unless essential for clarity, no further reference will be made to these.

Lemma 4.1: $\underline{m} + N\underline{\beta}$ is linearly estimable if and only if rows of NK belong to the row space of AK. In this case, any linear estimator of the form $\underline{m} + (N - QA)J\underline{\Omega}_1 + Q\underline{Y}$, where $QAK = NK$, is unbiased for $\underline{m} + N\underline{\beta}$.

Proof: $\underline{m} + N\underline{\beta}$ is linearly estimable if and only if there exist $\underline{p}$ and Q such that

$$\begin{aligned} E(\underline{p} + Q\underline{Y}) &= \underline{p} + QA\underline{\beta} = \underline{p} + QAJ\underline{\Omega}_1 + QAK\underline{\Omega}_2 \\ &= \underline{m} + N\underline{\beta} = \underline{m} + NJ\underline{\Omega}_1 + NK\underline{\Omega}_2 \end{aligned} \tag{4.1}$$

for every $\underline{\Omega}_2$ in R^{p-m}.

For the if part of the proof, let $NK = LAK$ and take for a linear estimator $\underline{m} + (N - LA)J\underline{\Omega}_1 + L\underline{Y}$, that is set $Q = L$. Then

$$\begin{aligned} E(\underline{m} + (N - LA)J\underline{\Omega}_1 + L\underline{Y}) &= \underline{m} + (N - LA)J\underline{\Omega}_1 + LAJ\underline{\Omega}_1 + LAK\underline{\Omega}_2 \\ &= \underline{m} + NJ\underline{\Omega}_1 + NK\underline{\Omega}_2 \\ &= \underline{m} + N\underline{\beta}. \end{aligned}$$

This completes the first part of the proof.

For the only if part of the proof, take the zero vector and then the columns of I_{p-m} in succession and substitute these for $\underline{\Omega}_2$ in (4.1). The result of the first substitution is

$$\underline{p} = \underline{m} + (N - QA)J\underline{\Omega}_1 \tag{4.2}$$

and of the second is

$$QAK = NK. \tag{4.3}$$

The necessity that rows of NK belong to the row space of AK is displayed in (4.3). This completes the second part of the proof.

Finally the cited form of a linear unbiased estimator of $\underline{m} + N\underline{\beta}$ can be obtained directly from (4.1) - (4.3) or by setting $Q = L$ in $\underline{m} + (N - LA)J\underline{\Omega}_1 + L\underline{Y}$ as suggested. This completes the proof of the lemma.

The next two lemmas are statements of equivalent necessary and sufficient conditions for a linear unbiased estimator $\underline{p} + Q\underline{Y}$ of $\underline{m} + N\underline{\beta}$ to be best. The first of these, which is due to Fisher (1925) and Rao (1952) for the estimation of a scalar parameter, is the more instructive one. It is a statement of the essential property of a Best Linear Unbiased Estimator, namely that there is no covariate estimator of $\underline{0}$ that can be used in conjunction with $\underline{p} + Q\underline{Y}$ to reduce dispersion of the errors of estimation. The criterion however is not always easy to apply, and therefore it is convenient to have an equivalent form. This, which is given in the second of the two lemmas, is a full generalization of special results due to Grenander and Rosenblatt (1957 for a summary of early work in stationary time series), McElroy (1967), Williams (1967, 1970), and Watson (1967).

Lemma 4.2: A linear unbiased estimator $\underline{p} + Q\underline{Y}$ of $\underline{m} + N\underline{\beta}$ is best if and only if $QCA' = 0$ for every $\underline{\delta}$, A such that $\underline{\delta} + A\underline{Y}$ is an unbiased estimator of $\underline{0}$.
Proof: The difference between two linear unbiased estimators of $\underline{m} + N\underline{\beta}$ is a linear unbiased estimator of $\underline{0}$. Therefore, $\underline{p} + Q\underline{Y}$ and $(\underline{p} + \underline{\delta}) + (Q + A)\underline{Y}$ are linear unbiased estimators of $\underline{m} + N\underline{\beta}$.

If $QCA' = 0$ for every A, then the dispersion matrix of $(\underline{p} + \underline{\delta}) + (Q + A)\underline{Y}$ is proportional in σ^2 to

$$QCQ' + ACA',$$

and the difference

$$(QCQ' + ACA') - QCQ' = ACA'$$

is nonnegative definite for all A and is 0 if and only if $A = 0$ because C is positive definite. If $A = 0$, then $\underline{\delta} = \underline{0}$. Therefore $\underline{p} + Q\underline{Y}$ is best. This completes the first part of the proof.

If $\underline{p} + Q\underline{Y}$ is best linear unbiased, then the difference between the dispersion matrix of

$$(\underline{p} + Q\underline{Y}) - QCA'(ACA')^*(\underline{\delta} + A\underline{Y}),$$

which is also linear unbiased, and of $\underline{p} + Q\underline{Y}$ must be nonnegative definite, that is

$$[QCQ' - QCA'(ACA')^*ACQ'] - QCQ' = - QCA'(ACA')^*ACQ'$$

must be 0. The remainder of the proof is now completed when it is noted that the matrix on the right-hand side of this final equation is invariant to the choice of the conditional inverse and that every nonnegative definite matrix, such as ACA', has at least one positive definite conditional inverse. Therefore, the equation can obtain if and only if $QCA' = 0$.
Lemma 4.3: A linear unbiased estimator $\underline{p} + Q\underline{Y}$ of $\underline{m} + N\underline{\beta}$ is best if and only if there exists M such that $QC = MA'$ where $MF' = 0$.
Proof: The necessary and sufficient conditions set out in Lemma 4.1 for $\underline{\delta} + A\underline{Y}$ to be an unbiased estimator of $\underline{0}$ are that $\underline{\delta} = -AAJ\underline{\Omega}_1$; $AAK = 0$. Both of these are needed here. The first part of the condition is contained in Lemma 4.2, namely that $\underline{p} + Q\underline{Y}$ is Best Linear Unbiased for $\underline{m} + N\underline{\beta}$ if and only if $QCA' = 0$ for every A such that $AAK = 0$. It is also part of the statement here in Lemma 4.3, namely that $\underline{p} + Q\underline{Y}$ is best linear unbiased for $\underline{m} + N\underline{\beta}$ if and only if $QC = MA'$, $MF' = 0$ where $AAK = 0$ for every A such that $\underline{\delta} + A\underline{Y}$ is a linear unbiased estimator of $\underline{0}$.

For the if part of the proof, observe that $MF' = 0$ implies that $M = LK'$ because columns of F' are linearly independent. It then follows from this and $QC = MA'$, that $QCA' = (LK'A')A' = L(K'A'A') = L0 = 0$ for every A such that $AAK = 0$. Therefore, $\underline{p} + Q\underline{Y}$ is a Best Linear Unbiased Estimator of $\underline{m} + N\underline{\beta}$ (Lemma 4.2). This completes the first part of the proof.

For the only if part of the proof, consider first the trivial case when rank $(AK) = n$. Then, the $t \times n$ matrix product QC must have rows that are linear combinations of the $p - m$ rows of $K'A'$. That is $QC = L(K'A') = (LK')A' = MA'$ (say). In this, $MF' = (LK')F' = L(K'F') = L0 = 0$. Consider next the nontrivial case when rank $(AK) < n$, and let rows of A_0 be a basis of the null space of columns of AK. Then, consider the $t \times n$ matrices $A_j = \underline{1}\underline{u}_j'A_0$, $j = 1(1)(n-\pi)$, where $\underline{1}$ is a vector of t ones, $\underline{u}_j$ is a vector with one in the jth position and zeros in the remaining positions, if any, and π = rank (AK). Clearly, the $-\Lambda_j AJ\underline{\Omega}_1 + \Lambda_j \underline{Y}$ are linear unbiased estimators of $\underline{0}$ (Lemma 4.1), and if $\underline{p} + Q\underline{Y}$ is a Best Linear Unbiased Estimator of $\underline{m} + N\underline{\beta}$, then $QC\Lambda_j' = 0$, $j = 1(1)(n - \pi)$ (Lemma 4.2). From this, we obtain

$QC\Lambda_j'=0,\ j=1(1)(n-\pi) \Rightarrow QC\Lambda_0 \underline{u}_j' = \underline{0},\ j=1(1)(n-\pi) \Rightarrow QC\Lambda_0'=0 \Rightarrow QC=L(K'A')=(LK')A'=MA'$ (say)

where

$$MF' = (LK')F' = L(K'F') = L0 = 0.$$

This completes the proof of the Lemma.

Lemma 4.4: There is at most one Best Linear Unbiased Estimator of $\underline{m} + N\underline{\beta}$.

Proof: Were there more than one Best Linear Unbiased Estimator of $\underline{m} + N\underline{\beta}$, then any two could be denoted by $\underline{p} + Q\underline{Y}$ and $(\underline{p} + \underline{\delta}) + (Q + \Lambda)\underline{Y}$ where $QC\Lambda' = 0$ (Lemma 4.2). The difference in dispersion matrices of the estimators therefore would be $-\Lambda C\Lambda' = 0$, for both to be best. From this, it would follow that $\Lambda = 0$, because C is a positive definite matrix, and then that $\underline{\delta} = -\Lambda AJ\underline{\Omega}_1 = \underline{0}$. Therefore, $(\underline{p} + \underline{\delta}) + (Q + \Lambda)\underline{Y} = \underline{p} + Q\underline{Y}$, so that any two Best Linear Unbiased Estimators would be one and the same. This completes the proof of the lemma.

The following lemma applies only to the multivariate Normal linear model (2.3). Since the likelihood function in this case is quadratic in $\underline{\beta}$, proofs based on simple algebraic methods can be found. One is presented here in order to maintain the form and level of mathematics used for the preceding proofs.

Lemma 4.5: The linear estimator $\underline{m} + NJ\underline{\Omega}_1 + NK\hat{\underline{\Omega}}_2$ is maximum likelihood for $\underline{m} + N\underline{\beta}$ if and only if $\hat{\underline{\Omega}}_2$ is a solution of

$$K'A'C^{-1}(\underline{Y} - AJ\underline{\Omega}_1 - AK\underline{\Omega}_2) = \underline{0}. \tag{4.4}$$

Furthermore, the estimator is unbiased and unique if $\underline{m} + N\underline{\beta}$ is linearly estimable.

Proof: Firstly, observe that the set of linear equations represented by (4.4) is consistent. Therefore, there exists a set of one or more solutions for $\underline{\Omega}_2$ that are linear in the elements of $\underline{Y} - AJ\underline{\Omega}_1$.

Secondly, $\underline{m} + N\underline{\beta} = \underline{m} + NJ\underline{\Omega}_1 + NK\underline{\Omega}_2$ where only $\underline{\Omega}_2$ is part of the index for the multivariate Normal model for $\underline{Y}$. Therefore, a Maximum Likelihood Estimator of $\underline{m} + N\underline{\beta}$, if one exists, is $\underline{m} + NJ\underline{\Omega}_1 + NK\hat{\underline{\Omega}}_2$ where $\hat{\underline{\Omega}}_2$ is a Maximum Likelihood Estimator of $\underline{\Omega}_2$.

Thirdly, the natural logarithm of the likelihood function of $(\underline{\Omega}_2, \sigma^2)$, aside from an additive constant, is

$$-\frac{n}{2}\ell n\sigma^2 - \frac{1}{2\sigma^2}(\underline{Y} - AJ\underline{\Omega}_1 - AK\underline{\Omega}_2)'C^{-1}(\underline{Y} - AJ\underline{\Omega}_1 - AK\underline{\Omega}_2) = -\frac{n}{2}\ell n\sigma^2 - \frac{1}{2\sigma^2}Q \text{ (say)}.$$

Since $(\underline{\Omega}_2, \sigma^2) \in R^{p-m} \times R^+$, this can be maximized in two steps: first with respect to $\underline{\Omega}_2$ to produce a set of local maxima with elements

$$-\frac{n}{2}\ell n\sigma^2 - \frac{1}{2\sigma^2}\hat{Q} \text{ (say)},$$

and then with respect to σ^2 to produce a global maximum

$$-\frac{n}{2}\ell n(\frac{\hat{Q}}{n}) - \frac{n}{2}.$$

Therefore, the problem that remains is to find $\underline{\Omega}_2$ for which Q (which is not a function of σ^2) is a minimum.

To show the if and only if property of (4.4), expand Q in $\underline{\Omega}_2$ about any solution $\hat{\underline{\Omega}}_2$ of (4.4). The result of this is

$$Q = \hat{Q} + (\hat{\underline{\Omega}}_2 - \underline{\Omega}_2)'K'A'C^{-1}AK(\hat{\underline{\Omega}}_2 - \underline{\Omega}_2) \geq \hat{Q}$$

where $\hat{Q}$ is Q evaluated at $\hat{\underline{\Omega}}_2$. Clearly the lower bound is achieved if $\underline{\Omega}_2 = \hat{\underline{\Omega}}_2$, and therefore, $\hat{\underline{\Omega}}_2$ is a Maximum Likelihood Estimator. This completes the if part of the proof. The lower bound is also achieved if

$$(\hat{\underline{\Omega}}_2 - \underline{\Omega}_2)'K'A'C^{-1}AK(\hat{\underline{\Omega}}_2 - \underline{\Omega}_2) = 0$$

for any other $\underline{\Omega}_2$. For any such solution, observe that

$$K'A'C^{-1}(\underline{Y} - AJ\underline{\Omega}_1 - AK\underline{\Omega}_2 + AK\underline{\Omega}_2 - AK\hat{\underline{\Omega}}_2) = \underline{0}$$

$$\Rightarrow K'A'C^{-1}(\underline{Y} - AJ\underline{\Omega}_1 - AK\underline{\Omega}_2) = K'A'C^{-1}AK(\hat{\underline{\Omega}}_2 - \underline{\Omega}_2)$$

$$\Rightarrow AK(K'A'C^{-1}AK)^*K'A'C^{-1}(\underline{Y} - AJ\underline{\Omega}_1 - AK\underline{\Omega}_2) = AK(\hat{\underline{\Omega}}_2 - \underline{\Omega}_2)$$

$$\Rightarrow (\hat{\underline{\Omega}}_2 - \underline{\Omega}_2)'K'A'C^{-1}AK(\hat{\underline{\Omega}}_2 - \underline{\Omega}_2)$$

$$= (\underline{Y} - AJ\underline{\Omega}_1 - AK\underline{\Omega}_2)'C^{-1}AK(K'A'C^{-1}AK)^*K'A'C^{-1}(\underline{Y} - AJ\underline{\Omega}_1 - AK\underline{\Omega}_2).$$

The right-hand side of the final equation displayed here is invariant to choice of the conditional inverse, and every nonnegative definite matrix has at least one positive definite conditional inverse. Therefore the quadratic on the right-hand side is zero if and only if $\underline{\Omega}_2$ is also a solution of (4.4). This completes the only if part of the proof.

To show that a Maximum Likelihood Estimator of linearly estimable $\underline{m} + N\underline{\beta}$ is unbiased, refer first to <u>Lemma</u> 4.1 for the estimability condition that there exists L such that LAK = NK and then obtain

$$K'A'C^{-1}AK\ E(\hat{\underline{\Omega}}_2) = K'A'C^{-1}AK\underline{\Omega}_2,\ \underline{\Omega}_2 \in R^{p-m}, \tag{4.5}$$

for any solution of (4.4) by taking expectations of the two terms in parenthesis in

$$K'A'C^{-1}[(\underline{Y} - AJ\underline{\Omega}_1 - AK\underline{\Omega}_2) + (AK\underline{\Omega}_2 - AK\hat{\underline{\Omega}}_2)] = \underline{0}.$$

The equation

$$AKE(\hat{\underline{\Omega}}_2) = AK\underline{\Omega}_2,\ \underline{\Omega}_2 \in R^{p-m},$$

can be obtained from (4.5) by premultiplication on both sides by $AK(K'A'C^{-1}AK)^*$, and from this, the required result follows in one step, <u>viz.</u>

$$E(NK\hat{\underline{\Omega}}_2) = E(LAK\hat{\underline{\Omega}}_2) = LAKE(\hat{\underline{\Omega}}_2) = LAK\underline{\Omega}_2 = NK\underline{\Omega}_2,\ \underline{\Omega}_2 \in R^{p-m}.$$

To show that the estimator is unique, take any two solutions of (4.4), say $\hat{\underline{\Omega}}_{21}$ and $\hat{\underline{\Omega}}_{22}$, and recall that

$$(\hat{\underline{\Omega}}_{21} - \hat{\underline{\Omega}}_{22})K'A'C^{-1}AK(\hat{\underline{\Omega}}_{21} - \hat{\underline{\Omega}}_{22}) = 0.$$

The matrix C is positive definite, so that every solution of the equation satisfies

$$AK\hat{\underline{\Omega}}_{21} = AK\hat{\underline{\Omega}}_{22}.$$

Also $\underline{m} + N\beta$ is linearly estimable so that there exists L such that LAK = NK (Lemma 4.1). Therefore

$$NK\hat{\underline{\Omega}}_{21} = NK\hat{\underline{\Omega}}_{22}.$$

This completes the proof of the lemma.

The final result that brings together Best Linear Unbiased and Maximum Likelihood Estimation is the following. When $\underline{m} + N\beta$ is linearly estimable, then there is a unique Best Linear Unbiased Estimator when (2.1) and (2.2) obtain and a unique Maximum Likelihood Estimator when (2.1) and (2.3) obtain. Furthermore, the two estimators are one and the same linear function of $\underline{Y}$. This we summarize in the following way.

Theorem 4.1: If $\underline{m} + N\beta$ is linearly estimable, then the Best Linear Unbiased and Maximum Likelihood Estimators of $\underline{m} + N\beta$ are one and the same. Furthermore the estimator is

$$\underline{m} + [N - NK(K'A'C^{-1}AK)^{*}K'A'C^{-1}A]J\underline{\Omega}_1 + NK(K'A'C^{-1}AK)^{*}K'A'C^{-1}\underline{Y} \quad (4.7)$$

$$= \underline{m} + NJ\underline{\Omega}_1 + NK(K'A'C^{-1}AK)^{*}K'A'C^{-1}(\underline{Y} - AJ\underline{\Omega}_1). \quad (4.8)$$

Remark: (4.7) is the Best Linear Unbiased Estimator arranged in the form given in Lemma 4.1, and (4.8) is a solution of estimation equation (4.4) for the Maximum Likelihood Estimator (Lemma 4.5).

Proof: Lemmas 4.1 - 4.5 will be used to establish the theorem.

Firstly, NK belongs to the row space of AK because $\underline{m} + N\beta$ is linearly estimable (Lemma 4.1). Therefore write LAK = NK and set $Q = NK(K'A'C^{-1}AK)^{*}K'A'C^{-1}$. Clearly then QAK = LAK = NK so that (4.7) has the form of the linear unbiased estimator set out in Lemma 4.1 for linearly estimable $\underline{m} + N\beta$.

Secondly, take Q as in the preceding paragraph and let Δ be any matrix that satisfies $\Delta AK = 0$. Then

$$QC\Delta' = NK(K'A'C^{-1}AK)^{*}(K'A'\Delta') = NK(K'A'C^{-1}AK)^{*}(0) = 0.$$

This is the necessary and sufficient condition in Lemma 4.2 for (4.7) to be a Best Linear Unbiased Estimator of linearly estimable $\underline{m} + N\beta$.

Thirdly, take Q as before. Then

$$QC = [NK(K'A'C^{-1}AK)^{*}K']A' = MA' \text{ (say)}$$

where

$$MF' = [NK(K'A'C^{-1}AK)^{*}K']F' = NK(K'A'C^{-1}AK)^{*}(K'F') = NK(K'A'C^{-1}AK)^{*}(0) = 0.$$

These are the necessary and sufficient conditions in Lemma 4.3 that (4.7) be a Best Linear Unbiased Estimator of linearly estimable $\underline{m} + N\beta$.

Fourthly, there is at most one Best Linear Unbiased Estimator of linearly estimable $\underline{m} + N\beta$ (Lemma 4.4). From the results in the preceding two paragraphs, that estimator must be (4.7).

Finally, take $\hat{\underline{\Omega}}_2 = (K'A'C^{-1}AK)^{*}K'A'C^{-1}(\underline{Y} - AJ\underline{\Omega}_1)$. Then

$$K'A'C^{-1}[\underline{Y} - AJ\underline{\Omega}_1 - AK(K'A'C^{-1}A)^{*}K'A'C^{-1}(\underline{Y} - AJ\underline{\Omega}_1)]$$

$$= K'A'C^{-1}(\underline{Y} - AJ\underline{\Omega}_1) - K'A'C^{-1}(\underline{Y} - AJ\underline{\Omega}_1) = \underline{0}.$$

This is the necessary and sufficient condition in Lemma 4.4 that (4.8) be the Maximum Likelihood Estimator of linearly estimable $\underline{m} + N\underline{\beta}$. This completes the proof of the theorem.

5. APPLICATION TO A MODEL FOR LONGITUDINAL ANALYSIS.

Four different classes of applications in multivariate analysis are collections of methods for the solutions of common problems. These are: growth-curve, longitudinal, profile, and repeated-measures analyses for which the following model is often appropriate:

$$Y = U\beta X' + \Sigma^{1/2} Z \Gamma^{1/2\prime} \tag{5.1}$$

(see for example Grizzle and Allen (1969) for a detailed discussion of uses of this model). Here rows of Y are indexed by variates, columns of Y are indexed by units ordered by randomization and/or probability sampling, columns of U are regressors for the description of variate effects, columns of X are regressors for the description of unit effects, $\Sigma^{1/2}\Sigma^{1/2\prime} = \Sigma$, and $\Gamma^{1/2}\Gamma^{1/2\prime} = \Gamma$ are two positive definite matrices of real constants. Elements of β represent real variables that are constrained in the following way:

$$L\beta R' = \Omega_1,$$

and elements of Z are uncorrelated $(0, \sigma^2)$ random variables with arbitrary variance, i.e. $\sigma^2 \in R^+$.

The relationship of this model to the ones in the forgoing sections is displayed in the following: successive columns of Y are successive partitions of $\underline{Y}$, $(U \otimes X) = A$, $(\Sigma \otimes \Gamma) = C$, $(L \otimes R) = F$, and successive columns of Ω_1 are successive partions of $\underline{\Omega}_1$. The model will be less than general in one aspect in order to simplify the estimation problem. The rows of L and R will be restricted to the row spaces of U and X respectively.

The first problem that will be considered is Best Linear Unbiased Estimation of

$$M + N_l \beta N_r' \tag{5.2}$$

where rows of N_l and N_r belong to the row spaces of U and X respectively. The second problem will be the same except that the linear combination of β will not be restricted to the $N_l \beta N_r'$ form. The final problem that will be considered concerns limiting properties of the Maximum Likelihood Estimator of the more general linear form when rank (U) = p and Σ is an arbitrary element of the class of positive definite matrices.

The model set out in (5.1) and the linear function of β given in (5.2) are of considerable practical importance. A single example can be used to illustrate this point. One may want to study operating characteristics of an industrial longitudinal process that develops in space and/or time (factors that determine U regressors) in an experiment based on a blocked multiple regression plan (factors that determine X regressors). Operating conditions may restrict effects of factors ($L\beta R' = \Omega_1$). Subject to these, the important operating characteristics are linear combinations of effects ($N_l \beta N_r'$) that are linearly modified by observable extrinsic effects (**M**).

It is not difficult to guess that the Best Linear Unbiased Estimator of **M** + $N_l \beta N_r'$ is the appropriate linear combination of

$$(U'\Sigma^{-1}U)^* U'\Sigma^{-1} Y \Gamma^{-1} X (X'\Gamma^{-1}X)^*$$

that has been adjusted by subtracting off the linear zero estimator

$$(U'\Sigma^{-1}U)^{*}L'[L(U'\Sigma^{-1}U)^{*}L']^{*}L(U'\Sigma^{-1}U)^{*}U'\Sigma^{-1}Y\Gamma^{-1}X(X'\Gamma^{-1}X)^{*}R'[R(X'\Gamma^{-1}X)^{*}R']^{*}R(X'\Gamma^{-1}X)^{*}$$
$$- (U'\Sigma^{-1}U)^{*}L'[L(U'\Sigma^{-1}U)^{*}L']^{*}\Omega_1[R(X'\Gamma^{-1}X)^{*}R']^{*}R(X'\Gamma^{-1}X)^{*} =$$
$$(U'\Sigma^{-1}U)^{*}L'[L(U'\Sigma^{-1}U)^{*}L']^{*}L(U'\Sigma^{-1}U)^{*}U'\Sigma^{-1}(Y-U\beta X')\Gamma^{-1}X(X'\Gamma^{-1}X)^{*}R'[R(X'\Gamma^{-1}X)^{*}R']^{*}R(X'\Gamma^{-1}X$$

for a covariate adjustment to reduce dispersion. The result of this, after correction for bias and rearrangement of terms, is

$$M + N_{\ell}(U'\Sigma^{-1}U)^{*}L'[L(U'\Sigma^{-1}U)^{*}L']^{*}\Omega_1[R(X'\Gamma^{-1}X)^{*}R']^{*}R(X'\Gamma^{-1}X)^{*}N_r' +$$
$$N_{\ell}\{(U'\Sigma^{-1}U)^{*}U'\Sigma^{-1}Y\Gamma^{-1}X(X'\Gamma^{-1}X)^{*} -$$
$$(U'\Sigma^{-1}U)^{*}L'[L(U'\Sigma^{-1}U)^{*}L']^{*}L(U'\Sigma^{-1}U)^{*}U'\Sigma^{-1}Y\Gamma^{-1}X(X'\Gamma^{-1}X)^{*}R'[R(X'\Gamma^{-1}X)^{*}R']^{*}R(X'\Gamma^{-1}X)^{*}\}N_r'. \quad (5.3)$$

However a derivation of this starting from (4.7) and (5.1) falls just short of an algebraic tour de force.

It is easier to use vector notation based on Kroneker products than the matrix notation displayed in (5.3) to show that the estimator is best linear unbiased because all of the conditions in Lemmas (4.1) - (4.5) are expressed in vector form. For example $(U'\Sigma^{-1}U)^{*}U'\Sigma^{-1}Y\Gamma^{-1}X(X'\Gamma^{-1}X)^{*}$ becomes $[(U'\Sigma^{-1}U)^{*}U'\Sigma^{-1} \otimes (X'\Gamma^{-1}X)^{*}X'\Gamma^{-1}]\underline{Y}$.

The first problem to solve is to demonstrate that the estimator is unbiased for $\underline{m} + N\underline{\beta}$ where successive columns of **M** form successive partitions of $\underline{m}$ and $N = (N_{\ell} \otimes N_r)$. For this, it is necessary and sufficient to verify that

$$QAK = NK$$

(Lemma 4.1). We have

$$(N_{\ell} \otimes N_r)\{(U'\Sigma^{-1}U)^{*}U'\Sigma^{-1} \otimes (X'\Gamma^{-1}X)^{*}X'\Gamma^{-1}$$
$$-(U'\Sigma^{-1}U)^{*}L'[L(U'\Sigma^{-1}U)^{*}L']^{*}L(U'\Sigma^{-1}U)^{*}U'\Sigma^{-1}\otimes(X'\Gamma^{-1}X)^{*}R'[R(X'\Gamma^{-1}X)^{*}R']^{*}R(X'\Gamma^{-1}X)^{*}X'\Gamma^{-1}\}$$
$$(U \otimes X)K$$
$$= (N_{\ell} \otimes N_r)K - N\{(U'\Sigma^{-1}U)^{*}L'[L(U'\Sigma^{-1}U)^{*}L']^{*} \otimes (X'\Gamma^{-1}X)^{*}R'[R(X'\Gamma^{-1}X)^{*}R']^{*}\}(L \otimes R)K$$
$$= NK. \quad (5.4)$$

This is the required result. The first reductions shown here are consequences of rows of N_{ℓ} and L belonging to the row space of U and rows of N_r and R belonging to the row space of X. The final reduction is a consequence of $(L \otimes R) = F$ and $FK = 0$.

Next it is necessary to verify that the constant term has been written out properly, namely that

$$(N - QA)J\underline{\Omega}_1 = (N_{\ell} \otimes N_r)\{(U'\Sigma^{-1}U)^{*}L'[L(U'\Sigma^{-1}U)L']^{*}\otimes (X'\Gamma^{-1}X)^{*}R'[R(X'\Gamma^{-1}X)^{*}R']^{*}\}\underline{\Omega}_1.$$

To do this, follow the development of the two equations in (5.4) when K is replaced by $J\underline{\Omega}_1$ and replace $(L \otimes R)J\underline{\Omega}_1 = FJ\underline{\Omega}_1$ in the first of these by $\underline{\Omega}_1$ since $FJ = I_m$. The required result is the difference of the reduced forms in the final equality.

The last problem to solve is to demonstrate that the unbiased estimator is best for $\underline{m} + N\underline{\beta}$. For this, it is necessary and sufficient to verify that

$$QC = MA' \text{ where } MF' = 0$$

(Lemma 4.3) and to apply Lemma 4.4. The product on the left-hand side of the first of the equalities is

$$(N_l \otimes N_r)\{(U'\Sigma^{-1}U)^* \otimes (X'\Gamma^{-1}X)^* - (U'\Sigma^{-1}U)^*L'[L(U'\Sigma^{-1}U)^*L']^*L(U'\Sigma^{-1}U)^*$$
$$\otimes (X'\Gamma^{-1}X)^*R'[R(X'\Gamma^{-1}X)^*R']^*R(X'\Gamma^{-1}X)^*\}(U' \otimes X')$$

where $(U' \otimes X') = A'$ as required. Finally, observe that

$$\{(U'\Sigma^{-1}U)^*L'[L(U'\Sigma^{-1}U)^*L']^*L(U'\Sigma^{-1}U)^*$$
$$\otimes (X'\Gamma^{-1}X)^*R'[R(X'\Gamma^{-1}X)^*R']^*R(X'\Gamma^{-1}X)\}(L'\otimes R')$$
$$= (U'\Sigma^{-1}U)^*L' \otimes (X'\Gamma^{-1}X)^*R'$$

so that $MF' = 0$ as required.

One useful modification of the forgoing results is to observe that the development does not depend on $N = (N_l \otimes N_r)$ where rows of N_l belong to the row space of U and rows of N_r belong to the row space of X, that is $N = (T_l \otimes T_r)(U \otimes X)$. It can be carried through with no more than $N = T(U \otimes X)$.

A second useful modification is to consider a Normal linear model indexed by $(\underline{\beta}, \sigma^2C) \in B \times D$ where D is the class of positive definite, real, symmetric matrices. This represents the case in model (5.1) when $\sigma^2\Sigma$ is unknown. When rank $(U) = p$, then the Maximum Likelihood Estimator of $M + N_l\beta N_r'$ (or the modified form) can be obtained from (5.3) by substitution of S for $\sigma^2\Sigma$, where

$$S = \frac{1}{n - \upsilon}Y[\Gamma^{-1} - \Gamma^{-1}X(X'\Gamma^{-1}X)^*X'\Gamma^{-1}]Y', \quad \upsilon = \text{rank}(X).$$

This result has been proved by us and appears in a companion to this paper (1987). (For a similar result when β is not linearly constrained and $M = 0$, $\Gamma = I_n$ see Khatri, 1966, or more generally Srivastava and Khatri, 1979). Clearly when S converges with probability one to $\sigma^2\Sigma$, then for any scaling and rotation for which the difference between (5.3) and $M + N_l\beta N_r'$ (or their modified forms) converges to a random matrix, so will the difference between the Maximum Likelihood Estimator (when $(\underline{\beta}, \sigma^2C) \in B \times D$) and $M + N_l\beta N_r'$ (or it's modified form) converge to the same matrix. The Maximum Likelihood Estimator is asymptotically best linear unbiased with an error of substitution of S for $\sigma^2\Sigma$ that, with probability one, vanishes at the same rate that S converges to $\sigma^2\Sigma$.

REFERENCES

Bellman, R.E. (1970). Introduction to Matrix Analysis, Second Edition. McGraw Hill, New York, pp. 114-123.

Courant, R. and Hilbert, O. (1931). Methoden der mathematischen Physik. Berlin, Reprinted by Interscience Publishers, Inc., New York.

Fan, Ky (1949). On a theorem of Weyl concerning eigen values of linear transformations I. Proceedings of the National Academy of Science, U.S.A., 35, 625 - 655.

Fischer, E. (1905). Uber quadratische Formen mit reelen Koeffizienten. Monatshefte Mathematika und Physik, 16, 234 - 249.

Fisher, R.A. (1925). Theory of statistical estimation. Proceedings of the Cambridge Philosophical Society, 22, 700 - 725.

Grenander, U. and Rosenblatt, M. (1957). Statistical Analysis of Stationary Time Series. John Wiley and Sons, New York, 226 - 258.

Grizzle, J.D. and Allen, D.M. (1969). Analysis of growth and dose response curves. Biometrics, 25, 357 - 382.

Khatri, C.G. (1966). A note on a MANOVA modèl applied to problems in growth curve. Annals Institute of Statistical Mathematics, 18, 75 - 86.

McElroy, F.W. (1967). A necessary and sufficient condition that ordinary least squares be best linear unbiased. Journal of the American Statistical Association, 62, 1302 - 1304.

Rao, C.R. (1945). Markoff's theorem with linear restrictions on parameters. Sankhyā, 7, 16 - 19.

Rao, C.R. (1952). Some theorems on minimum variance estimation. Sankhyā, 12, 12 - 42.

Srivastava, M.S. and Khatri, C.G. (1979). An Introduction to Multivariate Statistics. North Holland, New York, 191 - 192, 196 - 197.

Watson, G.S. (1967). Linear least squares regression. Annals of Mathematical Statistics, 38, 1679 - 1699.

Williams, J.S. (1967). The variance of weighted regression estimators. Journal of the American Statistical Asssociation, 62, 1290 - 1301.

Williams, J.S. (1970). The choice and use of tests for the independence of several sets of variates. Biometrics, 26, 613 - 624.

Zaatar, M.K. and Williams, J.S. (1987). Mathematical background for studies of minimal sufficiency and the method of Maximum Likelihood for estimation and testing with a multivariate normal linear model. Technical Report #87-17, Department of Statistics, Colorado State University.

Original paper received: 01.07.87
Final paper received: 05.11.87

Paper recommended by P.W. Mielke